Liquidated Damages and Extensions of Time

In Construction Contracts

Liquidated Damages and Extensions of Time

In Construction Contracts

Third Edition

Brian Eggleston
CEng, FICE, FIStructE, FCIArb

WILEY-BLACKWELL

A John Wiley & Sons, Ltd., Publication

This edition first published 2009
© 1992, 1997, 2009 Brian Eggleston

Blackwell Publishing was acquired by John Wiley & Sons in February 2007. Blackwell's publishing programme has been merged with Wiley's global Scientific, Technical, and Medical business to form Wiley-Blackwell.

Registered office
John Wiley & Sons Ltd, The Atrium, Southern Gate, Chichester, West Sussex, PO19 8SQ, United Kingdom

Editorial offices
9600 Garsington Road, Oxford, OX4 2DQ, United Kingdom
2121 State Avenue, Ames, Iowa 50014-8300, USA

First published 1992
Second edition 1997
Third edition 2009

For details of our global editorial offices, for customer services and for information about how to apply for permission to reuse the copyright material in this book please see our website at www.wiley.com/wiley-blackwell.

Library of Congress Cataloging-in-Publication Data

Eggleston, Brian, CEng.
    Liquidated damages and extensions of time in construction contracts /
Brian Eggleston. — 3rd ed.
            p. cm.
    Includes bibliographical references and index.
    ISBN 978-1-4051-1815-6 (hardback : alk. paper). 1. Construction contracts—
England. 2. Performance (Law)—England. 3. Breach of contract—England.
4. Damages—England. 5. Time (Law)—England. 6. Construction contracts—
Wales. 7. Performance (Law)—Wales. 8. Breach of contract—Wales.
9. Damages—Wales. 10. Time (Law)—Wales. I. Title.

    KD1641.E34 2009
    343.41'078624—dc22

                                                            2008044760

A catalogue record for this book is available from the British Library.

Set in 10/12 pt Palatino by SNP Best-set Typesetter Ltd., Hong Kong
Printed in Great Britain by TJ International Ltd, Padstow, Cornwall

1   2009

Contents

Preface to Third Edition

The first edition of this book, intended as a construction industry guide to the purposes and perils of liquidated damages and extensions of time, was published in 1992. The second edition followed in 1997, by which time it had become evident that the book was also of interest to lawyers as a compendium of cases. In this third edition I have endeavoured to retain the original intent whilst at the same time covering in some detail new cases, of which there are many, reflecting current approaches of the courts to various legalistic problems. For that reason, and recognising that many readers of this book may not have ready access to court judgments or the time or inclination to study them in depth, the extracts I have included from the more important judgments are sometimes lengthy but hopefully no more so than sufficient to explain their purpose.

I have been surprised at the large amount of new material. The flow of cases seems to have risen dramatically over the past ten years or so. One likely reason for this is that the right to adjudicate disputes has increased their volume with a corresponding increase of the follow-on work of the courts. Another likely reason is the present high standing of the Technology and Construction Court.

One change to the book I have been obliged to make has been to abandon the earlier inclusion of provisions for liquidated damages and extensions of time from all well-used standard forms of construction, process and plant contracts. Such has been the multiplicity of new and amended forms in recent years that all that can be done now is to include a selection of the leading standard forms.

That change apart, the structure of the third edition is broadly the same as in previous editions. There is, however, a new chapter on delay analysis and the chapters covering penalty clauses, the effects of conditions precedent and time-bars, and the complexities of causation have been significantly expanded.

I am indebted to many friendly lawyers for bringing cases to my attention and for providing me with transcripts of the judgments but special mention must go to Mr Neil Kelly of MacRoberts for compiling three enormous volumes of cases without which I would have struggled to produce this third edition.

Brian Eggleston
July 2008

References

The following books are referred to in various parts of the text:

Anson – Anson's Law of Contract, 26th edition, A.G. Guest (ed.), The Clarendon Press, Oxford.

BLR – Building Law Reports, Volumes 1–79, H. Lloyd, N. Baatz, D. Streatfield-James, P. Fraser and R. Clay (eds), F.T. Law and Finance, London.

Emden – Emden's Construction Law, 8th edition, A.J. Anderson, S. Bickford-Smith, N.E. Palmer and R. Redmond-Cooper (eds), Butterworth, London.

Hudson – Hudson's Building and Civil Engineering Contracts, 11th edition, I.N. Duncan-Wallace (ed.), Sweet & Maxwell, London.

Keating – Keating on Construction Contracts, 8th edition, V. Ramsay and S. Furst (eds), Sweet & Maxwell, London.

Acknowledgements

The clauses in Chapters 17, 18, 19 and 20 are reproduced by kind permission of the copyright holders:

Standard Form of Building Contract
- The Joint Contracts Tribunal

ICE Conditions of Contract

ICE Conditions of Contract for Minor Works
- Institution of Civil Engineers/Association of Consulting Engineers/ Civil Engineering Contractors Association

The New Engineering Contract
- Institution of Civil Engineers

FIDIC Conditions of Contract
- Fédération Internationale des Ingénieurs-Conseils

CECA Form of Sub-Contract
- Civil Engineering Contractors Association

Model Form – MF/1
- Institution of Electrical Engineers

The IChemE Red Book
- Institution of Chemical Engineers

Chapter 1
Introduction

1.1 *General overview*

There is an old saying that time is money. It certainly is in the world of business and no more so than when a construction contract overruns its allotted time for completion. When that happens, both parties can expect to suffer. This book is essentially about how contracts deal with the financial consequences of late completion and how, within the scope of the law and the scope of the contractual provisions, the parties endeavour to protect their respective interests.

To those not deeply involved with such matters it might seem extraordinary that it requires a book of some magnitude to explore the full complexities of the subject. To explain this conundrum this first section of the book provides a general overview of the law relating to liquidated damages and extensions of time. This review was, for the most part, the opening chapter of previous editions of the book.

Breaches of contract

Every breach of contract carries with it the potential for dispute. There may be those who thrive on dispute but they rarely include the parties to the contract. Not surprisingly it has long been accepted as good commercial practice for the parties to include in their contracts provisions for dealing with the most likely breaches. This is how standard forms of contract and the use of liquidated damages began to develop.

In the construction industry, breaches of contract are commonplace to the point of being routine. Did any employer ever wholly avoid impeding the contractor in the performance of his obligations and did any contractor ever wholly fulfil his obligations without fault? Not often. This is reflected in the standard forms and most contain clauses detailing the procedures to be applied and the recovery permitted in the event of those specified breaches identified and described with the benefit of centuries of experience.

When the employer is in breach by way of interference or prevention arising from late supply of information, failure to give full possession of the site and the like, the result for the contractor is delay, disruption and involvement in loss and expense or extra cost. The contractual remedy gives the contractor recovery of his provable loss and expense or extra cost and, in appropriate circumstances, an extension of time for completion. In some contracts certain breaches by the employer, such as failure to make payment on an interim certificate, entitle the contractor to determine his employment under the contract but such remedies are few and as a general rule the contractor's remedy for employer's breach is the recovery of general or unliquidated damages. That is to say, damages which are assessed after the breach.

The contractor's breaches of contract are most commonly failure to proceed with due diligence, failure to meet specified standards and failure to complete on time. Only in respect of the last does the employer have a solely financial contractual remedy. For other breaches he may have the right to terminate the contractor's employment or order reconstruction but he will rarely have an entitlement to deduct moneys from sums due to the contractor. The employer may, of course, sue for latent or patent damage but this is a common law remedy rather than a contractual one.

The employer's position is, therefore, significantly different from the contractor's. Whereas the contractor has a financial remedy for numerous and various breaches, the employer has his for only one breach of common occurrence – failure by the contractor to complete on time. And whereas the financial effects of the employer's breach on the contractor can rarely be estimated in advance of the breach, not least because of the involvement of sub-contractors, the financial effects of the contractor's late completion can usually be estimated with some certainty.

Consequently most standard forms of construction contract are drafted to permit the parties to fix in advance the damages payable for late completion. When these damages are a genuine pre-estimate of the loss likely to be suffered or a lesser sum they can rightly be termed liquidated damages.

In short, liquidated damages are fixed in advance of the breach, whereas general or unliquidated damages are proven damages assessed after the breach.

The practice of liquidating damages is by no means of recent origin. The *Shorter Oxford English Dictionary* gives 1574 as the earliest known date for 'ascertained and fixed in amount' as the meaning of the word 'liquidated'. And from the courts there are numerous law reports dating back to the early 19th century of cases concerned with liquidated damages in construction contracts.

Indeed this is a subject on which Kipling's famous lines seem to be particularly relevant:

'How very little since things were made
Things have altered in the building trade.'

Hudson's Building and Engineering Contracts gives the 1838 case of *Holme v. Guppy* in which carpentry contractors at a brewery finished late and sought relief from deduction of liquidated damages from the contract price on the grounds that the employer had prevented them from finishing on time by delay in giving possession of the site and by delays on the part of his own workmen. A familiar story which could have come from any modern day contract. And, as it happened, nearly a century and a half later, Lord

Justice Salmon referred to *Holme* v. *Guppy* in his judgment in *Peak* v. *McKinney* (1970) when saying:

> 'I cannot see how, in the ordinary course, the employer can insist on compliance with a condition if it is partly his own fault that it cannot be fulfilled.'

Peak v. *McKinney* is the modern authority on prevention – a subject to be considered in some detail in later chapters.

Liquidated damages and penalties

The association between liquidated damages and penalties lies in the nature of the remedy – an agreed price to be paid for breach or non-performance. The parties may agree any price they wish; they are not bound by any rules and if the price they agree is clearly intended to penalise the defaulting party rather than to compensate and restore the position of the innocent party, that is a matter for the parties. When Antonio in Shakespeare's *The Merchant of Venice* agreed to give a pound of his flesh if he defaulted on his bond he did so willingly – albeit in the mistaken belief that his default would never occur.

The question which is of prime importance to the parties in agreeing their price is, will the courts assist them in enforcing payment? In *The Merchant of Venice* Portia could find nothing in Venetian law to prevent the application of the penalty and Antonio was saved only by the impossible precision of the penalty – a pound of flesh, no more nor less; and not a drop of blood to be included.

In fact many legal systems do allow the recovery of penalties and it is something of a peculiarity of English law that the courts will look at the price irrespective of whether it is called liquidated damages or a penalty, and, if it is found to be a penalty, will limit damages to the amount flowing from the breach.

The origin of this lies in the branch of justice named equity, which traditionally relieved against penalties but for the last two centuries the doctrine has been taken up and applied by the common law. The logic of the position seems to be that since a penalty is designed to secure performance, the promisee is sufficiently compensated by recovery of his actual loss and he is not entitled to demand a sum which although fixed by agreement is disproportionate to the actual loss suffered.

An early example of the thinking of the courts is *Kemble* v. *Farren* (1829) where it was held that a sum of £1000 to be paid for any breach, and said by the parties to be liquidated and ascertained damages and not a penalty, was held nonetheless by the court to be a penalty.

However, the courts show a sensible reluctance to go too far in interfering in the commercial bargains struck by the parties. In the case of *Elsley* v. *J.G. Collins Insurance Agencies* (1978) Judge Dickson, delivering the judgment of the Supreme Court of Canada, said:

'It is now evident that the power to strike down a penalty clause is blatant interference with freedom of contract and is designed for the sole purpose of providing relief against oppression for the party having to pay the stipulated sum. It has no place where there is no oppression.'

Later thinking of the courts was given by Lord Roskill in the case of *Export Credits Guarantee Dept* v. *Universal Oil Products Company* (1983), where he said:

'My Lords, one purpose, perhaps the main purpose, of the law relating to penalty clauses is to prevent a plaintiff recovering a sum of money in respect of a breach of contract committed by a defendant which bears little or no relationship to the loss actually suffered by the plaintiff as a result of the breach by the defendant. But it is not and never has been for the courts to relieve a party from the consequences of what may in the event prove to be an onerous or possibly even a commercially imprudent bargain.'

And in *The Angelic Star* (1987) Lord Justice Gibson said that the doctrine of penalties was not a rule of illegality. It was a rule of public policy by which the courts refused to sanction legal proceedings for recovery of a penalty. The rule was not designed to strike down any more of a lawful contract than was necessary to apply public policy. It should interfere as little as possible with proper enforcement of a lawful contract.

The case of *Kemble* v. *Farren* was one of many considered by Lord Dunedin in his judgment in *Dunlop Pneumatic Tyre Company Ltd* v. *New Garage* (1915) which stands to this day as providing the principal tests for distinguishing liquidated damages from penalties.

Because of its continuing importance the case is examined in detail in Chapter 4 but the following short extract from Lord Dunedin's judgment is given here to sum up the point:

'The essence of a penalty is a payment of money stipulated as "in terrorem" of the offending party: the essence of liquidated damages is a genuine covenanted pre-estimate of damage.'

Contentions that liquidated damages are, in law, penalties rank highly amongst the defences put up to avoid payment of liquidated damages. This is not so much that the stipulated sums are patently extravagant and evidently intended as threats but more because of the ingenuity of lawyers in making arguable cases from discrepancies and oddities in contract documents.

The subject is undoubtedly complex and it offers an excuse perhaps for the common misconception that damage must be suffered before liquidated damages become payable. Usually this line of thought applies to public sector projects or non-commercial buildings such as churches. However, those who harbour such thoughts, mostly contractors it must be said, eventually learn to their dismay that the test for enforcement of liquidated damages is: were they a genuine pre-estimate of loss at the time the contract was made? If it is not, can loss be proved after the breach? Indeed as later chapters will reveal, providing the liquidated damages are a genuine pre-estimate or a lesser sum, not only need there be no proof of loss, there need be no loss at all for the damages to become enforceable.

Since the first edition of this book was published in 1992 the law on liquidated damages and penalties has been examined and clarified by the English and Commonwealth courts in a surprising number of cases. The more noteworthy are considered in later chapters.

Amongst the most important of these cases is that of *Philips Hong Kong Ltd* v. *Attorney General of Hong Kong* which travelled through three tiers of the courts in 1990, 1991 and 1993. The concluding decision of the Judicial Committee of the Privy Council given in 1993 provides the basis of the current approach of the courts on the question of when in law liquidated damages are to be regarded as penalties.

The decision of the Privy Council restated the principles that the courts should not adopt an approach to provisions for liquidated damages which could defeat their purpose and that the test for determining whether a provision for liquidated damages is a penalty is whether or not it is a genuine pre-estimate of loss.

Genuine pre-estimate of loss

The relationship between a pre-estimate of loss and liquidated damages raises some difficult questions, not least how can there be a genuine pre-estimate of loss for a non-commercial project? This was an argument put forward by the shipbuilders in *Clydebank Engineering Co.* v. *Yzquierdo y Castaneda* (1905) where the contract for the building of four warships provided that 'the penalty for late delivery shall be at the rate of £500 per week'. It was said by Clydebank in opposing an action to enforce the penalty clause that there can be no genuine pre-estimate of loss as a warship does not earn money. But Lord Halsbury refuted the argument and held the stated sum to be liquidated damages, establishing that difficulty in ascertainment is no barrier to an estimate being made.

The ruling has been followed by the courts on many occasions and not infrequently the point has been made that the very difficulty in ascertaining damages for late completion is a good reason why such damages should be liquidated. For example, Lord Dunedin in his judgment in the *Dunlop* case restated the point made by Lord Halsbury in the *Clydebank* case that:

> 'It is no obstacle to the sum stipulated being a genuine pre-estimate of damage, that the consequences of the breach are such as to make precise pre-estimation almost an impossibility. On the contrary, that is just the situation when it is probable that pre-estimated damage was the true bargain between the parties.'

There is no bar therefore to a genuine pre-estimate of loss in non-commercial projects and various methods and formulae have been devised for application in the construction industry which meet satisfactorily the test of a genuine pre-estimate.

If one party wishes to challenge the sum so calculated and stated in the contract as liquidated damages, the best time to do so is before the contract is signed. Post contract challenges to liquidated damages on the grounds that they are not a genuine pre-estimate of loss have had a poor record of success in the courts in the past and are likely to have even less in the future.

Agreed nature of damages

There is a view that too much attention is sometimes given to the 'pre-estimate' aspect of liquidated damages and not enough to the agreed nature of such damages. This is not a plea for the enforcement of penalties but a plea for greater recognition of the fact that liquidated damages are frequently not a genuine pre-estimate of loss but are simply a sum agreed by the parties as part of their commercial bargain. In short, the parties, using their commercial judgments, agree a sum which serves as compensation for the employer and limitation of liability for the contractor. Where that sum is less than a genuine pre-estimate it will not be in law a penalty – although that has not always been without doubt – but this does raise the question, who can say whether or not a sum is more or less than a genuine pre-estimate if there was no pre-estimate to start with? The implications of this are considered later.

Exhaustive remedy

There is sometimes the question of whether or not liquidated damages provide an exhaustive and exclusive remedy.

At first sight there would seem to be no doubt whatsoever on this. Why introduce liquidated damages into a contract to give both parties the benefit of certainty of knowledge of the consequences of the relevant breach if the liquidated damages clause can be avoided and general or unliquidated damages can be claimed?

The Court of Appeal in the case of *Temloc Ltd* v. *Errill Properties Ltd* (1987), where liquidated damages had been stated as £nil, firmly supported the exhaustive remedy principle. Lord Justice Nourse said in the course of his judgment:

> 'I think it clear, both as a matter of construction and as one of common sense, that if . . . the parties complete the relevant parts of the appendix, . . . then that constitutes an exhaustive agreement as to the damages which are . . . payable by the contractor in the event of his failure to complete the works on time.'

Why, then, should there be a question of whether or not liquidated damages provide an exhaustive and exclusive remedy? Firstly because it is not uncommon for employers who find their actual losses to be greater than their liquidated damages to argue that they have retained, and are entitled to pursue, their common law rights to sue for the damages they can prove to have been incurred. And secondly because both Lord Justice Bingham and Lord Justice Parker in the Court of Appeal in *E. Turner & Sons* v. *Mathind Ltd* (1986) expressed forceful views, albeit obiter and therefore not binding authority, to the effect that the employer could have both liquidated damages for failure to complete the whole works on time and unliquidated damages for failure to meet phased handover dates. Furthermore a New Zealand Court in the case of *Baese* v. *Bracken* (1989) declined to follow the ruling in *Temloc* v. *Errill* on 'nil' damages and advanced an interesting argument on why, given certain forms of wording, a liquidated damages clause is an alternative to unliquidated damages. More will be said on both cases later but it will be a bad day for the construction industry if the certainty brought to contracts by liquidated damages is ever lost.

Extensions of time

Just as there are many misunderstandings on the purpose and principles of liquidated damages, there are many on extensions of time.

It is a common belief that liquidated damage provisions are solely for the benefit of the employer and extensions of time provisions solely for the benefit of the contractor. Both views are not only wrong but almost the reverse of true intentions. Liquidated damages provisions are beneficial to contractors for they not only limit the contractor's liability for late completion to the sums stipulated, but they also indicate to the contractor at the time of his tender the extent of his risk.

Thus, if a contractor believes that he cannot complete within the time allowed he can always build into his tender price his estimated liability for liquidated damages.

All that the employer gets out of liquidated damages is relief from the burden of proving his loss and usually, in construction contracts, the right to deduct liquidated damages from sums due to the contractor. To the extent that the employer's true losses may be greater than the stipulated level of liquidated damages he is disadvantaged by agreeing to a restrictive remedy. Indeed during the property booms of the 1980s many employer / developers preferred to enter contracts without liquidated damages because rental values were rising so quickly that liquidated damages would almost invariably understate true losses. But not surprisingly few contractors were prepared to operate under such terms since their liability for late completion was not only uncertain but potentially crippling.

Similarly with extensions of time provisions, the fact that the contractor is the obvious recipient of benefit in gaining relief from liquidated damages obscures the primary purpose of such provisions. That is, they preserve the contractor's obligation to complete within a specified time and in doing so they preserve the employer's right to liquidated damages when, by prevention, he has delayed the contractor and is responsible in part for late completion. That was the point at issue in *Holme* v. *Guppy* (1838) and *Peak* v. *McKinney* (1970) mentioned earlier.

The amount of detail in standard forms of contract on procedural rules for making extensions of time varies enormously. Building contracts tend to have elaborate schemes whereas process and plant contracts say very little. Much of the case law relating to the procedures for extending time concerns the alleged non-observance of particular rules and is of only limited assistance in setting general guidelines. That may account for the fact that there is a great deal of variability and unpredictability in awards of extensions of time. In later chapters of this book an attempt is made to derive from case law the basic rules which should be followed. Of particular note are a clutch of recent cases reported since the first edition of this book was published.

John Barker Construction Ltd v. *London Portman Hotel Ltd* (1996) is important because it emphasises the need for an analytical approach to the investigation of delay and the corresponding extension due. *Balfour Beatty Civil Engineering Ltd* v. *Docklands Light Railway Ltd* (1996) confirms the need for discretion to be exercised with fairness when extensions are being considered. And, *Balfour Beatty Building Ltd* v. *Chestermount Properties Ltd* (1993) confirms that extensions granted for entitlements to extensions arising after the completion date has passed should be on a 'nett' rather than a 'gross' basis.

Prevention

If the extension of time clause fails to cover the employer's fault then usually the right to liquidated damages is lost; the liquidated damage clause fails; and the employer is left to sue for general damages which must be proved. Lord Salmon in *Peak* made the following statement:

> 'If the failure to complete on time is due to the fault of both the employer and the contractor, in my view, the clause does not bite.'

Consequently, extension of time clauses are drafted to include the likely range of events for which the employer is responsible, although as will be seen later, not all are successful and there remain some remarkable gaps in well used building and civil engineering forms.

Most standard forms do, of course, also provide for extensions of time for a range of neutral events associated with bad weather, industrial disputes and the like and it is understandable that in respect of these matters the benefit should seem to be wholly for the contractor. On a narrow view it is, but on a broader view the inclusion of neutral events for extensions of time is part of the give and take of the consultative drafting process and the establishment of an acceptable balance of risk between the parties.

Relationship to claims

The purpose of extension of time provisions is further complicated and widely confused by the linkage in the industry of extensions of time and claims for loss and expense or extra cost. This is not a legal link, nor is it in most standard forms a contractual link, since the extension of time clauses and the financial claim clauses usually stand alone. But what has happened over the years is that contractors have developed the maxim: get time first and the money will follow, and contract administrators have also found extensions of time a useful peg on which to hang claims when justification for approval of payment has to be made to the employer or his auditors.

From this has grown the operational practice of separating extensions of time and events giving rise thereto into 'reimbursable' and 'non-reimbursable' categories; the first set covering employer's fault and the second set covering neutral events. 'Operational' is used in the sense that those seeking and those granting extensions of time play their cards to suit the circumstances of their situation, and though there may be no distinctions in the contract between reimbursable and non-reimbursable extensions and probably none in the correspondence exchanged on the applications, both parties are aware of the other's intentions.

The court in *Fairweather & Co. Ltd* v. *London Borough of Wandsworth* (1987) took a practical view of the relationship between claims and extensions of time when considering the problem of concurrent delays, notwithstanding the well established dominant event approach which takes a detached view of the relationship.

Although this book is not about claims, the importance of extensions of time in claim submissions has to be recognised and consequently attention is given to the subject in later chapters. In connection with claims the phrase 'loss and expense or extra cost' is used. This is not a new form of contractual entitlement. It is simply that this book is intended to cover both building and civil engineering contracts and it so happens that building contracts usually refer to 'loss and expense' and civil engineering contracts to 'extra cost'.

Additional time for the employer

Because the obligation to complete the works of a contract on time rests with the contractor and because the essential purpose of an extension of time clause is to maintain a fixed time for completion it is understandable that most extension of time clauses are drafted so as to be applicable only to extending the time for the contractor's obligations. To the extent that extensions of time are available for delays for which the employer is responsible, it could not be otherwise. However, it is a fact that most extension of time clauses permit extensions for delays caused by events beyond the control of the contractor as well as for acts of prevention by the employer. The effect of this, as a judge once put it, is that the loss lies where it falls. The contractor obtains relief from his liability for damages but has no claim for delay, whilst the employer by losing his right to damages for delays stands his own costs of the delay. This is generally taken to be a fair and reasonable approach to the problem of delays caused by neutral events.

However, what of the position where the employer is delayed by circumstances beyond his control in the performance of his obligations to the contractor? What relief, if any, is to be afforded to the employer against claims for delay by the contractor? In construction contracts the answer is usually none – unless there is exceptionally a force majeure clause which expressly applies to both parties. However, in process and plant contracts such a clause is normal and the IChemE Red Book even goes so far as to formalise the procedure for granting extensions of time to the employer.

Liquidated damages other than for delay

It is not intended in this book to examine in any detail the provisions which appear in some forms of contract for liquidated damages for breaches other than delay in completion. The most common are, of course, liquidated damages for low performance. These are standard in process and plant contracts.

However, the point is worth making that liquidated damages, whether for delay or for some other default in the contractor's performance, are, as is stated often in this book, an exhaustive and exclusive remedy for the particular breach. It is usually not too difficult for an employer to see that by opting for liquidated damages for delay he is forgoing his right to have his damages assessed under his common law remedy. However, when liquidated damages are applied to other matters such as low performance it is easier to make the mistake of thinking that the liquidated damages give an additional remedy and there is a danger of failing to recognise the exclusive and exhaustive nature of such damages.

Caution does need to be exercised, therefore, in adding into contracts liquidated damages clauses for matters other than delay.

1.2 Legal developments

As is evident from the large number of new cases considered in this third edition, the courts remain busy with matters concerning liquidated damages and extensions of time. It is difficult to confirm whether or not the number of judgments released in the last ten years on such matters exceeds the number in any previous ten-year period but it seems to be a possibility.

Some of the new cases have been generated by adjudications and many of the judgments on these relate to procedural and / or jurisdictional disputes which do little to advance the law on substantive matters. These are not covered in this book except to the extent that they concern rights on the deduction of liquidated damages.

Most of the new cases of interest can be categorised as relating to:

- penalty clauses
- prevention / conditions precedent / time at large

- causation and / or concurrency
- apportionment / global claims
- delay analysis.

An interesting aspect of the judgments in some of these cases is their length – with a few running into hundreds of pages. The amount of detail covered would, by past standards, be considered extraordinary. It suggests that a preference for litigation over arbitration may be developing.

Penalty clauses

It is not stating anything new to say that present law on penalties clauses is less than satisfactory. The English Law Commission published a working paper on the subject in 1975. The Scottish Law Commission produced a report and a draft Bill in 1999. They both address the underlying issue of whether the courts should have any powers to strike down as penalties certain types of commercial agreements such as liquidated damages. They both look in detail at the problems created by the legal distinction between penalties for breach and penalties for exercising contractual rights and at the problems caused by the 'genuine pre-estimate of loss' rule for liquidated damages.

The recommendations of the Scottish report make interesting reading:

'1. (1) There should continue to be judicial control over contractual penalties.
   (2) The criterion for the exercise of that control should be whether the penalty is "manifestly excessive".
   (3) Penalties which are not manifestly excessive should be enforceable even if they cannot be regarded as based on a genuine pre-estimate of loss.
2. Judicial control over contractual penalties should not be confined to cases where the penalty is due when the promisor is in breach of contract. It should extend to cases where the penalty is due if the promisor fails to perform, or to perform in a particular way, under a contract or when there is an early termination of a contract.
3. (1) Judicial control over contractual penalties should apply whatever form the penalty takes. It should, in particular, apply whether the penalty takes the form of a payment of money, a forfeiture of money, a transfer of property, or a forfeiture of property.
   (2) Without prejudice to the possibility of a systematic review of the law on irritancies of leases of land, the recommended judicial control over contractual penalties should not apply to such irritancies.
4. In deciding whether a clause comes within the scope of the new law on penalty clauses regard should be had to the substance of the clause rather than to its form.

5. The enforceability of a penalty should be judged according to all the circumstances, including circumstances arising since the contract was entered into.

6. A court, or a tribunal or arbiter adjudicating on a penalty clause, should have power to modify a manifestly excessive penalty so as to make it enforceable – by for example, reducing its amount or attaching conditions to the exercise of the relevant right.

7. In any new law on penalty clauses it should be made clear that parties cannot contract out of the application of that law.

8. The onus of showing that a penalty is unenforceable should lie on the party so contending.

9. The proposed rules on penalty clauses should apply to penalty clauses in bonds and other unilateral voluntary obligations in the same way as to penalty clauses in contracts.

10. Any new legislation should apply only to penalty clauses agreed after it comes into force.'

And it will certainly be interesting if the Scottish draft Bill incorporating the above recommendations comes into force – thereby moving Scots law closer to continental law but creating a significant legal gulf between Scotland and the rest of the United Kingdom.

Most of the new cases on penalty clauses covered in this book show the difficulties of the courts in maintaining compliance with the genuine pre-estimate of loss rule and in using the *Dunlop Tyre* case criteria to distinguish between liquidated damages and penalties. Some of the cases, *Cine, Jeancharm*, and *Leisureplay* have Court of Appeal rulings but perhaps the most influential judgment is that of Mr Justice Coleman in *Lordsvale Finance*. He addressed head-on the difficulties of applying old English rules on penalties in a modern financial world and in doing so illuminated a path for others to follow.

Prevention / conditions precedent / time at large

A debate has been going on for some years as to how to reconcile the principle of prevention with conditions precedent to entitlement to extension of time such as notice requirements and time-bars. The key issue is to what extent the legal rule that a party cannot benefit from its own breach operates in circumstances where the employer has prevented completion on time but the contractor has not complied with the contractual requirements for obtaining extension of time. From that comes the questions – can the employer claim liquidated damages for a delay he has caused or is time put at large by lack of entitlement to an extension?

One aspect of the debate is whether the principle of prevention is a rule of law or a rule of construction. Another is whether a distinction should be made when considering the effects of conditions precedent between preventive acts which amount to breach of contract and preventive acts such as the ordering of extra works which are permitted by the contract.

Until recently there was little guidance from the courts on these matters save for a batch of conflicting decisions from the Australian and United States courts. We now have judgments in the Scottish case of *City Inn* v. *Shepherd*, the English case of *Multiplex* v. *Honeywell*, and other following cases. All incline towards upholding the operation of conditions precedent and rejecting time at large claims.

However, it is unlikely that the debate is concluded. The contractual provisions in *City Inn* were somewhat unusual and *Multiplex* provides observations rather than final decisions.

Causation and / or concurrency

Disputes on construction contracts, particularly those relating to claims for delay, extensions of time and liquidated damages are frequently beset by arguments on what events have caused delays and on how the concurrency of their effects should be treated. Such are the complications that it seems to be recognised by the courts and legal authorities that there is no single approach which fits all situations. This leads *Keating* to suggest a number of propositions which might apply: the *Devlin* approach, the dominant cause approach, the burden of proof approach, and the tortious solution.

As might be expected therefore the guidance to be derived from recent construction industry cases is of limited general effect. Nevertheless some useful pointers to the current thinking of the courts on particular situations can be gained from cases such as *Plant Construction* (2000), *John Doyle* (2004), *Great Eastern Hotel* (2005), *City Inn* (2007) and others.

Apportionment / global claims

Ever since the 1967 judgment in the *Crosby* v. *Portland* case contractors have endeavoured to extend the permissible boundaries of global claims. Generally such claims are for money but in complex delay situations they are sometimes made for extensions of time.

One of the difficulties faced by contractors making global claims is that until recently such claims, under English law, only stood if it was possible to exclude the effects of causes other than those relied on for entitlement. Apportionment was not permitted. Understandably therefore the judgment in the Scottish case of *John Doyle* v. *Laing* (2004) created some excitement. In that case the judges of the Inner House (equivalent to the English Court of Appeal), having reviewed decisions in overseas cases, concluded that the facility to undertake apportionment exercises as carried out in the United States should be available under Scots law.

Approval to that approach has since been given in a number of English cases. However, as to whether that approval extends to time claims as well as to money claims remains in some doubt.

Delay analysis

Until comparatively recently the process of calculating amounts due to contractors and to sub-contractors as extensions of time was not regarded as an exact science. Providing that a fair and reasonable award of extension was given there was no great concern as to the method of calculation. The courts were rarely troubled with the details of calculations – that burden fell upon arbitrators.

But with the advent of computers and sophisticated logic-linked programs there came increased interest in the techniques of delay analysis. Even so it came as something of a shock when in 1996 the judge in *John Barker* v. *London Portman Hotel* rejected an architect's assessment on grounds that it was impressionistic rather than calculated and that there was no logical analysis. That was a wake-up call to many architects and engineers.

Since the *John Barker* case the courts have become increasingly involved in the details of delay analysis. See, for example:

- *Ascon* v. *McAlpine* (1999)
- *Royal Brompton Hospital* v. *Hammond* (2002)
- *Great Eastern Hotel* v. *Laing* (2005)
- *Skanska* v. *Egger* (2004)
- *Mirant* v. *Ove Arup* (2007)
- *London Underground* v. *Citylink* (2007)
- *City Inn* v. *Shepherd* (2007).

If any common message can be taken from these cases it is that some form of methodical delay analysis is essential but over-elaboration is no substitute for common sense.

In an attempt to improve understanding of delay analysis and to establish a measure of conformity in its practice the Society of Construction Law published in 2002 its 'Delay and Disruption Protocol'. It was hoped that its recommendations would be adopted in standard forms of contract but as yet there is little evidence of that happening.

1.3 Contractual developments

In 1964 the government-sponsored Banwell Report 'The placing and management of contracts for building and civil engineering works' recommended the joint production of a single form of contract for the construction industry. That never happened. Instead of combining their efforts, the various professional and other bodies producing standard forms expanded their outputs into families of forms thereby largely dashing hopes that the construction industry would move towards rationalisation of its conditions of contract. With that went hope of industry-wide standard provisions for extensions of time and liquidated damages.

The 1994 government sponsored Latham Report 'Constructing the Team', without going so far as to recommend integration of building and civil engineering forms, did recommend that public and private sector clients should begin to use the New Engineering Contract. This did happen but there is still a long way to go before it can be said to be the construction industry's standard form.

However, if the New Engineering Contract ever does achieve that role there will be, so far as provisions for extension of time and liquidated damages are concerned, a measure of irony in the situation. As things presently stand there is a degree of uniformity between the main body of building and civil engineering forms on such provisions despite the variety of their titles. In contrast the provisions in the New Engineering Contract are not only significantly different to those in common use but also they have not been tested in the courts.

Chapter 2
Time in contracts

2.1 Problems with terminology

Contractual requirements on time differ greatly in status and on the consequences which flow from their breach. Failure to meet times for performance may attract sanctions ranging from repudiation to damages, liquidated or unliquidated; failure to meet times for payment may result in determination or payment of interest; failure to give notices on time most commonly leads to a loss of entitlements; and failure to undertake administrative duties frequently attracts no sanction at all. Much depends upon the intentions of the parties, their conduct in connection with the contract and the particular terms and conditions of the contract.

A problem for the non-lawyer is that legal terminology provides little assistance in pointing to the consequences of breach. Thus in *Wickman Machine Tools* v. *Schuler* (1972) it was said:

> 'If a term is described as a "condition" there is a strong indication that the parties intended any breach, however small, to be repudiatory, but the description is not conclusive and yields to the discovery of the parties' intentions as disclosed by the contract read as a whole. Conversely the use of the word "warranty" to describe a term is not conclusive that that term is not a condition.'

Lord Denning, in the same case, suggested that the word 'condition' has three possible meanings and he concluded that the words 'it shall be a condition of this agreement,' used in connection with the number of visits to be made by a sales representative, had an ordinary meaning as a term of the contract and breach did not free the other party from its obligations.

This reveals another problem, that everyday language endows some legal phrases with far wider meaning than the law recognises. Although well-worn phrases such as 'time of the essence' and 'time at large' may safely be used in ordinary dialogue without undue concern for precision, the application of these phrases to contractual situations needs to be handled with care. There is always a danger that the assumed meaning of the phrases will be translated into action which is incompatible with, or at odds with contractual provisions. This is easily done. If there is any common thread in construction disputes it is that one or both parties misunderstands or misreads the factual position and the legal remedies. This failing extends to interpretation of the contract itself as well as to phrases imported into the contractual relationship by common usage.

The ultimate dispute on a construction contract is for an employer to assert that time is of the essence and to determine without paying whilst the contractor is claiming time to be at large and determining for non-payment. This may be extreme, but is not so far removed from the circumstances of two cases: *J M Hill & Sons Ltd* v. *London Borough of Camden* (1980) and *Lubenham Fidelities and Investments Co. Ltd* v. *South Pembrokeshire District Council* (1986), where the contractor in each case determined his own employment for alleged non-payment and the employer concurrently determined the contractor's employment for failure to proceed regularly and diligently. As the contractor in the latter case was to learn to his cost, what seemed to him, taking a common sense approach to the meaning of non-payment, an obvious breach of contract was in law no breach at all. This is a story repeated time and again on many matters throughout the law reports. See, as a recent example, the case of *Shawton Engineering Ltd* v. *DGP International Ltd* (2005) examined later in this chapter.

It is clear that the first place for an injured party to look for a description of his remedy in the event of breach must be in the terms of his contract, and if a standard form has been used he will frequently find the remedy expressed with clarity and certainty: for example, the remedy for late completion will be stated as liquidated damages. However, if the contract is less than clear on matters such as remedies for late completion, the safe course for the employer is to rely on an implied term that the contractor will complete within a reasonable time and to use his common law rights to sue for general and provable damages. The unsafe and often fatal course is to assume that conditions of contract are conditions in the legal sense and that the parties intend any breach, however small, to be repudiatory.

Before going on, therefore, to consider the meaning of the phrases 'time of the essence', 'time at large', and 'reasonable time', some thought needs to be given to the legal meaning of 'conditions' and to the extent to which repudiation, rescission, or determination, however expressed, can be a proper remedy for failure to meet the time requirements of a contract.

2.2 Conditions and warranties

There is a line of thought that concepts such as conditions and warranties are outmoded. It is beyond the scope of this book to enter too far into that particular debate or to offer a detailed analysis of this complex subject but it is necessary for practical reasons to find definitions which still carry, at least in the construction industry, a broad level of agreement on their application.

Anson's Law of Contract, in considering the traditional approach of the courts to the terms of a contract, describes conditions and warranties in this way:

> 'If the parties regarded the term as essential, it is a condition: any breach of a condition entitles the innocent party, if he so chooses, to treat himself as discharged from further performance of the contract. He can also claim

damages for any loss sustained by the fact that the contract has not been performed. If the parties did not regard the term as essential, but as subsidiary or collateral, it is a warranty; its failure gives rise to a claim for such damages as have been sustained by the breach of that particular term, but the innocent party is not entitled to treat himself as discharged. The classification of a term as being either a "condition" or "warranty" will therefore determine the legal remedies available to the innocent party in the event of its breach.'

These definitions, even if old fashioned, do offer some assistance in that construction contracts although frequently inappropriately called 'conditions of contract' apply the above principles in their express terms. Thus the contractor's remedy for the employer's acts of prevention is usually given as loss and expense or extra cost and the employer's remedy for the contractor's time and quality faults is similarly to be found in damages.

Where the parties do intend a condition of the contract to be a condition in law and not merely a warranty they will incorporate a determination or forfeiture clause in the contract setting out those matters which they regard as essential. In JCT contracts, for instance, the employer can determine the contractor's employment for various defaults – suspending the works without reasonable cause; failing to proceed regularly and diligently; refusing or neglecting to comply with the architect's instructions; failing to comply with nomination procedures – and for other issues involving bankruptcy or corruption. Similar provisions apply in ICE contracts.

JCT and ICE forms differ, however, in their approach to the contractor's right to determine his own employment. JCT gives the contractor rights to determine where there is: failure to pay on any certificate by a set time; interference with the issue of any certificate; suspension of work exceeding a specified period or bankruptcy or similar by the employer. ICE forms have no provisions giving the contractor rights to determine his own employment.

It may seem surprising that on such an important matter as payment of certificates the building and civil engineering forms should be so far apart, with JCT forms making failure to pay on time a breach of a condition and ICE forms making failure to pay on time a breach of a warranty and allowing only the payment of interest as a remedy.

This illustrates the difficulty in distinguishing between conditions and warranties, particularly where there are no express terms in the contract to settle the matter. It also points to an area of overlap. Clearly there must come a time when continuing or repeated breach of a warranty exhibits an intention no longer to be bound by the contract or an inability to continue. With regard to interim payments, for example, single non-payment would not give grounds for determination in the absence of an express term but continuing non-payment could do so at common law in appropriate circumstances.

This, perhaps, is where the modern approach of the courts in moving away from the rigid distinction between conditions and warranties towards a third category of 'intermediate' terms is more helpful. The need for such flexibility was shown by Lord Justice Diplock in *Hongkong Fir Shipping Co. Ltd* v. *Kawasaki Kisen Kaisha* (1962) when he said:

'There are, however, many contractual undertakings of a more complex character which cannot be categorised as being "conditions" or "warranties"' . . . Of such undertakings all that can be predicated is that some breaches will and others will not give rise to an event which will deprive the party not in default of substantially the whole benefit which it was intended he should obtain from the contract; and the legal consequences of such a breach of such undertaking, unless provided for expressly in the contract, depend upon the nature of the event to which the breach gives rise and do not follow automatically from a prior classification of the undertaking, as a "condition" or a "warranty".'

Anson says of 'intermediate' terms:

'A term is most likely to be classified as "intermediate" if, as in the *Hongkong Fir* case, it is capable of being broken either in a manner that is trivial and capable of remedy by an award of damages or in a way that is so fundamental as to undermine the whole contract.'

The danger for the practical man is that flexibility in the law introduces uncertainty into the decision making process.

Fundamental breach

When faced with problems on a contract and the need to decide on a course of action because none is clearly defined in the contract, there is sometimes talk of a 'fundamental breach'. This is a dangerous phrase which attracts a number of meanings.

The principle of fundamental terms and fundamental breach developed to provide some relief against carefully drafted exemption clauses. *Anson* explains it as follows:

'There were, it was said in every contract certain terms which were fundamental, the breach of which amounted to a complete non-performance of the contract. A fundamental term was conceived to be something more basic than a warranty or even a condition. It formed the "core" of the contract, and therefore could not be affected by any exemption clause.'

Most legal cases on fundamental terms related to the sale of goods – the wrong goods altogether, thus peas instead of beans; goods of the wrong specification; or goods without legal title. In construction, the wrong building; or the right building in the wrong place; or a building in the wrong materials might similarly have qualified.

On the sale of goods, the Unfair Contract Terms Act 1977 largely obviated the need for the principle of fundamental breach. On a wider front it was said by Lord Justice Diplock in *Photo Production Ltd* v. *Securicor Transport Ltd*

(1980) that if the expression 'fundamental breach' was to be retained it should, in the interests of clarity, be confined to the ordinary case of breach where the consequences entitle the innocent party to elect to put an end to all remaining unfulfilled primary obligations of both parties.

The use of the phrase 'fundamental breach' to describe a breach of condition or an intermediate term is best avoided by all but lawyers.

Application to time

The relationships of these various matters, conditions, warranties and intermediate breaches, to the time requirements of construction contracts are therefore diverse and to some extent unpredictable. The express terms are not everything; implied terms are far from certain; and the facts of each case and the conduct of the parties have an important bearing on the outcome of any dispute.

In cases relating to time for completion, as distinct from time for payment or other issues, the question of whether the relevant terms of the contract are conditions or warranties can take on immense importance, particularly when there is doubt as to whether or not completion can ever be achieved within a satisfactory time or at all. In some such cases termination of the contract may be the only sensible course of action, notwithstanding other express contractual remedies. But this is very much a matter where the advice of lawyers is essential.

2.3 Termination

Termination of a contract can occur in various ways – by agreement between the parties; by novation; by acceptance of repudiation; or by application of the determination provisions of the contract.

Novation

Novation is a tripartite agreement whereby a contract is rescinded in consideration of a new contract being entered into, on the same or similar terms as the old contract, by one of the original parties and a third party. This frequently occurs when one of the original parties changes its legal status or goes into receivership.

Repudiation

Repudiation is an act or omission by one party which indicates that he does not intend to fulfil his obligations under the contract. In construction a builder who abandoned the site or an employer who refused to give possession of the site would be obvious examples. However, repudiation by one party does not in itself terminate the contract. It requires the other party to accept the repudiation.

It is theoretically possible for the innocent party to refuse to accept the repudiation and to press for specific performance of contractual obligations, but in construction contracts where there are no practical means of enforcement this would be exceptional.

Determination

The most common method of termination of construction contracts is by application of the determination provisions of the contract. Insolvency of the contractor accounts for the majority of cases but failure by the contractor to proceed with due diligence, and failure by the employer to pay on time, are not uncommon.

Not all construction contracts contain determination clauses – ICE Minor Works is one that does not – and the clauses of those that do differ widely. As mentioned earlier, JCT contracts generally give rights of determination to both the employer and the contractor, whereas ICE contracts generally give right only to the employer. It is therefore necessary for the innocent party where he has no express contractual rights of determination to establish repudiation by the other party or breach of a legal condition if he wishes to terminate the contract at common law.

The point is sometimes made in respect of determination under contractual provisions that this is not determination of the contract but only determination of the contractor's employment under the contract. This is a fine legal point of some interest to academics but it does not appear to have any practical effect on the outcome of proceedings.

Common law rights

Where there are no contractual provisions for determination and the innocent party relies wholly on his common law rights, the grounds for determination will not necessarily correspond with those in standard forms of construction contracts. Thus it would be hard to establish at common law that failure by the contractor to proceed with regular and due diligence was evidence of either repudiation or breach of a condition. The case of *GLC* v. *Cleveland Bridge & Engineering Ltd* (1986) shows the difficulty. The case itself did not concern determination but whether a term requiring the contractor to proceed with due diligence should be implied in connection with his entitlement to payments under a variation of price clause. One of the arguments for the GLC was that in the determination clause of the contract failure by the contractor to proceed with due diligence and expedition was ground for taking the work out of his hands. Notwithstanding this, the Court of Appeal refused to imply a term of due diligence into the contract since no such term was necessary to give it business efficacy. Moreover, the obligation on the contractor was to finish on time and he was entitled to proceed at his own pace in doing so.

Similarly it would be unlikely that an application for appointment of an administrative receiver – one of the grounds for determination under JCT contracts – would be accepted as a common law entitlement to determine since the purpose of the application would be an attempt to stay in business and not evidence of an intention to repudiate. But whilst the grounds for determination under the contract might be wider than for determination at common law, there can be disadvantages with determination under the contract in the need to follow precisely the procedures expressed in the determination provisions. This raises the question: when the procedures have not been followed sufficiently do the contractual provisions for determination exclude common law rights? The point came up in *Architectural Installation Services Ltd* v. *James Gibbons Windows Ltd* (1989) where a determination was held to have failed under the contractual provisions but to be rightful under common law rights which were not expressly excluded by the contract.

However, in a more recent case, *Lockland Builders Ltd* v. *Rickwood* (1995), the Court of Appeal took a different approach. Mr Rickwood, dissatisfied with the performance of Lockland, had excluded the firm from the site without following the determination provisions of the contract. It was held that where a contract provided machinery for determination of a dispute, and it was not expressed to be without prejudice to the parties' common law rights, then the contractual machinery and the common law rights could co-exist only when one party displayed a clear intention not to be bound by the contract.

Similarly in *Balfour Beatty Civil Engineering Ltd* v. *Docklands Light Railway Ltd* (1996) the Court of Appeal on the question of whether certificates should be challengeable in the courts took the view that it had no power to review decisions which were stated by the contract to be made by one of the parties.

In short, contrary to long-held belief that common law rights can only be excluded by express terms, the *Lockland* and the *Balfour Beatty* judgments suggest that common law rights can be excluded by implication if there is an alternative contractual machinery or remedy and there is no qualifying phrase such as 'without prejudice to other rights and remedies' applicable to or attached to such machinery or remedy.

The relevance of determination to time in contracts is that late completion, or the prospect of it, may be wholly unacceptable notwithstanding the remedies of damages, liquidated or unliquidated. There may be times when an employer can rightly consider determination and can rightly claim that time is of the essence.

2.4 *Time for performance*

Most construction contracts specify time for performance in achieving completion of the whole of the works and many have additional requirements for phased or sectional handovers.

Time may be fixed either by reference to specified dates or by reference to a construction period, but if the latter method is used it is essential that a precise completion date can be established. This means there must be an identifiable commencement date from which time runs and there must be no uncertainty on whether the construction period takes in or excludes holiday periods. These may seem obvious matters but it is extraordinary how often in construction industry disputes it is found that the intentions of the parties in respect of time have not been clearly expressed or have been misapplied.

Fixing time

Such troubles may arise in part from the varying approaches taken to time in the commonly used standard forms of contract. Building forms usually specify a date for completion in the appendix whereas civil engineering forms usually specify a time for completion, leaving the date for completion to be calculated from a date for commencement given by the engineer. But in both cases procedural variations are frequently introduced and sight can be lost of the objective, which is to establish precise dates for commencement and completion.

It is, for example, quite common to allow tenderers to give their own preferred times for completion or to allow them to offer an alternative to that specified in the tender documents. The practice is used more in building than in civil engineering although it has a great deal to recommend it whenever it is used. Firstly, it enables the tendering contractor to exercise his commercial and technical judgment in arriving at his best price; secondly, it gives the employer the opportunity to compare tenders on both price and time; and thirdly, if the contractor's own time for completion is used as the contractual time it eliminates the contentious and claims oriented business of shortened programmes – a subject considered in more detail in Chapter 15.

It is necessary, however, if the contractor is allowed to fix his own time that this time is linked in the contract documentation to either a start date or a completion date. Without one or the other there will be no firm date for completion.

Fixing the completion date

Similar problems in fixing the date for completion with certainty can arise when extensions of time are granted. Again, differences of approach in various standard forms of contract may be in part responsible. JCT contracts for instance require the architect to fix a new completion date but ICE contracts require the engineer to grant a period of time. Often in practice, however, architects grant periods of time and engineers fix new completion dates.

The danger in granting periods of time instead of fixing new dates is that uncertainty can be created as to whether such periods, particularly where they are expressed in days, cover working days only or include weekends and holidays. If the contractor has applied for an extension of ten working days and has been granted ten days, he may well assume that he has been granted a two-week extension of time, whereas his application may have been scaled down and ten days is the full amount including weekends. Unless and until a new date is fixed the misunderstanding may not come to light but if the application of liquidated damages becomes an issue the contractor may well feel that he has been misled.

Finishing early

On many construction projects the problem is not always one of late completion. Contractors strive to finish early and often do so to the embarrassment of their employers. It may be that the employer has no use for a building or engineering project before a particular date and has no wish to accept premature responsibilities of care and insurance. It may be that the employer has inflexibility in his funding arrangements and is unprepared for early payment. Generally, however, under most standard forms such matters are of no concern to the contractor and he is entitled to finish early if he can. Under JCT contracts the contractor's obligation is generally to complete 'on or before the Completion Date' and under ICE contracts the obligation is generally to complete 'within the time prescribed'.

Obligation to proceed 'regularly and diligently'

In *West Faulkner Associates* v. *London Borough of Newham* (1994) the Court of Appeal had to consider the obligation imposed on a contractor by a clause requiring the contractor to proceed 'regularly and diligently'. The court held that literal interpretation, commercial logic and common sense required the contractor to proceed both regularly and diligently and he could be dismissed from the site if he failed to do either. For further comment on similar obligations such as requirement to use best endeavours see Chapter 15.

The last hour

When it does come to a close finish contractors can be relieved to know that they have until the last hour of the last day to complete their work – a ruling established in the very old case of *Startup* v. *McDonald* (1843).

When no time for performance is specified the contractor has a reasonable time in which to complete the work. The question of what is reasonable is considered later in this chapter.

2.5 *Time of the essence*

The phrase 'time of the essence' has the ordinary meaning that unless something is done quickly it will be done too late: too late to be effective in itself
or too late to facilitate or prevent some other endeavour. When used in a
contractual context the phrase takes on a more precise meaning. It is not
then a matter of completing as soon as possible; it is a matter of completing
by a specified date.

In short when time is of the essence in a contract failure to complete by
the specified date is a breach of a condition entitling the innocent party to
treat the contract as repudiated.

If the contract is a supply contract and the goods are offered late, acceptance of the goods can be refused. If the contract is a construction contract
and the contractor fails to finish on time, the employer is entitled to dismiss
the contractor from the site and has no liability for payment for the unfinished work.

Clearly this is not the usual position in a construction contract. Finishing
late does not normally entitle the employer to dismiss the contractor from
the site; it is a breach of warranty and damages are the employer's remedy.
Nor does finishing late normally excuse the employer from payment for
unfinished work; even where determination is made under contractual provisions or at common law the employer must pay for any benefit he has
received.

The question then is: what governs whether or not time is of the essence
in contracts? Is it the use of the phrase in the contract; is it the specification
of fixed time; or must other circumstances be taken into account? To answer
this it is necessary to examine the courts' approach to time.

Rules of equity

Common law originally held that when time for performance was specified
then time was of the essence. Equity took a different view and inquired
whether by fixing time the parties intended anything more than to secure
performance within a reasonable time.

However, there were three situations where equity was of no assistance:

(i)   where the contract expressly stated time to be of the essence;
(ii) where time, not originally of the essence, was made so by one party
      giving reasonable notice to the other;
(iii) when from the nature of the contract or its subject matter time must
      obviously be intended to be of the essence.

Rules of common law

The rules of equity are now also the rules of common law, and the general
rules as set out in Halsbury's Laws of England, and approved by the House

of Lords in *United Scientific Holdings Ltd* v. *Burnley Council* (1977), are that time will not be considered of the essence unless:

(i)   the parties expressly stipulate that conditions as to time must be strictly complied with;

or

(ii) the nature of the subject matter of the contract or the surrounding circumstances show that time should be considered to be of the essence;

or

(iii) a party who has been subjected to unreasonable delay gives notice to the party in default making time of the essence.

Note, however, that under this ruling the stipulation or statement in a contract that time is of the essence is not in itself sufficient for the law to hold time to be of the essence.

Cases in contract

It is not difficult to envisage contracts for the sale of goods or the provision of services where these rules will apply and time will be of the essence by the very nature and circumstances of the purchaser's requirements.

Thus in *Rickards (Charles) Ltd* v. *Oppenheim* (1950) a Rolls Royce car was to be delivered 'at the most' within six or seven months. When the car was not delivered on time the purchaser did not cancel the contract but continued to press for delivery. After three months the purchaser wrote saying that unless the car was delivered within a month in time for him to take it abroad on holiday he would have to buy another car. Again the car was not delivered, the purchaser made other arrangements and when three months or so later the original car was completed it was not accepted. The Court of Appeal found in favour of the purchaser, holding that the original order made time of the essence and although this had been waived by the purchaser after the first failure of delivery, and the supplier's obligation then became to complete within a reasonable time, the final written notice reinstated time of the essence and that the supplier had failed to comply.

Regarding the service of notices making time of the essence Lord Justice Denning said:

'If the defendant, as he did, led the plaintiff to believe that he would not insist on the stipulation as to time and that if they carried out the work, he would accept it, and they did it, he could not afterwards set up the stipulation as to time against them. Whether it be called waiver or forbearance on his part or an agreed variation or substituted performance does not matter. It is a kind of estoppel. By his conduct he evinced an intention to affect their legal relations. He made in effect a promise not to insist upon his strict legal rights. That promise was intended to be acted upon and was in fact acted upon. He cannot afterwards go back on it.

However,

'It would be most unreasonable if the defendant having been lenient and waived the initial expressed time, should, by so doing, have prevented himself from ever thereafter insisting on reasonably quick delivery. In my judgment, he was entitled to give a reasonable notice making time of the essence of the matter.'

Cases in construction contracts

The application of the rules on time of the essence to construction contracts shows how unusual it is for time to be of the essence in such contracts. It is possible to envisage circumstances when late completion of the contract works would render them valueless to the employer: for example, a marquee for a wedding; a stand for a show; a car park for a festival, and in such circumstances time could properly be of the essence. But more commonly, the inclusion of extension of time provisions; the express remedy of liquidated damages; and the value to the employer of goods fixed on his land make the proposition that time is of the essence in a construction contract difficult to sustain.

The courts have historically been reluctant to hold time to be of the essence in construction contracts. Thus in *Lucas* v. *Godwin* (1837) it was said:

'It never could have been the understanding of the parties that if the house were not done by the precise day the plaintiff would have no remuneration; at all events if so unreasonable an engagement had been entered into the parties should have expressed their meaning with a precision which could not be mistaken.'

In *Lamprell* v. *Billericay Union* (1849) it was stated that:

'We are of the opinion that time for completion was not an essential part of the contract; first because there is an expressed provision made for a weekly sum to be paid for every week during which the work should be delayed . . . and secondly, because the deed clearly meant to exempt the plaintiff from the obligation . . . should he be prevented by fire or other circumstances satisfactory to the architect.'

In *Felton* v. *Wharrie* (1906) a demolition contractor failed to clear the site by the specified date and when asked when he would do so declined to say. The employer, without notice and without express right under the contract, took possession and put in another contractor. It was held that notwithstanding the contractor's answer there was no evidence of any repudiation on the part of the contractor entitling the employer to determine.

This case illustrates not only the difficulty of claiming time to be of the essence but also the danger mentioned earlier of termination on an assumption that late completion is repudiation.

As to the actual incorporation of the phrase 'time of the essence' in construction contracts, this is something of a rarity.

Peak v. McKinney (1970)

The phrase time of the essence was used in the case of *Peak Construction (Liverpool) Ltd* v. *McKinney Foundations Ltd* (1970) in a clause which read:

> 'Time shall be considered as of the essence of the contract on the part of the contractor, and in case the contractor shall fail in the due performance ... shall be liable to pay the corporation, as and for liquidated damages, the sum of ...'

There then followed an extension of time clause. This was the case in which Lord Justice Salmon said:

> 'The form of this contract has been much criticised during the course of the argument – and not without justification. Indeed if a prize were to be offered for the form of a building contract which contained the most one-sided, obscurely and ineptly drafted clauses in the United Kingdom, the claim of this contract could hardly be ignored even if the RIBA form of contract was among the competitors.'

When considering later in his judgment the significance of the phrase 'time shall be considered as of the essence', Lord Justice Salmon said:

> 'No doubt this gave the corporation the right to determine the contract at the end of the 24 months period as extended by the architect. Had they done so, some other contractors might have been called in to complete the work, or the plaintiffs might have completed it on freshly negotiated terms. But the corporation did not determine the contract. They elected to leave the plaintiffs to complete the work. ...'

This would seem to suggest that the corporation had as its option alternative remedies – determination or liquidated damages – but the point was not considered by the other members of the Court of Appeal.

McAlpine Humberoak (1992)

The phrase time of the essence was also found in *McAlpine Humberoak Ltd* v. *McDermott International Inc.* (1992), a case concerning the construction of part of the deck structure for an off-shore drilling rig. Clause 2 of the contract read:

> '2. Commencement and Completion
> Time is of the essence of this contract. Contractor shall commence the work after receipt of notice from McDermott and shall complete the work in accordance with the dates set out in Exhibit B SC 5.'

Disputes arose on the effects of variations on contract price which was on a lump sum basis. The judge at first instance held that the contract had been frustrated saying, amongst other things:

> 'Time went out of the window of this contract with the first two issues of additional drawings in December 1981. The effect of its departure and the introduction of the additional drawings was to put paid to the lump sum constituent of the contract as well. From that time, time was at large.'

The Court of Appeal soundly rejected the judge's decision. In examining the proposition put forward for the contractor that 'Since time was of the essence of the contract, and since the [employer] had no power to fix a new completion date, time became at large', Lord Justice Lloyd said this:

> 'It is worth pausing here to consider the consequences of Mr Thomas's argument, if it is correct. In its extreme form it comes to this. If, in a contract which provides for a lump sum price and a firm delivery date, the employer causes the contractor to miss the delivery date by one day, as he might, for example, by ordering extra work, both the lump sum and the delivery date are displaced. Otherwise the contract remains intact. So the contractor can take as long as he likes, provided only he is not guilty of culpable delay, and can at the end recalculate his price based on the time actually taken.
>
> The only authority cited in support of this novel doctrine of quasi-frustration is *Wells* v. *Army & Navy Co-operative Society* (1902) 86 LT 764. In that case there was a contractual date for completion, a provision enabling the employer to extend the time for completion in certain defined circumstances, and a liquidated damages clause. The contractor was fifteen months late in completing the contract. The employer purported to extend the completion date by three months, and then claimed liquidated damages for the remainder. It was held that the extension clause did not apply, and that since the employer had contributed to the delay, thereby preventing the contractor from completing by the contractual completion date, he could not rely on the liquidated damages clause.
>
> The principle enunciated in *Wells* v. *Army & Navy Co-operative Society* was not new. It is as old as *Holme* v. *Guppy* (1831) 3 M & W 387, where Baron Parke first used the phrase, often since repeated, of the contractor being "left at large". In recent times the principle has been applied in such cases as *Peak Construction (Liverpool) Ltd* v. *McKinney Foundations Ltd* (1970) 1 BLR 114, *The Cape Hatteras* [1982] 1 Lloyd's Rep 518 and *SMK Cabinets* v. *Hili Modern Electrics Pty Ltd* [1984] VR 391. In all these cases the employer was claiming liquidated damages. In all of them it was held that the claim for liquidated damages must fail since the employer could not rely on the original date of completion, nor on a power to extend the date of completion. In the absence of such a power, there could be no fixed date from which the liquidated damages could run.
>
> In the present case the defendants are not seeking liquidated damages, since there is no liquidated damages clause. So the line of cases has no

direct application. It is true that the defendants have a modest counter-claim for unliquidated damages. We will discuss what effect, if any, *Wells v. Army & Navy Co-operative Society* has on a claim for unliquidated damages when we come to discuss the counterclaim. But one thing is quite clear. The principle on which Mr Thomas relies cannot possibly help him establish his claim. Even if time is "at large" (whatever that may mean) there is nothing in the quoted line of authorities to suggest that the price is at large.'

Time of the essence generally in construction contracts

Perhaps all that can be said of time of the essence generally is that time will not normally be of the essence in a construction contract which contains extension of time and liquidated damages provisions. However, if the contract additionally contains an express statement that time is of the essence, the employer may have a stronger defence to any legal challenge on action taken arising from determination of the contract.

In any event, there are practical problems for the employer in treating time to be of the essence in a construction contract, since the right to terminate does not arise until the completion date, whilst the obligation to make interim payments continues up to that date notwithstanding the employer's eventual likely loss.

2.6 Notice making time of the essence

The realisation by one party that time is of the essence, or needs to be made of the essence, is not always evident at the start of a contract. Frustration or concern is likely to develop progressively as completion recedes to an uncertain date in the distant future.

The remedy of liquidated damages may seem inadequate, whilst the course of determining the contract under express provisions or at common law may seem too risky.

At what stage is it possible for one party to say enough is enough and, by notice, make time of the essence in the contract?

In *Rickards v. Oppenheim* (1950), mentioned earlier in this chapter, the exasperated car purchaser succeeding in reinstating time of the essence in a contract where by his previous conduct the provision had been waived. But what of the situation where time was not originally of the essence? Can one party by notice unilaterally change the terms of the contract and make time of the essence?

This is what Lord Simon had to say in *United Scientific Holdings Ltd* v. *Burnley Council* (1977):

'The notice operates as evidence that the promisee considers that a reasonable time for performance has elapsed by the date of the notice and as evidence of the date by which the promisee now considers it reasonable for the contractual obligation to be performed. The promisor is put on notice of these matters. It is only in this sense that time is made of the essence of a contract in which it was previously non-essential. The promisee is really saying unless you perform by such and such a date I shall treat your failure as repudiation of the contract. To say that time can be made of the essence of a contract by notice except in the limited sense alone would be to permit one party to the contract unilaterally by notice to introduce a new term into it.'

In *Felton* v. *Wharrie* (1906) where the employer expelled the contractor from the site without notice, it was said:

'If he were going to act upon the plaintiff's conduct as being evidence of his not going on he ought to have told him of it, and to have said "I treat that as a refusal".'

Notice making time of the essence can therefore be effective but it operates with and not against the principle of a reasonable time for completion.

Contractual provisions

Shawton Engineering v. *DGP*

Some of the difficulties of making time of the essence were considered by the Court of Appeal in the case of *Shawton Engineering Ltd* v. *DGP International Ltd* (2005). The case concerned the circumstances in which a contracting party may lawfully terminate the contract for delay in performance by the other party when that party's obligation is to complete its work within a reasonable time.

DGP was employed on a sub-subcontract basis to produce designs for a design and manufacture subcontract undertaken by Shawton at British Nuclear Fuels Sellafield plant. Variations were issued but there was no provision in the sub-contract for extending time. It was accepted that the original contractual date for completion was lost and that DGP's obligation was to complete within a reasonable time. Shawton was not satisfied with DGP's performance and terminated the sub-subcontract contending that DGP was in breach of its obligation to complete within a reasonable time and, in any event it had made time of the essence.

The Court of Appeal upheld the findings of the judge at first instance that Shawton had not made time of the essence and that no breach of DGP's obligation to complete within a reasonable time had been established. Lord Justice May said:

'71. Mr Thomas was unable to show us any evidence or correspondence earlier than 7th November 2000 to show that Shawton were complaining of delay by DGP. Letters from DGP of 19th May and 2nd

October 2000, to which Mr Thomas did refer, contain explanations by DGP relevant to their own progress. But he showed us nothing to the effect that Shawton rejected their explanations. There was nothing to gainsay the judge's findings to the effect that up to November 2000 Shawton were simply not insisting on early completion by DGP. The original completion dates, and, indeed, the original completion periods had ceased to be of any relevance. Shawton were, in the language of Denning LJ in *Rickards v Oppenheim*, not insisting on the stipulations as to time, nor were they insisting on any times or periods for completion. This circumstance, in my view, overlaid to extinction any question of calculating time periods by reference to the original dates for completion and the work content of variations. In the strange circumstances of this case, a reasonable time for completion was literally at large, in the sense of being undefined, until Shawton took steps, as they did, on 7th November 2000 to start negotiating for its better definition.

72. The judge was, in my judgment, accordingly right to hold that Shawton had not established what was a reasonable time for completion. He was right to hold that DGP were not in breach for delay on 7th November 2000. He was right to hold, as he implicitly did, that on 7th November 2000 the reasonable time for completion was to be assessed afresh, mainly with reference to the outstanding work content including variations. That was not solely or mainly because Shawton had issued variation instructions, but because, until 7th November 2000, Shawton had not insisted on completion by any particular date or within any particular period.

73. Since DGP were not in breach for delay on 7th November 2000, Shawton were not then able to give notice making time of the essence. Mr Thomas accepts this, if DGP were not in breach for delay. Shawton's appeal accordingly fails in so far as they contend that time was of the essence on 26th March 2001.

and

76. I have already considered and rejected the alternative submission that, even if time was not of the essence, DGP were in repudiatory breach of contract on 26th March 2001, such that Shawton were entitled to accept the repudiation by determining the contracts. The judge was right to hold that the main case here was the delay claim. There were other allegations, some of them insubstantial, but the main significant consequence of them was delay. I accept that, even if time is not of the essence, it is theoretically possible for a party to show that another party's delay is so profound as to be repudiatory. But what has to be shown is, not mere breach, but a breach of such gravity as to deprive the other party of substantially the whole benefit which it was the intention of the parties that they should obtain from the contract. Mr Thomas accepted that the judge correctly articulated the law – see paragraphs 165 and 169, 4th sentence.'

In short, the Court of Appeal ruling confirms that before a valid notice making time of the essence can be given in circumstances where the obligation is to complete within a reasonable time it must first be established that there has been breach of an obligation to complete within a reasonable time. The ruling also confirms that what was a reasonable time had to be judged at the time the question arose, in the light of all the relevant circumstances.

Contractual provision making time of the essence

Although it is not unusual to find the phrase 'time of the essence' in the non-standard parts of construction contracts, more often than not this is mere exhortation and it is not intended in its true legal sense. There is, however, one standard form of contract, MF/1, the model form conditions for electrical and mechanical plant, which does have a provision for notice to be given making time of the essence – although it does not actually use those words. By clause 34.2 of MF/1, if the contractor is in prolonged delay and the specified maximum amount of liquidated damages for late completion has been exhausted, then the purchaser (employer) can give notice requiring the contractor to complete within a stated reasonable time and can terminate if that is not achieved.

2.7 Time at large

The phrase 'time at large' is much loved by contractors. It has about it the ring of plenty; the suggestion that the contractor has as much time as he wants to finish the works.

This is not what it means.

Time becomes at large when the obligation to complete within the specified time for completion of a contract is lost. The obligation then becomes to complete within a reasonable time. The question of what is a reasonable time will be considered in the next section but it is most certainly not 'as and when the contractor sees fit'.

The circumstances of time becoming at large are usually where an act of prevention by the employer creates delay and that delay is not covered by an extension of time provision; and, to a lesser extent:

(i)   where there is no stated time or date for completion;
(ii) where there is lack of clarity in the provisions for extending time;
(iii) where the provisions for extension of time have not been properly administered, have been misapplied; or have not been utilised;
(iv) where there has been waiver of the original time requirements;
(v)   where there has been interference by the employer in the certifying process.

All of these matters will be considered in greater detail in later chapters (particularly Chapter 5), but what is generally at stake in the matter of whether or not time is at large is the employer's right to deduct liquidated damages for late completion. This right is lost completely if time becomes at large – the employer can still sue for general or unliquidated damages for late completion – but regard will then be had to the contractor's entitlement to a reasonable time. As noted in Section 2.5 above, the Court of Appeal in the *McAlpine Humberoak* case disposed of the proposition that if time becomes at large that has the effect of putting the contract price at large.

2.8 Reasonable time

The question of what is a reasonable time for completion is a matter of fact to be decided in the light of the circumstances of each case.

Guidance on this can be had from the House of Lords' ruling in the case of *Hick* v. *Raymond and Reid* (1893) where it was said that where the law implies a contract shall be performed within a reasonable time it has:

> 'invariably been held to mean that the party upon whom it is incumbent duly fulfils his obligations, notwithstanding protracted delay, so long as such delay is attributable to causes beyond his control and he has neither acted negligently nor unreasonably.'

This principle has been the foundation for many subsequent decisions. Thus, in *British Steel Corporation* v. *Cleveland Bridge and Engineering Co. Ltd* (1984) the judge said:

> 'It was common ground between the parties that the principles I had to apply in this connection were those stated by the House of Lords in *Hick* v. *Raymond & Reid, viz.* that the question what constituted a reasonable time had to be considered in relation to the circumstances which existed at the time when the contractual services were performed, but excluding circumstances which were under the control of the party performing those services. As I understand it, I have first to consider what would, in ordinary circumstances, be a reasonable time for the performance of the relevant services; and I have then to consider to what extent that time for performance . . . was in effect extended by extraordinary circumstances outside their control.'

There remains, however, some uncertainty as to whether the assessment of reasonable time should put the time actually taken under examination or whether the task is to build up a theoretical time allowance having regard to all the circumstances. The answer to this may depend upon whether the issue is whether or not the contractor is entitled to a reasonable time for completion and, if so, how much or whether the issue is whether or not the contractor has failed to complete within a reasonable time. The burden of proof rests on the asserting party and in practical terms that means that the contractor has to prove that the time taken was reasonable whereas the employer has to prove that the time taken was not reasonable.

No time for completion specified

The principles of reasonable time apply not only to contracts where specified time has been lost and reasonable time substituted, but also to contracts where no time for completion has been specified in the first place. Such contracts are very common and, surprisingly even in the construction industry, they may form the majority. This is because when the average householder employs a builder to fit new windows, erect a porch, or decorate the bedrooms, he pays most attention to the price and the written quotation. Beyond that the contractual details are frequently left open – including the time for performance.

In Machenair Ltd v. *Gill & Wilkinson Ltd* (2005), Mr Justice Jackson, in considering a sub-subcontract with sparse express terms said:

'In my judgment, on a proper construction of this sub-sub-contract, alternatively by implication, Gill's obligation was to complete the mechanical works within a reasonable time. In determining what constitutes a reasonable time it is necessary to have regard to the main contractor's programme, and also to all the other circumstances.'

Factors relevant to reasonable time

Where there is a formal contract and time is at large the defunct extension of time provisions may well serve as some guide as to what is reasonable time. Thus, extra works, exceptional weather, strikes etc., might all be taken into account. With or without a formal contract it might be appropriate to look at the production capability of the contractor, his management and financial resources, and his other contractual commitments – particularly if known to both parties.

Late instructed variations

One point of general interest is how late instructed variations affect the reasonable time for completion. This point came up in the *Shawton* case with Lord Justice May saying:

'69. I am not convinced that the judge was entirely correct in what he said about DGP's misapprehension of the work content, nor about the effect of Shawton instructing variations after the original completion dates. What is a reasonable time has to be judged as at the time when the question arises in the light of all relevant circumstances. One such circumstance was that DGP had originally agreed fixed time periods, although they did so upon a misapprehension as to the work content. It was a relevant factor that Shawton originally had the contractual benefit of these time periods, and that fact was not to be entirely ignored simply because DGP's obligation became to complete within a reasonable time. Equally, the true work content was a relevant circumstance. If these two factors had been the only relevant circumstances, judging what was a reasonable time may have presented something of a conundrum, since the two factors worked in opposite directions. But they were not the only relevant circumstances. The mere instructing of a (perhaps quite modest) variation after the original date for completion would not by itself necessarily mean that a reasonable time had to be assessed afresh by reference only to the variation and whatever work happened to remain at the date of the variation instruction – which is what the judge appears to say in the final sentences of paragraphs 101 and 108 of his judgment. Mr Thomas may well be right that a modest variation instruction given after an original completion date has passed could, depending on all the circumstances, result in an obligation to complete within a reasonable time whose assessment would produce a date which was in the past. But I accept Mr Friedman QC's submission that the question is a composite one. The circumstances in the present case included that the variations were significant in scope and, importantly, that, throughout most of the year 2000, Shawton were not insisting on, nor particularly concerned about, early completion of DGP's drawing work.'

Urgency and expedition

Another point of general interest is to what extent the contractor has an obligation to recognise the urgency of the situation faced by the employer. This point was considered in the *Astea* case detailed below.

Astea v. *Time Group* (2003)

In the case of *Astea (UK) Ltd* v. *Time Group Ltd* (2003) Judge Seymour sitting in the Technology and Construction Court had to consider whether Astea, a supplier of computer software, was liable for damages for failure to complete within a reasonable time in a contract where there was no set time for completion. Astea contended that its obligation was to complete within a time which was reasonable in the circumstances – although conceding that it could not rely on its own failings as extending such time. Time contended that a reasonable time should be assessed by reference not so much as the time actually taken but more by reference to the time which could have been achieved with due expedition.

The judge put it this way:

'141. The distinction between these two approaches seemed to be that Mr Hossain [for Time] in effect was contending that Astea was bound to complete the Services as fast as humanly or technically possible, subject only to being excused in respect of delays over which it had no control, while Mr Kinsky [for *Astea*] sought to persuade me that the question was not so much how fast the Services could have been performed by Astea had it chosen to allocate to doing so the greatest possible resources and to maintain them for as long as necessary, but rather, considering all of the circumstances, how long, as things turned out, it was reasonable for Astea to take.'

The judge went on to say:

'142. Both Mr Hossain and Mr Kinsky endeavoured to seek support for their respective emphases from the well-known decision of the House of Lords in *Pantland Hick* v. *Raymond & Reid* [1893] AC 22. The leading speech in that case was that of the Lord Chancellor, Lord Herschell. At pages 28 and 29 of the report Lord Herschell said this:

"The bills of lading in the present case contained no such stipulation [as to time for performance], and, therefore, in accordance with ordinary and well-known principles the obligation of the respondents was that they should take discharge of the cargo within a reasonable time. The question is, has the appellant proved that this reasonable time has been exceeded? This depends upon what circumstances may be taken into consideration in determining whether more than a reasonable time was occupied.

The appellant's contention is, that inasmuch as the obligation to take discharge of the cargo, and to provide the necessary labour for that purpose, rested upon the respondents, the test is what time would have been required for the discharge of the vessel under ordinary circumstances, and that, inasmuch as they have to provide the labour, they must be responsible if the discharge is delayed beyond that period.

The respondents on the other hand contend that the question is not what time would have been necessary or what time would have been reasonable under ordinary circumstances, but what time was reasonable under existing circumstances, assuming that, in so far as the existing circumstances were extraordinary, they were not due to any act or default on the part of the respondents.

My Lords, there appears to me to be no direct authority upon the point, although there are judgments bearing on the subject to which I will presently call attention. I would observe, in the first place, that there is of course no such thing as a reasonable time in the abstract. It must always depend upon circumstances. Upon 'the ordinary circumstances' say the learned counsel for the appellant. But what may without impropriety be termed the ordinary circumstances differ in particular ports at different times of the year. As regards the practicability of discharging a vessel they may differ in summer and winter. Again, weather increasing the difficulty of, though not preventing, the discharge of a vessel may continue for so long a period that it may justly be termed extraordinary. Could it be contended that in so far as it lasted beyond the ordinary period the delay caused by it was to be excluded in determining whether the cargo had been discharged within a reasonable time? It appears to me that the appellant's contention would involve constant difficulty and dispute, and that the only sound principle is that the "reasonable time" should depend on the circumstances which actually exist. If the cargo has been taken with all reasonable despatch under those circumstances I think the obligation of the consignee has been fulfilled. When I say the circumstances which actually exist, I, of course, imply that those circumstances, in so far as they involve delay, have not been caused or contributed to by the consignee. I think the balance of authority, both as regards the cases which relate to contracts by a consignee to take discharge, and those in which the question what is a reasonable time has had to be answered when analogous obligations were under consideration, is distinctly in favour of the view taken by the Court below."

143. I was also referred by Mr Kinsky to the speech of Lord Watson at pages 32 and 33 of the report, where he said:
"When the language of a contract does not expressly, or by necessary implication, fix any time for the performance of the contractual obligation, the law implies that it shall be performed within a reasonable time. The rule is of general application, and is not confined to contracts for the carriage of goods by sea. In the case of other contracts the condition of reasonable time has been frequently interpreted; and has invariably been held to mean that the party upon whom it is incumbent duly fulfils his obligation, notwithstanding protracted delay, so long as such delay is attributable to causes beyond his control, and he has neither acted negligently nor unreasonably."

and

'144. . . . What it seems to me the application of the test formulated by the House of Lords in *Pantland Hick* v. *Raymond & Reid* involves in a case such as the present is a broad consideration, with the benefit of hindsight, and viewed from the time as at which one party contends that a reasonable time for performance has been exceeded, of what would, in all the circumstances which are by then known to have happened, have been a reasonable time for performance. That broad consideration is likely to include taking into account any estimate given by the performing party of how long it would take him to perform; whether that estimate has been exceeded and, if so, in what circumstances; whether the party for whose benefit the relevant obligation was to be performed needed to participate in the performance, actively, in the sense of collaborating in what was needed to be done, or passively, in the sense of being in a position to receive performance, or not at all; whether it was necessary for third parties to collaborate with the performing party in order to enable it to perform; and what exactly was the cause, or were the causes of the delay to performance. This list is not intended to be exhaustive.'

The judge went on to hold that no breach of contract by Astea had been established.

Summary

As is evident from the closing words of the above passage, assessment of a reasonable time for completion is not an exact science bound by rigid rules. However, the general principles for assessment of a reasonable time for completion remain as stated in *Hick* v. *Raymond and Reid* (1893) – sometimes named *Pantland Hick* v. *Raymond Reid*. Generally the assessment is a retrospective exercise since a reasonable time includes delays beyond the control of the contractor – and such delays are unlikely to be capable of prospective assessment. The burden of proof rests on the party which seeks to rely on reasonable time or on the party which asserts that the obligation to complete within a reasonable time has been breached.

2.9 *Fixing time by reference to correspondence*

It is not unusual, particularly in sub-contracts, for important terms on time to be less than clearly stated. The details of the particular contract may be set out in exchanges of correspondence but they are not always transferred into a formal contract document. In such cases there is, of course, considerable scope for dispute as to what terms apply.

One party, for example, might argue that in the absence of a clearly or formally stated time for completion then a reasonable time should apply.

The point came up in the case of *J and J Fee Ltd* v. *The Express Lift Company Ltd* (1993), where the parties agreed to use DOM/2 conditions but never went beyond an extensive exchange of correspondence on what completion date was to apply. The judge, after reviewing the correspondence, held that the operative completion date was that stated in the final non-contested letter and that the date applied as if it had been written into a DOM/2 contract signed by the parties.

2.10 *The effect of time at large on the contract price*

As noted in Section 2.5 above, the Court of Appeal in *McAlpine Humberoak* soundly rejected the simple proposition when time is put at large the contract price is also put at large. But that does not dispose of such questions as how prolongation costs should be assessed when time is at large or how other costs should be assessed if the effect of time at large is to render the contract price inapplicable.

For example, it might be said in relation to prolongation costs that since time at large requires the contractor to complete within a reasonable time then recoverable prolongation costs should be those which are likewise reasonable. Or put another way, since the burden falls on the employer to show that the contractor failed to complete within a reasonable time if the employer wishes to recover delay damages then, applying the same principle to prolongation costs, it should be for the employer to prove that the contractor's incurred prolongation costs are not reasonable.

There seems to be little legal authority on this matter and there is probably no single rule to fit all cases. The terms of the contract may be relevant and it may depend upon the circumstances which put time at large. Thus for breach of contract the contractor may be able to recover prolongation costs as damages whereas for additional works not covered by any extension of time provision the valuation of variation rules of the contract may take effect.

Similar points to these came up in the case of *Wiltshier Construction (Scotland) Ltd* v. *Drumchapel Housing Co-operative Ltd* (2003) where it was argued by the contractor that as a result of excessive vandalism the contract time and contract price were superseded by a reasonable time for completion and reasonable remuneration. The court ruled, however, that the original contract provisions were not displaced.

Chapter 3
Damages for late completion

3.1 *Liquidated and general damages distinguished*

As explained in Chapter 1, liquidated damages are fixed in advance of the breach and can be recovered without proof of loss; whereas general damages are assessed only after the breach and can only be recovered upon proof of loss.

Reasons for use

There are sound commercial reasons for using liquidated damages whenever possible. Firstly because of the certainty they bring to the consequences of breach; and secondly because they avoid the expense and dispute involved in proving loss. As Lord Justice Diplock said in the case of *Robophone Facilities Ltd* v. *Blank* (1966) when summing up the balance between the parties:

> 'The court should not be astute to descry a penalty clause in every provision of a contract which stipulates a sum to be payable by one party to the other in the event of a breach by the former. Such a stipulation reflects good business sense and is advantageous to both parties. It enables them to envisage the financial consequences of a breach; and if litigation proves inevitable it avoids the difficulty and the legal costs, often heavy, of proving what loss has in fact been suffered by the innocent party.'

Pre-estimates of loss

It is clearly not easy to estimate in advance the financial consequences of the various breaches of a construction contract which the contractor might allege, such as the damages arising from late instructions, prevention, and the like. Consequently most claims from contractors come to be settled by way of general damages. Some contracts have been put out where the contractor is required to state a sum per week for reimbursable delay but the practice is not widespread. It was considered but not used in edition 3 of GC / Works / 1 /.

For the employer, however, the most common breach suffered is late completion by the contractor and here it is possible to make genuine pre-estimate of the loss and to incorporate the same into the contract as liquidated damages.

There are clear advantages to the employer in this because he does not have to prove his loss and there will probably be a mechanism in the contract for deduction of the damages from sums due to the contractor. There are also corresponding benefits to the contractor in that he knows in advance what damages he is liable for in the event of late completion. At tendering stage this is often an important factor in the contractor's bid. If he feels that he cannot risk the level of damages stated for late completion, he can withdraw or bid high. If he thinks that he cannot complete in the time allowed he knows how much to add to his bid for anticipated late completion and then, during construction, when there may be a balance to be struck between spending more money to complete on time or facing damages for late completion, the contractor knows what these damages will be and can calculate accordingly.

Mutuality

The mutuality of liquidating damages is not always recognised by the courts or by contractors. Thus, in *Peak* v. *McKinney* (1970), Lord Justice Salmon said:

> 'The liquidated damages clause contemplates a failure to complete on time due to the fault of the contractor. It is inserted by the employer for his own protection; for it enables him to recover a fixed sum as compensation for delay instead of facing the difficulty and expense of proving the actual damage which the delay may have caused him.'

Indeed, many of the successful challenges to liquidated damages rely on the traditional hostility of the courts to such damages, often demonstrating a logic which now seems distinctly old fashioned.

As for contractors, those who dislike liquidated damages frequently misunderstand basic principles and believe either that such damages are penalties for late completion, or that such damages are imposed in circumstances where no other damages would be payable. The reality is very different. If liquidated damages can be shown to be penalties they cannot be enforced; and if liquidated damages cannot be enforced, for whatever reason, then general damages are payable. Thus omitting liquidated damages clauses from construction contracts would not relieve the contractor of liability for damages for late completion. The best perhaps that can be said for the hostile contractor's view is that proving damages for late completion in some projects, particularly those in the public sector, would be no easy matter and many employers would not consider it worth the effort.

However true this point may be and however different liquidated and general damages may appear, the principles which apply to the pre-estimation of liquidated damages cannot be divorced from the principles which apply to the calculation of general damages.

3.2 Principles of general damages

Two aspects of general damages need to be considered. First, remoteness of damage, which relates to liability and is in effect a defence against a claim for breach on the grounds that the consequences could not have been foreseen.

Secondly, measure of damages, which relates to the quantum of a claim once the principle of liability has been established.

Remoteness of damage

The law does not allow a claimant to succeed in every case where damage follows a breach but draws a practical line by excluding that which is too remote. Lord Wright in *Liesbosch Dredger* v. *Edison Steamship* (1933) put it this way:

'The law cannot take account of everything that follows a wrongful act; it regards some subsequent matters as outside the scope of its selection, because "it were infinite for the law to judge the cause of causes", or consequences of consequences. In the varied web of affairs the law must abstract some consequences as relevant, not perhaps on grounds of pure logic, but simply for practical reasons.'

Hadley v. Baxendale (1854)

The guiding principles of remoteness applying to cases of breach of contract derive from the judgment of Baron Alderson in the very old case of *Hadley* v. *Baxendale* (1854). In the course of his judgment he said:

'Where two parties have made a contract which one of them has broken, the damages which the other party ought to receive in respect of such breach of contract should be such as may fairly and reasonably be considered either arising naturally, i.e. according to the usual course of things, from such breach of contract itself, or such as may reasonably be supposed to have been in the contemplation of both parties, at the time they made the contract, as the probable result of the breach of it.'

The facts of the case were that the mill of the plaintiffs at Gloucester was brought to a standstill by a broken crank shaft and it became necessary to send the shaft to the makers at Greenwich as a pattern for a new one.

The defendant, a common carrier, promised to deliver it at Greenwich on the following day. Owing to his neglect it was unduly delayed in transit with the result that the mill remained idle for longer than it would have done had there been no breach of the contract of carriage. The plaintiffs therefore claimed to recover damages for the loss of profit caused by the delay.

In his judgment Baron Alderson demonstrated that in accordance with the principle that he had just expressed there were only two possible grounds upon which the plaintiffs could sustain their claim. First, that in the usual course of things the work of the mill would cease altogether for the want of the shaft. This, he said, would not be the normal occurrence for, to take only one reasonable possibility, the plaintiffs might well have had a spare shaft in reserve. Secondly, that the special circumstances were so fully disclosed that the inevitable loss of profit was made apparent to the defendant. This, however, was not the case since the only communication proved was that the article to be carried was the shaft of a mill and that the plaintiffs were the owners of the mill. The jury, therefore, should not have taken the loss of profit into consideration in their assessment of damages.

Rule in *Hadley* v. *Baxendale*

The rule in *Hadley* v. *Baxendale* is taken as having two branches and is commonly expressed as:

> 'Such losses as may fairly and reasonably be considered as either arising: (1st rule) "naturally", i.e. according to the usual course of things, or (2nd rule) "such as may reasonably be supposed to be in the contemplation of both parties at the time they made the contract, as the probable result of breach of it".'

Victoria Laundry v. *Newman* (1949)

The test of remoteness laid down by Baron Alderson was reformulated in the judgment of Lord Justice Asquith in the case of *Victoria Laundry (Windsor) Ltd* v. *Newman Industries Ltd* (1949).

The plaintiffs who wished to extend their business contracted to buy a second-hand boiler which was then damaged in dismantling and delivered five months late. They sued for loss of profit during the period of delay, which profit would have come from two sources; firstly the general extension of their business and secondly, highly lucrative contracts from the Ministry of Supply. The Court of Appeal allowed damages under the first heading but not the second.

Lord Justice Asquith, having reviewed the law as it then stood, gave six propositions in which he introduced the test of reasonable foreseeability:

> 'What propositions applicable to the present case emerge from the authorities as a whole, including those analysed above? We think they include the following:
>
> (1) It is well settled that the governing purpose of damages is to put the party whose rights have been violated in the same position, so far as

money can do so, as if his rights had been observed. This purpose, if relentlessly pursued, would provide him with a complete indemnity for all loss de facto resulting from a particular breach, however improbable, however unpredictable. This, in contract at least, is recognized as too harsh a rule. Hence,

(2) In cases of breach of contract the aggrieved party is only entitled to recover such part of the loss actually resulting as was at the time of the contract reasonably foreseeable as liable to result from the breach.

(3) What was at that time reasonably so foreseeable depends on the knowledge then possessed by the parties or, at all events, by the party who later commits the breach.

(4) For this purpose, knowledge "possessed" is of two kinds; one imputed, the other actual. Everyone, as a reasonable person, is taken to know the "ordinary course of things" and consequently what loss is liable to result from a breach of contract in that ordinary course. This is the subject matter of the "first rule" in *Hadley* v. *Baxendale*. But to this knowledge, which a contract-breaker is assumed to possess whether he actually possesses it or not, there may have to be added in a particular case knowledge which he actually possesses, of special circumstances outside the "ordinary course of things", of such a kind that a breach in those special circumstances would be liable to cause more loss. Such a case attracts the operation of the "second rule" so as to make additional loss also recoverable.

(5) In order to make the contract-breaker liable under either rule it is not necessary that he should actually have asked himself what loss is liable to result from a breach. As has often been pointed out, parties at the time of contracting contemplate not the breach of the contract but its performance. It suffices that if he had considered the question, he would as a reasonable man have concluded that the loss in question was liable to result.

(6) Nor, finally, to make a particular loss recoverable need it be proved that upon a given state of knowledge the defendant could, as a reasonable man, foresee that a breach must necessarily result in that loss. It is enough if he could foresee it was likely so to result. It is indeed enough if the loss (or some factor without which it would not have occurred) is a "serious possibility" or a "real danger". For short we have used the word "liable" to result. Possibly the colloquialism "on the cards" indicates the shade of meaning with some approach to accuracy.'

Current position

In *Czarnikow Ltd* v. *Koufos* (1969), known as *The Heron II*, the House of Lords moved away from the foreseeability test to one of assumed common knowledge. The effect of this on the law and a summary of the law as it now stands was admirably expressed by the Court of Appeal of New Zealand in *Bevan Investments* v. *Blackall & Struthers* (1977) as follows:

'(1) The aggrieved party is only entitled to recover such part of the loss actually resulting as may fairly and reasonably be considered as arising naturally, that is according to the usual course of things, from the breach of the contract.

(2) The question is to be judged as at the time of the contract.

(3) In order to make the contract-breaker liable it is not necessary that he should actually have asked himself what loss was liable to result from a breach of the kind which subsequently occurred. It suffices that if he had considered the question he would as a reasonable man have concluded that the loss in question was "liable to result".

(4) The words "liable to result" should be read in the sense conveyed by the expressions "a serious possibility" and "a real danger".'

Presumed knowledge

In *Balfour Beatty Construction (Ltd)* v. *Scottish Power plc* (1994) the House of Lords had to consider the extent to which one party to a contract is presumed to know about the business activities of the other. Scottish Power had provided a power supply to Balfour Beatty's concrete batching plant and as a result of a failure of the power supply during the continuous pour of an aqueduct structure Balfour Beatty had to demolish the partly finished pour. It was held that the need for demolition was not within the contemplation of Scottish Power and they were not liable for the resulting financial damages.

For further discussion on applicability of the rules of *Hadley* v. *Baxendale* to genuine pre-estimates of loss see Chapter 4 and, in particular, the case of *Multiplex* v. *Abgarus* (1992).

Measure of damages

The principles applied by the courts in measuring damages date back to the case of *Robinson* v. *Harman* (1848) where it was stated:

'The rule of common law is that where a party sustains a loss by reason of a breach of contract, he is, so far as money can do it, to be placed in the same situation, with respect to damages, as if the contract had been performed.'

This rule is, of course, subordinate to the rule on remoteness first considered. As Lord Esher in *The Argentino* (1888) said:

'This rule does not come into play with regard to any claimed head of damage until it has been determined by the rule as to remoteness whether that head of damage can be brought into consideration at all.'

The distinction between remoteness of damage and measure of damage is not always obvious. Thus, in *Parsons (Livestock) Ltd* v. *Uttley Ingham & Co. Ltd* (1977) a defective ventilator in a feed hopper led to mould in the pig feed and the death of 254 top-grade pigs. The pig farmer claimed £36,000 for loss of his herd; the hopper manufacturer offered £18 for replacement feed. The Court of Appeal held that the type of damage (the death of the pigs) was foreseeable from the consequences of the breach and the pig farmer's losses were recoverable.

Further complications arise when the measure of damages is to cover diminution in value and / or liability for rectification. In *Ruxley Electronics and Construction Ltd* v. *Forsyth* (1995), a case which concerned a swimming pool built to a depth of 6 feet 9 inches instead of the specified 7 feet 6 inches, the House of Lords held that the employer could not recover the full cost of a replacement pool and that, when such expenditure would be out of all proportion to the benefit to be obtained, the appropriate measure of damages was diminution in value. Or, put another way, the proper measure of damages was not the monetary equivalent of specific performance but the loss suffered as a result of the breach.

Wasted expenditure

Some judicial guidance on wasted expenditure can be gained from the case of *C & P Haulage* v. *Middleton* (1983).

In that case, a motor repairer executed certain works to premises he occupied for his business to render them suitable for his purpose and sued for wasted expenditure when his lease was terminated in breach of contract.

The Court of Appeal held that he could not succeed as he had suffered no loss of profit because he had found alternative accommodation and the earlier 'wasted' expenditure would have been spent anyway even if the contract had not been broken. The court held that the correct approach was that he should be put in the position he would have been in had the contract been performed. He could not have damages to put him in the position as if the contract had never been made.

The general point here is that a claim for damages is not intended to improve one's position on what it would have been without any breach. In short, a claim is not a device for turning loss into profit.

Mitigation of loss

It is sometimes said that a claimant has a duty to mitigate his loss. This is true to the extent that the claimant seeks to recover his loss as damages, but it does not follow that an injured party in a breach of contract situation should have his conduct determined by the breach.

The following extracts from legal judgments explain this.

Viscount Haldane in *British Westinghouse Electric & Manufacturing Co. Ltd v. Underground Electric Railways of London Ltd* (1912) said that:

'A plaintiff is under no duty to mitigate his loss, despite the habitual use by the lawyers of the phrase "duty to mitigate". He is completely free to act as he judges to be in his best interest. On the other hand, a defendant is not liable for all loss suffered by the plaintiff in consequence of his so acting. A defendant is only liable for such part of the plaintiff's loss as is properly to be regarded as caused by the defendant's breach of duty.'

Sir John Donaldson, Master of the Rolls, in *The Solholt* (1983) said:

'The fundamental basis is thus compensation for pecuniary loss naturally flowing from the breach; but this first principle is qualified by a second, which imposes on a plaintiff the duty of taking all reasonable steps to mitigate the loss consequent on the breach, and debars him from claiming any part of the damage which is due to his neglect to take such steps.'

There is clearly wide scope for debate on how far the concept of 'neglect to take such steps' should apply. It is doubtful, for example, that it extends to expenditure of further moneys which might or might not be recoverable but it probably does include taking-up reasonable offers and applying practical steps.

In *Pilkington* v. *Wood* (1953) it was held that a house purchaser was under no duty to sue the vendor for conveying a defective title in order to mitigate his loss in proceedings against his solicitor.

Against that, in *Brace* v. *Calder* (1895), an employee who sued for breach of his employment contract was awarded only nominal damages because he rejected an offer of a new contract.

See also the case of *Murray* v. *Leisureplay Plc* (2005), discussed in Chapter 4, on the question of whether foreseeable prospects for mitigation of loss need to be included in genuine pre-estimates of loss to avoid them being declared penalties.

Best endeavours and the like

Obligations to use best endeavours, reasonable endeavours and other terms of similar intent are generally related in the construction industry, to progressing the works. Often they amount to no more than exhortation and only rarely will breach lead directly to liability for damages. The usual situation is that breach impacts on entitlement to extension of time or may provide grounds for determination. These matters are discussed in later chapters.

However, note the computer software case of *Astea (UK) Ltd* v. *Time Group Ltd* (2003) discussed in Chapter 2 where an attempt was made to fix a reasonable time for completion by reference to due expedition.

3.3 *Alternative remedies*

The rule is well settled that when a liquidated damages clause fails to operate because it is successfully challenged as a penalty, or fails because of some defect in legal construction, act of prevention or other obstacle, then general damages can be sought as a substitute. Thus, Lord Justice Phillimore in *Peak* v. *McKinney* (1970) said:

> 'If the employer is in any way responsible for the failure to achieve the completion date, he can recover no liquidated damages at all and is left to prove such general damages as he may have suffered.'

Lord Justice Stephenson in *Rapid Building Group Ltd* v. *Ealing Family Housing Association* (1984) said:

> 'It is accepted that a party must elect whether to claim liquidated or unliquidated damages; but as it seems to me, where the claim for liquidated damages has been lost or has gone . . . the defendants are not precluded from pursuing their counterclaim for unliquidated damages.'

The statement in this latter quotation that a party must elect whether to claim liquidated or unliquidated damages requires some explanation. At first sight it implies that liquidated damages and general damages are alternative remedies at the option of the claiming party; but this cannot generally be the case since it would defeat the purpose of liquidating damages.

Clearly, at the outset before the contract is prepared there are genuine alternatives to consider because a decision has to be made on whether to rely on general damages or whether to include within the contract express provisions for liquidated damages. The point is fairly obvious, but it is worth stating that: general damages can follow implied terms; but liquidated damages can only follow express terms.

In practice it is normally solely the party who prepares the contract who makes the decision on whether or not to include liquidated damages and at what rates they should be, and that does leave open some scope for later dispute on whether the rates stated are truly liquidated damages or are penalties. But that argument apart, once liquidated damages are included in a contract they are deemed to be there by agreement between the parties. Their application can be challenged later on various grounds but the argument sometimes put forward by a party facing liquidated damages that they should not apply because they were set without consultation has no merit.

However, where there is a liquidated damage clause in a contract, a major point to consider is to what extent is it possible for either party to avoid liquidated damages and substitute general damages?

The position of the party facing liability for liquidated damages is perhaps the most straightforward since that party always has a choice; to pay or accept the deduction of the liquidated damages due, or to challenge them and face general damages. The desire to avoid liquidated damages might arise from an attempt to defer payment or a belief that such general damages as could be proved would be less than the liquidated damages. There might well also be the belief, and not without some foundation, that avoidance of liquidated damages would in practice mean the avoidance of general damages since pursuit of the latter is a time consuming and expensive process.

This is a choice, however, on whether or not to challenge liquidated damages; it is not an election on whether or not they should apply. That power of election, if such a power exists, can only vest in the party seeking to apply damages. The point at issue then becomes: is a liquidated damages clause in the nature of an exclusion clause shutting out the alternative remedy of general damages?

Exclusion effect

This was one of the matters which came to be considered by the Court of Appeal in the unusual case of *Temloc Ltd* v. *Errill Properties Ltd* (1987) where the entry in the appendix to a JCT 80 contract was stated as £nil liquidated damages. The contract was finished late and the employer / developer who was liable to the property purchaser for damages sought to recover them as general damages from the contractor. On the argument that the employer had a choice of damages, Lord Justice Croom-Johnson had this to say:

> '[Counsel for Errills] submits that Errills had a choice as to which they should go for, whether for the liquidated damages or for damages at large. On the wording of Clause 25 there is no choice available. Any such claim for damages at large would have to be based on an implied term in the contract. If Clause 24 had been excluded from the contract altogether, as was submitted by [Counsel for Errills], it would have been necessary to imply such a term and give effect to it. But as Clause 24 is tied to dates certified by the architect and a method of calculation is provided in the appendix, there is no room for implying such a term. Clause 24 is headed "Damages for non-completion", and then lays down an agreed provision for calculating those by liquidated damages, which is covering all the damages for non-completion. There is every reason why parties to building contracts should agree to liquidated damages for non-completion. Proof of such loss is often difficult to achieve and agreement in advance is a saver of disputes.'

Exhaustive remedy

As to whether liquidated damages provide an exhaustive remedy, in *Temloc* v. *Errill,* Lord Justice Nourse, agreeing with Lord Justice Croom-Johnson, said:

> 'I think it clear, both as a matter of construction and as one of common sense, that if (1) Clause 24 is incorporated in the contract and (2) the

parties complete the relevant part of the appendix, either by stating a rate at which the sum is to be calculated or, as here, by stating that the sum is to be nil, then that constitutes an exhaustive agreement as to the damages which are or are not to be payable by the contractor in the event of his failure to complete the works on time.'

He went on to say:

'Viewing the clause in this way, I find it impossible to attribute to parties who complete the appendix in one way or the other an intention that the employer shall have the option of claiming damages of precisely the same character but in an unliquidated amount.'

Summary

All of the above can be summed up as:

(i) express terms on liquidated damages exclude the possibility of implied terms for general damages;
(ii) liquidated damages are an exhaustive remedy for the breach to which they apply;
(iii) the employer has no option of claiming general damages instead of liquidated damages.

Other cases

The exhaustive effect of a liquidated damages clause was also considered in the case of *Pigott Foundations Ltd* v. *Shepherd Construction Ltd* (1993). The sub-contract contained an agreement which read:

'With regard to B9, it was agreed that damages would only apply in the event of Pigott's not completing within 10 weeks and any sum would be limited to £40,000 (max) at the rate of £10,000 per week.'

In bringing a counterclaim against Pigott, Shepherd argued that the agreement applied only as a limitation of Pigott's liability in respect of any liquidated damages flowing down from the main contract and that it did not apply to the damages for delay, disruption and consequential loss and expense.

Rejecting that argument, Judge Gilliland QC said:

'The effect of a provision for the payment of liquidated damages for delay in a building contract has been considered in a number of recent authorities from which it is clear that not only does such a clause have the effect of imposing a liability upon the party who is responsible for the delay to pay damages at the stated rate but also it has the effect of precluding the other party to the contract from seeking to avoid the limitation on any amount of damages contained in a liquidated damages clause by claiming damages for delay or disruption arising from delay in completing the works as damages for the breach of some other provision of the contract. See for example *Temloc Ltd* v. *Errill Properties Ltd* (1988). In that case the amount of the liquidated damages was stated to be nil, but it was held by the Court of Appeal that the provisions constituted an exhaustive agreement as to the amount of the damages which were to be payable by the contractor in the event of his failure to complete the works on time. A similar conclusion was reached in *Surrey Heath Borough Council* v. *Lovell Construction Ltd* (1988).'

A recent thorough review of the law on liquidated damages as an exhaustive remedy is found in Mr Justice Ramsey's judgment in the case of *Biffa Waste Ltd* v. *Maschinfabrik Ernst Hese GMBH* (2008). The contract in that case contained a provision capping liquidated damages for delay at 7.5% of the contract price. The claimant sought to recover unliquidated damages relying on other provisions in the contract. The judge said:

'102. I consider first, the provisions of Clause 47.1 of the Design and Build Deed. This clause provides for liquidated damages for delay. Under Clause 43 there are a number of obligations relating to completion. Clause 43.1 states that the whole of the Works shall be completed in accordance with the provisions of Clause 48 (the Taking over Certificate) by the Time for Completion. Clause 43.2 provides that MEH shall complete any task specified in Part B of Schedule 3 by the date specified or such extended time as may be allowed under Clause 44.

103. Time for Completion is defined in Clause 1.1.49 as the time stated in Part A of Schedule 3 (or as extended under Clause 44 or reduced under Clause 44.3). Part A of Schedule 3 provides: "Time of completion for the whole of the Works (save for achieving biogas production and generated electricity tests at Wanlip described in Schedule 2.0 of the Plant Specification) 18 June 2004 (with commissioning operations to start no later than 18 April 2004)".

104. Part B of Schedule 3 provides dates or a period of time for different tasks under a heading of "time for sectional completion".

105. As a result, there are a number of contractual obligations as to time under Clauses 43.1 and 43.2 which, if breached, would ordinarily give rise to a claim for damages for breach of contract in an unliquidated amount.

106. Clause 47.1 deals both with the position where MEH "fails to comply with the Time for Completion in accordance with Clause 48 for the whole of the Works within the relevant time prescribed by Clause 43" and also where MEH fails "to comply with clause 43.2".

107. Clause 47.1 then provides that if there is such a failure then MEH shall pay Biffa Waste "the relevant sum stated in Schedule 12 as liquidated damages for such default and not as a penalty *(which sum shall be the only monies due from the Contractor for such Default)* for every week or part week which shall elapse between (a) the Time

for Completion and that date stated in a Taking Over Certificate of the whole of the Works or (b) the date specified in part B of Schedule 3 and the date the task specified in Part B of Schedule 3 is actually completed." *(emphasis added)*.

108. Biffa accepts, the phrase in parentheses, *"which sum shall be the only monies due from the Contractor for such Default"*, would have the effect of making Clause 47.1 the exclusive remedy for such delay. This would be consistent with the general position that a liquidated damages clause in a contract covers "all the damages for non-completion" or "constitutes an exhaustive agreement as to the damages which are or are not to be payable by the contractor in the event to the failure to complete the works on time": see *Temloc Ltd v. Errill Properties Ltd* (1987) 33 BLR 30 at 38 to 40.

109. Clause 47.1 then includes this provision: "The payment or deduction of such damages shall not relieve the Contractor from its obligation to complete the Works or from any other of its obligations and liabilities under the Contract and shall be without prejudice to any other right or remedy of the Employer."

110. Biffa submits that this wording opens up a claim by Biffa Waste against MEH for damages for delay where that delay is not simply a breach of the requirements of Clause 43, which it refers to as "simple" delay. Thus, in this case, on the basis of delay caused by the pleaded breaches of Clauses 8.1(a), 8.2, 15.1 and 36.1(g), Biffa submits that Biffa Waste is entitled to unliquidated damages which are not affected by Clause 47.1.

111. MEH submits that the distinction between a breach of Clause 43 and breaches of other terms of the contract leading to delay is not one which is properly made. MEH relies on *Piggott Foundations Ltd v. Shepherd Construction Ltd* (1993) 67 BLR 48 and *Surrey Heath Borough Council v. Lovell Construction Ltd* (1988) 42 BLR 25 and submits that the position is correctly stated in *Keating on Construction Contracts (8th Edition)* at para 9-006. It is submitted by MEH that the sentence in Clause 47.1 relied on by Biffa merely acts as a reminder that the obligation to pay liquidated damages does not relieve MEH of its other obligations under the Design and Build Deed. If it were read as Biffa contends, MEH submits that it would conflict with and deprive the earlier phrase in parentheses of any meaning.

112. I accept MEH's submission. The sentence relied on by Biffa commences by reminding MEH that the payment of liquidated damages does not relieve MEH of its obligation to complete the Works or from any other obligations or liabilities under the contract. When read in context, I do not consider that the other liabilities can include a liability to pay unliquidated damages for delay for breach of other provisions of the Design and Build Deed.

113. The phrase that the payment of liquidated damages *"shall be without prejudice to any other right or remedy of the Employer"* when read with

the words in parentheses must refer to a right or remedy which is not a monetary right or remedy. The words in parentheses make it clear that liquidated damages shall be "the only monies" due from MEH for such failure to complete. If a contractor fails to complete the employer has rights and remedies other than damages. In this case these include rights in relation to rates of progress in Clause 46.1 and termination under Clause 59.1 or at common law. It is those other rights which are not prejudiced. However if liquidated damages are the only monies payable for failure to complete, that must exclude other remedies for payment of damages.

114. Further I do not consider that the provisions of clause 47.1 can be construed to draw a distinction between a "simple" failure to complete and a failure to complete caused by the breach of another obligation under the Design and Build Deed. First, I do not consider that it is possible to draw a distinction between a "simple" failure to complete and a failure to complete caused by breach of another obligation. If there is a failure to complete then liquidated damages are "the only monies" due for such default. If there is a breach of another obligation and that breach causes a failure to complete then liquidated damages are still the only monies due for that default, that is a breach of contract causing a failure to complete on time.

115. Secondly, I do not accept that a liquidated damages clause which only applied to a case where there was simply a failure to complete on time without a breach of any other provision would make commercial sense. The purpose of the liquidated damages clause is, as Lord Upjohn said in the *Suisse Atlantique* case, for the benefit of both parties: "the party establishing breach by the other need prove no damage in fact; the other must pay that, no less and no more." Assessment of damages for delay is a difficult process as the expert evidence in this case has shown. The advantage of certainty in the sum payable as liquidated damages provides advantages to both sides. If that benefit were limited to cases of "simple" delay but not to cases where that "simple" delay had been caused by breach of another obligation, the commercial purpose would disappear. A party wishing to avoid liquidated damages and argue for no loss or a smaller sum would attempt to find some other breach of an implied or express term to hang the delay on. A party seeking to uphold the clause would be trying to disprove that another breach was the cause of the delay.

116. In the context of the Design and Build Deed, Clause 41.1 also makes the argument difficult because it provides that "the Contractor shall proceed with the works with due expedition and without delay" which would be capable of giving rise to another breach for "simple delay". If the liquidated damages provision applied only to those cases where there was no other breach then such a construction would neither be consistent with the phrase in parentheses nor

give sensible commercial meaning to the liquidated damages provision.

117. I consider that my view is consistent with the decision of His Honour Judge Gilliland QC in *Piggott Foundations Ltd* v. *Shepherd Construction Ltd* (1993) 67 BLR 48 at 68 where he held that there was a liquidated damages provision and that this provision "prevents the defendant from seeking to avoid the overall limitation of damages to £40,000 by claiming as a head of general damages for the breach of any other provisions or obligation under the contract such damages which have resulted from the failure of the plaintiff to complete the piling work within the period of 10 weeks." The same consistency is implicit in the decision of His Honour Judge Fox-Andrews QC in *Surrey Heath Borough Council* v. *Lovell Construction Ltd* (1988) 42 BLR 25 where at 37 he found that the liquidated damages were an exhaustive remedy for delay where a building had been damaged by a fire.

118. In *Keating on Construction Contracts (8th Edition)* at para 9-006 the issue of whether liquidated damages are an exhaustive remedy for delay caused by breach of an obligation other than the obligation to complete is dealt with. It is stated that "It is suggested that the solution is primarily a question of the construction of the contract in question. If, as in most (if not all) cases, the clause is clearly expressed to be or, as a matter of proper construction appears to be, a complete remedy for delayed completion then it matters not why the contractor failed to complete by the due date . . . The fact that the delay is due to a breach of contract by the contractor as opposed to merely going slow, cannot affect the nature or quality of the loss which the liquidated damages is intended to compensate. In reality, in such situations, there are two breaches: the carrying out of the defective work . . . and the failure to complete by the due date. Neither the employer nor the contractor can avoid liquidated damages by simply relying on the first breach."

119. I consider that this passage correctly sets out the position. In this case, on a true construction of the Design and Build Deed, Clause 47.1 provides a complete remedy in damages for delayed completion. As a result, in my judgment, Biffa Waste cannot recover from MEH in respect of delay caused by the breaches of the Design and Build Deed, other than liquidated damages under Clause 47.1.'

Exhaustive in contract and in tort

A further aspect of the exhaustive nature of liquidated damages provisions confirmed in the *Surrey Heath Borough Council* v. *Lovell Construction Ltd* case was that liquidated damages excluded any parallel remedy in tort for late completion. Shortly before completion of a new office building a fire destroyed the works, allegedly due to the negligence of a sub-contractor. Lovell obtained an extension of time for re-building but were sued in contract and in tort for various sums, some of which related to late completion. It was held by Judge Fox-Andrews following *Temloc* that the liquidated damages provisions were exhaustive of Surrey Heath's remedies in respect of any heads of claim relating to damages for late completion.

It does not follow, however, that claims other than for late completion would be similarly treated.

An Australian view on alternative remedies

The decision in the Australian case of *Baese Pty Ltd* v. *Bracken Building Pty Ltd* (1989) appears to show a different view than that taken by the English courts. This was another case of nil liquidated damages but the Supreme Court of New South Wales declined to follow the judgment in *Temloc* and held that:

(i) the liquidated damages clause was not an exhaustive statement of entitlement to damages in the event of late completion;
     and
(ii) the function of the clause was to provide a mechanism for invoking liquidated damages if the employer so wished but if he did not do so he was entitled to rely on his common law rights.

However, the decision rested on giving the phrase 'if such notice is given' (in relation to the architect's duty to issue a certificate on non-completion) a different effect from the phrase 'then the architect shall' which applied in *Temloc*. 'If' was taken to give the employer an option whereas 'shall' was said to be imperative.

Accordingly the decision may be of limited effect.

Mitigation costs

It is well established that acceleration costs incurred to relieve effects of breach are recoverable as damages. Thus it was said in *Great Eastern Hotel Company Ltd* v. *John Laing Construction Ltd* (2005):

> 'Any acceleration measures even if partially successful, were clearly measures adopted in order to mitigate GEH's losses and as such the cost of such measures are recoverable from the contract breaker, see *Lloyds and Scottish Finance Limited* v. *Modern Cars and Caravans* [1966] 1 QB 764 at page 782'

However, when the remedy for breach is expressed in the contract as liquidated damages operation of the above rule may well be excluded as shown by the judgment in the above-mentioned *Biffa Waste Services* case where having held that liquidated damages provided an exhaustive remedy the judge said:

'120. Biffa also claims the costs of running the plant with a temporary liner. This raises the question of whether the recovery of liquidated damages precludes the recovery of any costs incurred in reasonable mitigation of delay. Biffa contends that the costs of mitigation are recoverable. MEH submits that they are not.

121. The cost of taking reasonable mitigating steps is generally recoverable as part of the damages for the breach: see *The World Beauty* [1970] P 144 at 156 per Winn LJ. As stated above, liquidated damages are an exhaustive remedy for delay. That exhaustive remedy therefore includes any damages which could be recovered as damages for a failure to complete. Where, as here, Biffa took reasonable mitigating steps to avoid delay loss then I consider that the cost of taking such steps is treated as being included in the pre-estimate of loss which forms the basis of the liquidated damages clause. Clause 47.1 of the Design and Build Deed provides that liquidated damages "shall be the only monies due from the Contractor for such default" and to permit further damages to be recovered for the reasonable costs of steps to mitigating that default would, in my judgment, be contrary to the express terms of that provision.

122. As a result, Biffa cannot recover the cost of taking reasonable steps to mitigate delay as an extra head of damages because those damages are included within the exhaustive remedy of liquidated damages.'

3.4 Can general damages exceed liquidated damages?

There is no firm legal ruling in English law that liquidated damages invariably act as a limit on any general damages which may be awarded as a substitute and the courts take a cautious approach to the matter.

In *Widnes Foundry (1925) Ltd* v. *Cellulose Acetate Silk Co. Ltd* [1933] Lord Justice Scrutton said:

'I do not decide that a party is always bound by the figure mentioned from recovering a larger sum; it turns upon whether the sum mentioned can be said to be an estimate of the damage to be paid for the breach.'

The 'sum mentioned' in the *Widnes Foundry* case was for liquidated damages at the rate of £20 per week for late delivery and erection of an acetone recovery plant. When a dispute arose on final payment, after a 30 week delay in delivery and erection, the employer counterclaimed unsuccessfully not the £600 due as liquidated damages but the sum of £5850 as general damages using the argument that the liquidated damages clause was a penalty clause because it was described as such in the contract.

In *Rapid Building* v. *Ealing Family Housing* (1984), where the liquidated damages clause was held to have failed, neither Lord Justice Stephenson nor Lord Justice Lloyd would be drawn on the proposition that the quantum of general damages was limited to the quantum of liquidated damages.

Lord Justice Lloyd stated that:

> 'Counsel has argued that although the liquidated damages clause has ceased, for the reasons I have mentioned earlier, to be applicable, nevertheless the defendants will not be entitled to recover more than the amount they would have recovered under Clause 22 if Clause 22 had continued to be applicable. Even if that be right, as to which I say nothing, . . .'

Continuing uncertainty

It may well be that the cautious approach of the courts to whether or not general damages can ever exceed liquidated damages reflects the different and sometimes surprising ways in which the point can emerge but as the law stands at present it would seem:

(i) that an employer will not be successful in seeking general damages higher than liquidated damages on the grounds that the liquidated damages are a penalty;

(ii) that a contractor has no certainty that general damages will be limited to liquidated damages when he defeats such damages.

However, until an English court follows the example of the Supreme Court of Canada which held in *Elsley* v. *JG Collins Insurance Agencies Ltd* (1978) that where the stipulated sum is held to be a penalty any general damages awarded cannot exceed that sum, this area of the law will remain uncertain.

There is a further complication in that many standard forms of construction contract allow for a ceiling on the amount of liquidated damages – 10% of the contract sum or similar. It is difficult to assess how provisions such as these affect the general principles of limitation on damages but they would seem to add to the element of risk that a contractor takes in challenging liquidated damages on the assumption that general damages can never be greater. However, note the view expressed by the judge in the *Steria* v. *Sigma* (2007) case discussed in Chapter 5 that where a capped liquidated damages clause is held to be inoperable the cap disappears with the clause.

3.5 Under-liquidation of damages

The question of whether liquidated damages set at a lower level than the employer's pre-estimate of loss could invalidate a liquidated damages clause was one of the points considered in *Multiplex* v. *Abgarus* (1992). It was held that liquidated damages need not provide for the entirety of the employer's likely loss.

The following extract from the decision explains why an attack on under-liquidated damages as not being 'a genuine pre-estimate of loss' must fail.

> 'It is clear as a matter of principle, and established by authority, that if parties agree upon a quantum of damage as liquidated damages which is *less* than the damage which would be suffered from such breach, no attack can be made upon such a liquidated damages provision upon the basis that it is "extravagant or unconscionable". The attack upon a liquidated damages clause has traditionally been based upon it not being a "genuine pre-estimate of damage", but such attacks are grounded upon the concept of equity interfering to prevent a party imposing a penalty upon the other party for breach of contract in the sense that the sum designated, or to be determined as payable on breach, is greater, and unreasonably or inequitably so, than the true damage reasonably assessed at the time of contract as being the damage which the innocent party might suffer. It can never be inequitable so far as the defaulting party is concerned for an innocent proprietor to offer or agree to accept as liquidated damages a sum less than the damages which, at contract, it is reasonably assessed it will suffer resulting from breach. Thus a liquidated damages clause providing for such a lesser payment can never be a penalty. The true vice in a penal damages clause is not that it is not a genuine pre-estimate of damage, but rather that it yields a result which exceeds that which a genuine pre-estimate of damage would have yielded.'

The practice of under-liquidation, when it is done intentionally, is usually done for the sound commercial reasons of attracting competitive tenders or striking a deal with a particular contractor. In such situations both employer and contractor knowingly derive benefit from the arrangement. When, as sometimes happens, under-liquidation occurs as a result of an error by the employer in calculating his pre-estimate of loss or in his understanding on how the law on liquidated damages applies, it nevertheless remains the case that the employer is bound by the amounts specified in the contract. As the *Widnes Foundry* case shows and the *Temloc* case shows it is not open to employers to escape from a bad bargain by seeking to defeat their own liquidated damages clauses.

Limitations on liquidated damages

A practice, particularly common in process and plant contracts, is to under-liquidate damages by setting the specified amounts for liquidated damages by reference to a percentage of the contract sum rather than by reference to the employer's pre-estimate of loss. Typically, the amount will be 0.5% of the contract sum per week up to a maximum amount recoverable as liquidated damages of 10% of the contract sum.

This practice is fully in keeping with the general policy of limitations on the contractor's liability found in process and plant contracts and, because it is such an effective limitation of the contractor's liability for late completion, it would be exceptional if it were challenged by a contractor as departing from the basic principles for pre-estimation of loss. However, although it is probably safe in most instances for an employer to use a figure of 0.5% per week of the contract sum (or thereabouts) as the amount of liquidated damages – because such a figure is likely to be below a properly made preestimation of loss – there are obvious dangers in the practice if there are no back-up calculations to prove the under-liquidation.

A distinction should, however, be noted between the practice of limiting liability for damages by under-liquidation and the practice of limiting liability by imposing a ceiling on the total amount of damages which are recoverable by the employer. Many construction contracts include the facility for entering a ceiling on the amount of liquidated damages and, although this is often expressed as a percentage of the contract sum, the underlying daily or weekly rate is normally derived from a genuine preestimation of loss.

This later practice is a straightforward limitation of liability and it cannot in any way offend any of the principles of liquidated damages.

Effect of nil damages

The decision of the Court of Appeal in *Temloc Ltd* v. *Errill Properties Ltd* (1987) insofar as it relates to nil damages can be summarised as:

(i) the effect of a nil entry in the appendix to a liquidated damages clause is not that the clause is to be disregarded or ineffective, but that there should be no damages for late completion;

(ii) no claim for general damages can be sustained on an implied term since the express provisions of the liquidated damages clause leave no room for any such clause to be implied.

The case concerned a contract under JCT 80 conditions but it is relevant to most other standard forms of construction contracts.

The decision caused some surprise in the construction industry, if not in legal circles, because the practice of making a 'nil' entry is not uncommon; or was not prior to publicity of the *Temloc* decision.

Nil entries

Nil entries are made with a variety of intentions and beliefs. Where the employer and the contractor are on good terms with a long-standing relationship the stipulation of liquidated damages might appear unnecessary and inappropriate; alternatively the parties might be under the mistaken belief that liquidated damages for late completion are only effective when there are corresponding bonus provisions for early completion; and then it might be thought that stating nil damages is the proper way of leaving damages open. Whatever the reasons, it is unlikely that the parties apply their minds to whether 'nil' damages will exclude general damages and this will rarely be their intention.

In the *Temloc* case most of the above ingredients were present – a long standing business connection, with four or more contracts successfully completed earlier with nil damages and the misunderstanding over the need for bonus provisions. As it was said for the employer, Errill Ltd:

> 'By putting "nil" in . . . we would not expect to take the contractor to Court. We agreed, it was not practical to provide a bonus for finishing early and therefore no penalty. It was a tit for tat contract.'

Errill argued that the insertion of nil in the Appendix, meant that Clause 24 of JCT 80 dealing with liquidated damages was excluded from the contract altogether and they could, therefore, claim general damages for breach. In this they were unsuccessful. A harsh judgment, perhaps, in terms of natural justice but an inevitable one given the express terms of the contract. This is certainly not the first time that one or both contracting parties have been dismayed by the literal interpretation of their contract by the courts and the comments of Lord Justice Browne-Wilkinson in *Northern Regional Health Authority* v. *Derek Crouch Construction Co. Ltd* (1984), although on a different matter, should be taken as a warning by all:

> 'In principle, in an action based on contract, the court can only enforce the agreement between the parties; it has no power to modify that agreement in any way. Therefore, if the parties have agreed on a specified machinery for establishing their obligations, the court cannot substitute a different machinery.'

The *Temloc* case does not provide a firm ruling on contracts where a 'dash' is made in the Appendix or where the rate space is left blank and it is possible that in such cases the ruling in *Temloc* does not apply.

In regard to *Baese* v. *Bracken* (1989), it has been suggested that the employer wrote 'nil' damages out of ignorance of the level of damages that would be suffered and meant no more than a dash or a blank. However, the ruling did not hinge on this and the case is not particularly helpful.

3.6 Double damages

The law does not permit the recovery of double damages for the same breach and accordingly it is not permissible to claim both liquidated damages and general damages for late completion. This may appear to be so obvious that there should be no need for further comment but construction contracts are rarely straightforward and problems with double damages do occur.

Sub-contracting

The most common cause is the stepping down into sub-contracts of the provisions of the main contract for liquidated damages, whilst at the same time including in the sub-contract additional provisions for the recovery of loss and expense or extra cost if the sub-contractor finishes his work late.

This was the situation in *M J Gleeson plc* v. *Taylor Woodrow Construction Ltd* (1989) where Gleeson were sub-contractors under a management form of contract. Clause 11 of the sub-contract made the sub-contractor liable for the loss and expense of the management contractor in the event of failure to complete on time, and Clause 32 of the sub-contract provided for the payment of liquidated damages at the same rate as in the main contract. Taylor Woodrow deducted from amounts due to Gleeson in respect of both liquidated damages and what were termed 'set-off' claims under Clause 11. It was held by Judge Davies that:

> 'Taylor Woodrow Construction's course of action against Gleeson in respect of the set-offs is for delay in completion. It follows that it is included in the set-off for liquidated damages, and to allow it to stand would result in what can be metaphorically described as a "double" deduction.'

The message in this for contractors using any form of sub-contract is: do not state liquidated damages unless they are intended to cover all loss arising from late completion by the sub-contractor.

Phased completions

Another common cause of double damage problems in construction contracts is the practice of stipulating phased or sectional completion obligations in addition to an overall completion date. Most standard forms of contract endeavour to deal with this in a logical manner and contain provisions for scaling down rates of liquidated damages to correspond with the value of any work handed over, or they contain sectional completion dates and liquidated damages relating thereto. However, the scope for getting it wrong is considerable as the following cases illustrate.

In *M J Gleeson (Contractors) Ltd* v. *London Borough of Hillingdon* (1970) under a JCT 63 contract the bills of quantities set out detailed provisions for sectional completion each with damages but the contract conditions were unamended and stipulated an overall date for completion also with damages. Relying on a clause in the contract that nothing in the bills could override the conditions, Mr Justice Mocatta held that the provisions in the bills were to be ignored.

In *Bramall & Ogden Ltd* v. *Sheffield City Council* (1983) again under JCT 63, liquidated damages were expressed at the rate of £20 per week for each uncompleted dwelling, but the Appendix gave only one date for completion and the works covered not only dwellings but also communal areas. It was held that, in the absence of provisions for sectional completion, liquidated damages could not be claimed.

Double breach / double damages

On this matter there is the case of *E. Turner & Sons Ltd* v. *Mathind Ltd* (1986) where Lords Justice Parker and Bingham, when hearing an appeal against summary judgment, made some intriguing comment, albeit *obiter,* and not therefore binding authority, on double damages and phased completions. In that case, the bills gave phased handover dates but the Appendix gave only a final completion date with damages at the rate of £1000 per week. When late completion of the phases occurred, the employer tried various approaches to calculating damages. Firstly, he divided the £1000 per week by the number of phases to arrive at a rate per phase but this had no legal basis. Secondly, he claimed £1000 per week per phase but this contradicted the contract. Finally he claimed that the liquidated damages were a penalty and he was entitled to general damages. In considering the argument by the contractor that liquidated damages for the whole of the works excluded general damages for the phases, Lord Justice Bingham said:

> 'The plaintiffs may ultimately be held to be correct in advancing that construction; but it has this odd consequence. Even though ex hypothesi the earlier completion dates are binding on the plaintiffs, the defendants would have no remedy in damages for breach in those terms. To achieve that result one would look for a clause excluding any right to damages for a breach which would, in the ordinary course, give a right to damages. The plaintiffs say there is such an exclusion in clause 22. Again, that clause must be construed in the context of what is ultimately held to be the whole contract between the parties; but it does not seem to me, standing alone, to be an effective exclusion of any right to damages for earlier breaches.'

He went on to say:

> 'It may be that, on a true construction of clause 22, the provision for liquidated damages at the stipulated rate applies upon failure to complete the last sub-area by the final completion date, leaving the defendants to their right for general damages for breach of the earlier obligations.'

Lord Justice Parker was even more forthright in his view that liquidated damages for the whole of the works should not necessarily exclude general damages for late completion of phases. He said:

> 'It appears to me that there is a perfectly good business reason for applying the liquidated damages only to the whole works, but no reason at all for saying that liquidated damages for any breach in that respect constitutes a ceiling to what may be recovered for failure to meet the successive handover dates.

Suppose, for example, that the first handover date was sixteen weeks after start. Suppose further that it was not met and that, although all other dates were met, the first phase was not ready for handover until the same date as the last phase. There would, in such circumstances, be no liquidated damages recoverable at all for the whole of the works would have been completed on time. But the employer would have been deprived of the profit earning capacity of phase I for some forty six weeks. Why then it can be asked, should he be entitled to no damages merely because the whole of the works were completed on time? If the provisions were contractual, a clear exclusion would, as it seems to me, be required to produce such a strange result and it is, to say the least, arguable that there was no such exclusion.

Provisions for phased handover in the case of a large development are of prime importance and may, in many circumstances, lead to a higher contract price. There is every reason to suppose that the parties may well have intended those provisions to be contractual and, if they did, then, apart from a specific overriding provision, they are contractual, and breach of them sounds in damages.'

Comment

Their Lordships' comments were greeted with some surprise but nothing was said to suggest that double damages should be paid for the same breach. What was under consideration was whether separate damages should apply to separate breaches.

Even as the law stands it may not be necessary for all of the damages for a particular breach to be stipulated within the liquidated sum. There are some damages which can be pre-estimated with reasonable precision and others which cannot. It should be possible to draft a liquidated damages clause which would expressly liquidate only part of the loss; providing that part was clearly specified, and the liability for general damages for any non-specified part was apparent.

3.7 Liability for damages in tort

The possibility of an employer side-stepping the liquidated damages provisions of the contract and claiming damages in negligence for late completion was cautiously canvassed in the years when the law of torts was in the ascendancy. Lord Denning had said in *Photo Production Ltd* v. *Securicor Transport Ltd* (1980):

'If the facts disclose the self-same duty of care arising both in contract and in tort and a breach of that duty, then the plaintiff can sue in either contract or tort.'

This rule is subject to the limitation that it is not permissible to seek a wider remedy in tort than is available under the contract. Lord Justice Cumming

Bruce in *William Hill Organisation Ltd* v. *Bernard Sunley & Sons Ltd* (1982) said:

> 'The Plaintiffs are not entitled to claim a remedy in tort which is wider than the obligations assumed by the defendants under their contract.'

The Privy Council in *Tai Hing Cotton Mill* v. *Liu Chong Hing Bank* (1986) confirmed this:

> 'Their Lordships do not, however, accept that the parties mutual obligations in tort can be any greater than those to be found expressly or by necessary implication in their contract.'

However, the Privy Council in that case also threw doubt on the existence of a parallel obligation in tort. Lord Scarman said:

> 'Their Lordships do not believe that there is anything to the advantage of the law's development in searching for a liability in tort where the parties are in a contractual relationship.'

Both *William Hill* and *Tai Hing* were considered by Judge Fox-Andrews in *Surrey Heath* v. *Lovell* (1988) where it was conceded that if a contract expressly deals with the subject of a claim there is no room for a parallel duty in tort but, it was claimed, where the contract does not so deal there is room for a claim in tort. The judge, in finding that the contract made provision for all the claims, held that the claim in tort could not succeed.

Recent cases

However, it should be noted that the law of tort continues to develop and continues to produce surprises. In a series of cases since 1990 the courts both in England and the Commonwealth have shown an increasing tendency to accept concurrent duties in contract and tort – subject only to restrictions in a contract expressly excluding remedies in tort. Thus in one of the cases concerning the Lloyd's insurance loses of the late 1980s – *Henderson* v. *Merrett Syndicates Ltd* (1994) – Lord Goff had this to say on the subject:

> 'My own belief is that, in the present context, the common law is not antipathetic to concurrent liability, and that there is no sound basis for a rule which automatically restricts the claimant to either a tortious or a contractual remedy.
>
> The result may be untidy; but, given that the tortious duty is imposed by the general law, and the contractual duty is attributable to the will of the parties, I do not find it objectionable that the claimant may be entitled to take advantage of the remedy which is most advantageous to him, subject only to ascertaining whether the tortious duty is so inconsistent with the applicable contract that, in accordance with ordinary principle, the parties must be taken to have agreed that the tortious remedy is to be limited or excluded.'

The swing back towards concurrent liability would not in itself be suffi-cient to suggest that damages for late completion could be pursued in tort rather than in contract. The House of Lords ruling in *Murphy* v. *Brentwood District Council* (1990) had appeared to curtail claims in tort for purely eco-nomic loss excepting those falling within what is known as the *Hedley Byrne* v. *Heller* principle. That is where there is a special relationship from which reliance emerges. However, that principle which at first seemed to apply only to negligent advice is now being extended. Thus in *Barclays Bank plc* v. *Fairclough Building Ltd* (1995) the Court of Appeal held that a person who undertakes skilled work and who fails to exercise the skill and care reason-ably to be expected of one professing his calling can be held liable either in tort or in contract. It was said that a skilled contractor undertaking mainte-nance work assumes a responsibility which invites reliance no less than the professional adviser does.

The consequences of these developments on damages for late completion will only be seen in time. But two obvious points of importance are that actions in tort can often be brought when actions in contract would be out of time and actions in tort can be brought against persons or parties with whom there is no contract.

It will certainly revolutionise the whole of the law and practice of liqui-dated damages if it is ever held that an employer who has recovered less than the full economic losses he has suffered from late completion of the main contract can successfully sue a defaulting sub-contractor who is respon-sible for the delay.

3.8 *The* Panatown *problem*

As a general rule of English law the losses recoverable by a party suing for damages for breach of contract are restricted to losses the party has itself suffered. There are long standing exceptions to this rule, originating from shipping cases, most notably *Dunlop* v. *Lambert* (1839) and *The Albazero* (1977).

However, in a series of cases in the 1990s English law moved some way towards abolishing the general rule, or at least extending the scope of the exceptions to include construction cases. The cases culminated in two House of Lords' decisions which have been much debated as to their application, meaning and effect. The first was *St Martins Property Corporation Ltd* v. *Sir Robert McAlpine* (1994); the second was *Alfred McAlpine Construction Ltd* v. *Panatown Ltd* (2000).

In the *St Martins Property* case it was a sister company St Martins Invest-ment (the developer) which actually suffered the loss under the building contract for which St Martins Property was the employer. The House of Lords applied the *Dunlop* v. *Lambert* exception in holding that the property company could recover from the contractor, McAlpine, the losses suffered by its sister company. However, it was made clear that the exception would not have applied if there was a direct contract (e.g. a collateral warranty) between McAlpine and the sister company.

Panatown

In the *Panatown* case similar circumstances applied except that there was a collateral warranty between the contractor and the developer although it was restrictive in its terms as to what was recoverable as damages. Panatown's claim against McAlpine for damages was upheld by the Court of Appeal notwithstanding the collateral warranty. The case went on appeal to the House of Lords which upheld McAlpine's appeal by the narrow margin of three to two.

Unfortunately it cannot be said that the judgments in *Panatown* have entirely clarified the law on third party losses. Their Lordships considered two issues, one described as 'the narrow ground' (whether the general rule subject to its exceptions applied), the other described as 'the broader ground' (whether Panatown had not received the bargain it contracted for). The unanimous decision on the narrow ground was to the effect that the existence of a deed of care between McAlpine and Panatown's sister company prevented Panatown recovering its sister company's losses (the sister company had its own remedies). The decision on the broader ground was evenly split with the fifth judge apparently favouring Panatown's case in principle but not on the facts.

Doubts remain therefore on the circumstances in which claims can be brought in contract for third party losses. In *Panatown* and the cases which preceded it the judges struggled with the problem of damages falling into a legal black hole. That problem was examined in some depth in the *Biffa Waste* v. *MEH* (2008) case mentioned earlier in this chapter. Mr Justice Ramsey, having considered the relationship between Biffa Waste's sister company Biffa Leicester and MEH and warranties relating thereto said:

'128. The provision of warranties ensures that the beneficiary does not have to rely on an action in contract against the next party in the contractual chain who may have no assets or, particularly in the case of a PFI contract, might be an associated company. The warranty also avoids the difficulties of establishing a duty of care against the contractor and consultants following *Murphy* v. *Brentwood DC* [1991] 1 AC 598. It also avoids the "legal blackhole" which might arise if the project owner, lessor, occupier or user suffered the loss but another party retained the cause of action.

129. The result of the increased use of warranties is that in many cases a party may give a number of warranties to different participants in a project. When such legal rights are given to two or more parties in relation to the same transaction, the arrangement overcomes the prospect of the "legal blackhole" or other problems in obtaining a

remedy but, as Lord Millett said in his dissenting speech in
McAlpine Construction v. *Panatown* [2001] 1 AC 518 at 595, the
existence of rights by a number of parties raises the *"spectre of double
recovery"* or multiple recovery.

130. In this case, Biffa's case on quantum demonstrates that there has
been uncertainty as to whether Biffa Leicester or Biffa Waste suffered certain losses caused by delay. In principle, if Biffa Leicester
suffered losses caused by a delay in completion it could bring proceedings against MEH under the Direct Agreement or it could bring
proceedings against Biffa Waste for breach of the time obligations
under the Works Agreement between Biffa Leicester and Biffa
Waste. If Biffa Leicester pursued the latter course or Biffa Waste was
concerned that it might be liable to Biffa Leicester then Biffa Waste
could claim against MEH for both its own losses caused by delay
and any sums which it had paid or was liable to pay to Biffa
Leicester under the Works Agreement for that delay.

131. In such circumstances, MEH could face liability for overlapping
damages both to Biffa Leicester under the Direct Agreement and to
Biffa Waste under the Design and Build Deed. That was a problem
which Lord Goff of Chieveley and Lord Millett had to deal with in
Panatown having reached the conclusion that both the employer,
Panatown and the building owner, UIPL, had causes of action
against McAlpine, the contractor, and could recover the cost of
remedying the defects.

132. Lord Goff at 560 considered that UIPL would generally leave it to
Panatown "to enforce its more valuable rights under the building
contract, rather than have recourse to its more uncertain remedy
under the [Duty of Care Deed] under which it has to prove negligence on the part of McAlpine." However at 561A, he considered
the position where UIPL suffered damage distinct from that covered
by Panatown's claim. In those circumstances, as he pointed out at
561C "a successful claim by UIPL against McAlpine in respect of
such damage could not give rise to any double recovery." At 561D
he said that if any such possibility should exist, it could be disposed
of by joinder of the relevant party or parties to the proceedings in
the manner indicated by in the speech of Lord Millett.

133. Lord Millett at 595 A to C said this:
"By giving the third party a cause of action, it raises the spectre
of double recovery. Even though the plaintiff recovers for his
own loss, this obviously reflects the loss sustained by the third
party. The case is, therefore, an example, not unknown in other
contexts, where breach of a single obligation creates a liability to
two different parties. Since performance of the primary obligation
to do the work would have discharged the liability to both parties,
so must performance of the secondary obligation to pay damages.
Payment of damages to either must pro tanto discharge the liability
to both."

134. Lord Millett then considered the relationship between the proceedings commenced by Panatown and proceedings commenced by UIPL. He said at 595 D to E:

     "While, therefore, I do not accept that Panatown's claim to substantial damages is excluded by the existence of the [Duty of Care Deed], I think that an action like the present should normally be stayed in order to allow the building owner to bring his own proceedings. The court will need to be satisfied that the building owner is not proposing to make his own claim and is content to allow his claim to be discharged by payment to the building employer before allowing the building employer's action to proceed."

135. In the present case, the existence of a cause of action by both Biffa Leicester and Biffa Waste against MEH gives rise to similar problems of double recovery. In the absence of a liquidated damages provision, the court could stay proceedings or stay a judgment to ensure that there was not double recovery by both Biffa Leicester and Biffa Waste in relation to the same heads of damage.

136. The liquidated damages clause has two effects on this position. First, it defines the liquidated amount to be paid per week by MEH to Biffa Waste as an exhaustive remedy for delay. Secondly, it limits the overall liability of MEH to Biffa Waste for delay to 7.5% of the Contract Price.

137. If Biffa Waste were permitted to recover liquidated damages and Biffa Leicester were permitted to recover unliquidated damages there would, in my judgment, be double recovery. The fact that, as I have found, Biffa Waste and MEH have agreed that Clause 47.1 provides an exhaustive remedy for failure to complete in accordance with clause 43 means that any additional sum recovered by Biffa Leicester would amount to recovery of damages in addition to the exhaustive recovery. If Biffa Leicester recovered additional damages then MEH would have to pay twice for the breach of Clause 43: once for the exhaustive remedy and a second time for a sum in excess of the exhaustive recovery. Such a spectre of double recovery is no more acceptable than it would have been in *Panatown*.

138. In this context, clause 2.2 of the Direct Agreement contemplates that the liability of MEH to Biffa Leicester should be no greater than if Biffa Leicester had been named as provider under the supply contract. If Biffa Leicester had been named as the provider in the supply contract then Biffa Leicester would have been entitled to liquidated damages. Equally, if Biffa Leicester were named as provider, either instead of or together with Biffa Waste then the liability of MEH would be limited to paying liquidated damages to one party or jointly to the two parties.

139. Whilst Clause 2.2 is phrased as a limit on liability, liquidated damages are the exhaustive remedy and when read with clause 47.1, I consider that this amounts to an agreement both that damages

are limited to the value of the liquidated damages but also an acceptance that liquidated damages are an exhaustive remedy.

140. I therefore consider that clauses 2.1 and 2.2 of the Direct Warranty give Biffa Leicester an entitlement to damages but on the basis that any payment of liquidated damages is an exhaustive remedy for the particular delay. This does not mean that both Biffa Leicester and Biffa Waste can make a recovery for the same delay. By using the mechanism of a stay as indicated in *Panatown*, I consider that the Court can overcome the spectre of double recovery.

141. If MEH pays Biffa Waste the liquidated damages under Clause 47.1 then MEH has complied with the terms of the Design and Build Deed and has no remaining liability so that Biffa Leicester can recover nothing. Meanwhile, I consider that Biffa Leicester's claim should be stayed.'

The law in Scotland

It should be noted, however, that under the law of Scotland there is no corresponding black hole. See, by way of example and explanation, the case of *Clark Contracts Ltd v. The Burrell Co. (Construction Management) Ltd* (2003) mentioned in Chapter 4 where it was held that *Panatown* should not be followed in Scotland.

Breach of duty claims

Although the *Panatown* problem relates principally to claims made by an employer against a contractor it may also have application to claims made by contractors against sub-contractors, suppliers and against professionals engaged to provide design services and the like.

Thus if a contractor is the building arm of a development company its ability to include in breach of duty claims against the designer losses suffered as a result of late completion of the development may depend not only on whether or not it is a separately registered legal entity but also on whether or not any collateral warranty has been provided by the designer to the developer.

Chapter 4
Liquidated damages and penalties

4.1 Penalties – general introduction

For centuries the courts of the United Kingdom have toiled with cases concerning penalties. There is a Scottish case, *Home* v. *Hepburn*, dating from 1549; and an English case, *Sloman* v. *Walter*, dating from 1783. Overall the number of judgments on the subject runs into hundreds and yet it retains its mysteries. Judgments continue to flow from the English, Scottish, Commonwealth and USA courts on a regular basis. Not all relate to construction disputes. The range also covers property, shipping, commerce and employment. Most of the cases concern damages for breach of contract but some concern charges payable on the occurrence of specified events. There has long been debate on whether the distinction between the two which the law presently recognises should be maintained.

The underlying cause of much of this is tension between concepts of freedom of contract and of equitable relief. Common law courts are instinctively disposed to uphold the bargain the parties have made for themselves but are reluctant to enforce contractual terms which, on examination, can rightly be described as penal or penalty clauses. Inevitably this leads to disputes on the true nature of particular terms and clauses and the need for examination of their true purpose and the extent to which they are oppressive, extravagant or unconscionable. In dealing with such disputes the courts have produced a wealth of case law but there is continuing necessity to give modern interpretation to earlier decisions, particularly those of great importance but which are now nearly a hundred years old. Additionally new styles of business, new procurement methods and new forms of contract bring with them new problems for the parties and more disputes for the courts to solve. All of which suggests, as Lord Hailsham once said on the subject, 'the last word has not yet been spoken'.

Laws on penalties

The common law approach to penalties for breach is that a plaintiff who sues for enforcement of a penalty can recover only the loss he can prove. Or as Lord Ellenborough said in *Wilbeam* v. *Ashton* (1807):

'Beyond the penalty you shall not go; within it you are to give the party any compensation which he can prove himself entitled to.'

It is not wholly clear how the jurisdiction of the courts to give relief against penalty clauses developed. Lord Justice Kay in the case of *Law* v. *Local Board of Redditch* (1892) suggested that originally the courts of equity granted relief where a sum of money was agreed to be paid as a penalty for non-performance in circumstances where it was possible to ascertain the actual loss suffered.

Another explanation is that a penalty was originally held to be in the nature of a threat held over the other party 'in terrorem'. The courts of equity took the view that since a penalty was designed to secure performance, the promisee was sufficiently compensated by being indemnified for his actual loss and he was not entitled to demand a sum which, although fixed by agreement, might be disproportionate to the actual loss suffered. Whatever the background common law now applies the same approach, as did the courts of equity. Thus it was said in *Public Works Commissioner* v. *Hills* (1906) that a penalty covers but does not assess the damage.

The legal effect of this, as approved in *Watts, Watts & Co. Ltd* v. *Mitsui & Co. Ltd* (1917), is that when there is a penalty clause the plaintiff may sue either on the penalty clause, in which case he cannot recover more than the stipulated sum, or he can ignore the penalty clause and sue for breach of contract to recover damages in full. In either case damages can only be recovered to the extent they are proved.

The common law approach to charges which become due on the occurrence of specified events is significantly different from its approach to penalties for breach. Albeit that such charges may sometimes be called penalties, recovery is not dependent on proof of loss or damage and is subject only to observance of any statutory rules applicable to the particular type of transaction. Relief from enforcement requires the paying party to show that in truth the charge is related to breach.

Various attempts have been made over the years to bring the two approaches closer together and more into line with the civil laws of continental jurisdictions. Thus the Law Commission published in 1975 its Working Paper No. 61 on Penalty Clauses and Forfeiture of Monies Paid and the Scottish Law Commission published in 1999 its Report on Penalty Clauses.

European and international developments

The Scottish report has this to say of European and international developments in approaches to penalties:

> '1.7 There has been recent European and international activity in the area of penalty clauses. The Council of Europe published a report on penalty clauses in 1978. The Committee of Ministers recommended that the governments of the member states took into consideration the principles in the appendix to their Resolution when preparing new legislation on this subject. In 1983 UNCITRAL adopted *Uniform Rules on Liquidated Damages and Penalty Clauses*, for international

contracts and the General Assembly of the United Nations recommended that States should give serious consideration to the rules and, where appropriate, implement them in the form of either a model law or a convention. The *Principles of International Commercial Contracts,* produced by UNIDROIT and the *Principles of European Contract Law* prepared by the Commission on European Contract Law under the chairmanship of Professor Ole Lando both contain articles on 'Agreed payment for non-performance'. The issue has also been considered by law reform bodies in England, California, Canada and Australia.

1.8 There has been a convergence between the formerly disparate approaches of civil and common law countries. In countries whose law was heavily influenced by the English Common law, penalty clauses were once viewed as completely unenforceable. In countries whose law was heavily influenced by the Napoleonic code, however, penalty clauses were fully enforceable and were seen as an effective way to encourage performance and thus avoid litigation. However, most modern or recently revised civil codes now depart from the general principle of literal enforcement by allowing penalties to be modified where they are "disproportionately high" or "excessively high" or "excessive" or "unreasonable" or "manifestly excessive". In common law systems the distinction between penalties and liquidated damages can be used, or deliberately blurred, to allow recovery of many sums which the parties have agreed should be payable in the event of non-performance. Thus, in many systems it seems that a degree of compromise has been accepted in order to minimise the tension between the desire to enforce what was agreed between the parties and the injustice of enforcing an excessively penal provision.

1.9 This convergence of approaches is reflected in recent international instruments on the subject. The Council of Europe's Resolution on Penalty Clauses, for example, assumes that penalty clauses are, in general enforceable, but provides that

> "The sum stipulated may be reduced by the court when it is manifestly excessive".

The Principles of European Contract Law provide that

> "(1) Where the contract provides that a party who fails to perform is to pay a specified sum to the aggrieved party for such non-performance, the aggrieved party shall be awarded that sum irrespective of his actual loss.
>
> (2) However, despite any agreement to the contrary the specified sum may be reduced to a reasonable amount where it is grossly excessive in relation to the loss resulting from the non-performance and the other circumstances."

The Unidroit *Principles* have a virtually identical provision.'

Penalties arising other than on breach

The above-quoted extracts from the Scottish report are concerned mainly with whether or not amounts in contracts stipulated as penalties, or otherwise held to be penalties, are excessive. However, as noted at the start of this chapter, that is only part of the problem. There is also the problem of the different approaches in law to penalties for breach and penalties arising other than on breach.

Modern authority for the rule that the doctrine of penalties does not apply to stipulated payments enforceable on the happening of specified events is found in the House of Lords majority view in the case of *Export Credits Guarantee Dept Products Co.* v. *Universal Oil* (1983). There, Lord Roskill, agreeing with Lord Diplock, said:

> 'I, for my part, am not prepared to extend the law by relieving against an obligation in a contract entered into between two parties which does not fall within the well defined limits in which the Court has in the past shown itself willing to interfere.'

Lord Denning, in the minority famously disagreed.

Commenting on the existing law, the Scottish Report on Penalty Clauses said:

> '4.4 Because the law on penalty clauses applies only when there is a breach of contract, the law seems to favour the party who acts in breach rather than the party who complies with the terms of the contract. This is because the party in breach can seek judicial scrutiny of a penalty whilst the other party may not.
>
> "The hirer who honestly admits that he cannot keep up payments and terminates his agreement may have to pay a penalty; his less responsible neighbour, who simply goes on failing to pay the installments until the finance company is forced to take action, may escape . . . I have felt myself oppressed by that consideration. But the remedy is for the legislature." – *Mercantile Credit Co. Ltd* v. *McLachlan* (1962)
>
> 4.5 Take, for example, the instance of a contract terminated on an event such as insolvency or the appointment of an insolvency practitioner. As we have seen, the rules on penalty clauses do not apply. Potentially, claims made in the insolvency may therefore be out of all proportion to any loss. Indeed, they may be extravagant or unconscionable or excessive. This could severely prejudice other creditors and might provide an incentive to draft extortionate provisions, and to have a termination without a breach.
>
> 4.6 Indeed there exists general scope for avoiding the rules on penalties by drafting contracts so that, instead of providing for one method of performance with a penalty for breach, they provide options for

performing in different ways, some of which may attract heavy penal consequences.'

The report went on to recommend that:

'2 Judicial control over contractual penalties should not be confined to cases where the penalty is due if the promisor is in breach of contract. It should extend to cases where the penalty is due if the promisor fails to perform, or to perform in a particular way, under a contract or when there is an early termination of a contract.'

However, as things presently stand, it remains no more than a recommendation for future law reform and cases will continue to arise where there is debate as to whether a stipulated sum is payable on a breach or payable on occurrence of a specified event.

The Law Commission Working Paper (1975) considers the matter in some detail. It includes the following comments and example:

'17. In the present law, before a sum due can be struck down as an invalid penalty there must be a breach of contract. Thus, if a sum of money is payable on an event other than a breach of contract it is not open to the courts to hold that the sum is a penalty. This distinction between sums payable on breach and sums payable otherwise than on breach has emerged in a number of hire-purchase cases which will be referred to below, but it could arise in other types of agreement.

18. The distinction just described could arise in any contract which entitled one contracting party to perform in alternative ways. For example, a contract to clean the windows of a house might be drawn in one of two ways:

   (a) it might provide that "X hereby agrees to clean the windows on January 1 and in default of so doing to pay Y the sum of £100 as liquidated damages for the breach".

   or

   (b) it might provide that "X hereby agrees to clean the windows on January 1 or, at his option, to pay Y the sum of £100 instead, and X shall be taken to have exercised his option to pay the £100 if he does not clean the windows on January 1".

Existing authority clearly lays down that the contract in form (a) entitles the court to decide whether the sum of £100 is truly liquidated damages or is an unenforceable penalty, but suggests that in form (b) if X neither cleans the windows nor pays the £100 he is not in breach of any obligation to clean the windows; his only breach would be in respect of his promise to pay £100 and provided there was consideration for this promise he could be sued for the £100, the court having no power to grant relief. This example is deliberately an extreme one, but seems to follow from the authorities.'

4.2 *Liquidated damages*

'The essence of liquidated damages is a genuine covenanted pre-estimate of loss.' So said Lord Dunedin in *Dunlop Pneumatic Tyre Company Ltd* v. *New Garage* (1915). The characteristic of liquidated damages is that loss need not be proved.

Meaning of genuine pre-estimate

Interpretation of Lord Dunedin's definition has not always been without difficulty. In *Widnes Foundry (1925) Ltd* v. *Cellulose Acetate Silk Co. Ltd* which travelled through three layers of the courts in 1930, 1931 and 1932, a sum of £20 per week was stipulated as a penalty for late completion of an acetone recovery plant. The silk company sued for their full losses of £5850 for 30 weeks of late completion. The foundry company argued that their liability was limited to £600. It was not disputed that the silk company's losses were in the region of £5850. The judge at first instance, Mr Justice Wright, took Lord Dunedin's phrase literally and held that because the stipulated sum was obviously not a genuine pre-estimate of loss it could not be liquidated damages. He suggested that a sum could be a penalty under Lord Dunedin's rules by reason of its extravagant inadequacy as well as by reason of its exorbitance. He found favour of the silk company.

In the Court of Appeal his decision was overturned.

Lord Justice Scrutton said:

'I have come to the conclusion that this case comes within the rule laid down by Lord Dunedin, following earlier cases, that if the parties agree a figure fixing the amount for breach of one stipulation varying with the time of the delay, it is not a penalty but is liquidated damages.'

He went on to say:

'Here is one obligation – namely, to deliver something by a fixed date, and there is a clause fixing a payment which is commensurate with the amount of the delay in the delivery. The phrase "genuine covenanted pre-estimate of damage" does not state it in terms accurately.'

Lord Justice Slesser, in the same case, stated:

'To read the word "penalty" in this contract literally would lead to an absurdity, as it cannot be a penalty if the person is benefiting. In fact, this is a benefit to the person breaking the contract, and does not penalise him in any way. We may therefore properly read the expression as intended to be liquidated damage.'

The House of Lords confirmed the decision of the Court of Appeal on the basis that the stipulated sum was by way of compensation. Lord Atkin made the following pronouncement:

'Except that it is called a penalty, which on the cases is far from con-clusive, it appears to be an amount of compensation measured by the period of delay. I agree that it is not a pre-estimate of actual damage. I think it must have been obvious to both the parties that the actual damage would be much more than £20 a week; but it was intended to go towards the damage, and it was all that the sellers were prepared to pay.'

Australian view

In the Australian case of *W T Malouf Pty Ltd* v. *Brinds Ltd* (1981) it was said:

'A genuine pre-estimate means a pre-estimate which is objectively of that character: that is to say, a figure which may properly be called so in the light of the contract and the inherent circumstances. It will not be enough merely that the parties honestly believed it to be so.'

Genuine pre-estimate or lesser sum

It follows from the *Widnes Foundry* case that sums stipulated as liquidated damages may be either a genuine pre-estimate of loss or such smaller sums as the parties may agree. It will be no barrier to liquidated damages that the stipulated sums are patently inadequate as recovery of full loss.

More recently in *Multiplex Constructions Pty Ltd* v. *Abgarus Pty Ltd* (1992) it was said:

'It is clear as a matter of principle, and established by authority, that if parties agree upon a quantum of damage as liquidated damages which is less than the damage which would be suffered from such breach, no attack can be made upon such a liquidated damages provision upon the basis that it is extravagant or unconscionable.'

Liquidated damages need not be proved

When liquidated damages are a genuine pre-estimate of loss or a lesser sum it is fundamental that loss does not have to be proved to obtain recovery. Providing they are within the terms of a binding contract the courts will not allow a challenge on the basis that loss cannot be proved.

The parties to a contract are bound by its terms and the courts will enforce those terms except where they are penalties or they are caught by other legal impediments. The courts will not alter the contract the parties have made for themselves.

The basic difference between general damages which do have to be proved, and liquidated damages which do not have to be proved, is widely misunderstood. Perhaps it is because both types of damages can flow from breach of contract. But it comes down to this: general damages are not contemplated in the contract, whereas liquidated damages are not only contemplated but stipulated. There is no reason why sums for liquidated damages, save for the possibility that they are penalties, should suffer any more scrutiny than other sums in the contract.

Loss need not be suffered

It is only one step from the principle that liquidated damages can be recovered without proof of loss to the principle that they can also be recovered even when it is apparent there has been no loss. Why should the contract-breaker be excused his promise to pay damages by inquiry or reference into the circumstances of the innocent party?

The issue came up in the case of *BFI Group of Companies Ltd* v. *DCB Integration Systems Ltd* (1987). DCB, the contractor, carried out alteration and refurbishment work at BFI's transport depot. BFI were given possession on the extended date for completion but it was another six weeks before roller shutter doors were installed. BFI utilised this time to fit out the premises and did not suffer any delay in commissioning or any loss of revenue. Various disputes went to arbitration and on the matter of delay the arbitrator held there was a delay in completion but he declined to award liquidated damages on the grounds that BFI had suffered no loss. The case went to appeal where it was argued for the contractor that provisions for liquidated damages presupposed some loss and served only to quantify such loss. Where it was found there was no loss they should not apply. Judge Davies rejected this argument and accepted that it was irrelevant to consider whether there was any loss; the liquidated damages provisions worked automatically once breach was established.

4.3 Liquidated damages and penalties distinguished

Because payment of liquidated damages can be claimed without proof of loss, and payment of penalties claimed only with proof of loss, it is obviously a matter of considerable financial interest to the parties to a contract that they should know, preferably in advance of making their contract, whether the sums they have stipulated for breach are liquidated damages or penalties.

The terminology itself is not decisive and if the courts find, as a matter of construction, that liquidated damages are penalties or vice versa they will award accordingly. Thus, Lord Dunedin in *Dunlop Pneumatic Tyre Company* v. *New Garage & Motor Company Ltd* (1915) said:

> 'Though the parties to a contract who use the words "penalty" or "liquidated damages" may prima facie be supposed to mean what they say, yet

the expression is not conclusive. The Court must find out whether the payment stipulated is in truth a penalty or liquidated damages.'

This search for 'truth' has not always been without criticism. The Lord Chamberlain, Lord Cranworth, in *Ranger* v. *GWR* (1854) said:

'I am not sure that benefit has, on the whole, resulted from the struggles which courts, both of law and equity, have made to relieve contracting parties from payments which they have bound themselves to make by way of penalty. Such a course may have been very reasonable and useful where the damage resulting from the violation of the contract is capable of being exactly measured, but whenever the quantum of damage is in its nature uncertain and the due performance of it has been secured, or purports to have been secured, by a penalty, it might, perhaps, have been safer and more convenient to have always understood the parties as meaning what their language imports namely, that on failure to perform the contract, the stipulated penalty should be paid. But this has not always been the doctrine of the courts. The distinction between a penalty and a fixed sum as the conventional amount of damages is too well established to be now called in question, however difficult it may be to say in any particular case under which head the stipulation is to be classed.'

And over a century later, Lord Justice Diplock in *Robophone Facilities Ltd* v. *Blank* (1966) made the following statement:

'The court should not be astute to decry a penalty clause in every provision of a contract which stipulates a sum to be payable by one party to the other in the event of a breach by the former. Such stipulation reflects good business sense and is advantageous to both parties. It enables them to envisage the financial consequences of a breach; and if litigation proves inevitable it avoids the difficulty and the legal costs, often heavy, of proving what loss has in fact been suffered by the innocent party.'

How the courts distinguish between liquidated damages and penalties

In deciding whether, in the particular circumstances of the cases before them, stipulated sums are liquidated damages or penalties there can be many factors for the courts to consider, some of them typical, some of them unique to the case. Commonly, however, the courts may hear arguments on the true purpose of the payment provisions, the clarity or ambiguity of the provisions, the reasonableness or otherwise of the charges, the basis of calculation of the charges, the relative bargaining powers of the parties, and the virtue of upholding the agreement between the parties. Over the years the courts have developed sets of rules to facilitate consistency in the decision making process which for the most part do just that. Nevertheless, as times change, differing interpretations of the rules emerge and debate arises as to their order of precedence.

As will be seen from the cases which follow, the courts are not slow to express their views on the current standing of the law. The picture which emerges is that although attempts have been made to shift the focus of examination away from whether there is genuine pre-estimate of loss and towards examination of the commercial agreement between the parties there remains great respect for the House of Lord's ruling in *Dunlop Pneumatic Tyre Company* v. *New Garage and Motor Company Ltd* given in 1915. In hierarchical terms only the House of Lords ruling in the *Widnes Foundry* case and the Privy Council ruling in *Philips Hong Kong* v. *Attorney General of Hong Kong* (1993) approaches the same status. And although some interpretations of the Privy Council ruling suggested a shift away from the genuine pre-estimate of loss test in *Dunlop Tyre* the recent set of Court of Appeal cases examined later in this chapter indicates reluctance to move too far away from the ruling.

Early rulings

Kemble v. *Farren* (1829)

A contract for the defendant to appear as principal comedian at Covent Garden Theatre at the rate of £3.6s.8d per night for four seasons, contained a provision that if either party failed to fulfil the contract or any part thereof, such party should pay the other party by way of liquidated damages the sum of £1000. The defendant refused to act during the second season and was sued.

It was held that the sum of £1000 was a penalty because, had the plaintiff failed to make a single payment of £3.6s.8d, he would have been liable to pay £1000 and had the defendant contravened any regulations of the theatre, however minute, he would have been similarly liable.

That the payment of a very large sum in consequence of the non-payment of a very small sum should not be a penalty, was a contradiction in terms.

Lord Elphinstone v. *Monkland Iron and Coal Co.* (1886)

It was from the *Elphinstone* case that Lord Dunedin took his ruling that a single sum can be presumed to be a penalty in some circumstances. He said:

'I think *Elphinstone's* case, or rather the dicta in it, do go this length, that if there are various breaches to which one indiscriminate sum to be paid in breach is applied, then the strength of the chain must be taken at its weakest link. If you can clearly see that the loss on one particular breach could never amount to the stipulated sum, then you may come to the conclusion that the sum is penalty. But further than this it does not go.'

Law v. Redditch Local Board (1892)

'The distinction between penalties and liquidated damages depends on the intention of the parties to be gathered from the whole of the contract. If the intention is to secure performance of the contract by the imposition of a fine or penalty, then the sum specified is a penalty; but if, on the other hand, the intention is to assess the damages for breach of the contract, it is liquidated damages.' – Mr Justice Lopes

Public Works Commissioner v. Hills (1906)

A contract for construction of a railway provided that the contractor should forfeit the retention moneys under the contract 'as and for liquidated damages' for late completion.

It was held that since the amount of retention money would depend on progress with the works it was an indefinite sum and could not be a genuine pre-estimate of loss. It was, therefore, to be regarded as a penalty.

Ford Motor Company v. Armstrong (1915)

Armstrong, a retailer, agreed not to sell any Ford car, or any part thereof, below list price; and for every breach was to pay £250 as agreed damages.

It was held that the £250 was a penalty since it could become payable for a breach which would cause only trifling damage.

The *Dunlop Tyre* case (1915)

The historic definitive ruling on the distinction between liquidated damages and penalties comes from the case of *Dunlop Pneumatic Tyre Company Ltd* v. *New Garage & Motor Company Ltd* (1915). Lord Dunedin's classic judgment in that case remains the test on which subsequent judgments have relied. The point at issue was whether a price maintenance agreement which bound Dunlop's customer, New Garage & Motor Company, not to sell below list prices and concluded 'we agree to pay to Dunlop Pneumatic Tyre Company Ltd the sum of £5 for each and every tyre, cover or tube sold or offered in breach of this agreement, as and by way of liquidated damages and not as a penalty', was a liquidated damages or a penalty clause. The trial master found the stipulated sum to be liquidated damages but that finding was reversed by a majority in the Court of Appeal. The case then went to the House of Lords.

Lord Dunedin began by reviewing the law:

'I do not think it advisable to attempt any detailed review of the various cases, but I shall content myself with stating succinctly the various propositions which I think are deducible from the decisions which rank as authoritative:

(1) Though the parties to a contract who use the words "penalty" or "liquidated damages" may prima facie be supposed to mean what they say, yet the expression used is not conclusive. The court must find out whether the payment stipulated is in truth a penalty or liquidated damages.

(2) The essence of a penalty is a payment of money stipulated as in terrorem of the offending party; the essence of liquidated damages is a genuine covenanted pre-estimate of damage.

(3) The question whether a sum stipulated is penalty or liquidated damages is a question of construction to be decided upon the terms and inherent circumstances of each particular contract, judged of as at the time of making the contract, not as at the time of the breach.

(4) To assist this task of construction, various tests have been suggested, which if applicable to the case under consideration may prove helpful, or even conclusive. Such are:

   (a) It will be held to be a penalty if the sum stipulated for is extravagant and unconscionable in amount in comparison with the greatest loss that could conceivably be proved to have followed from the breach.

   (b) It will be held to be a penalty if the breach consists only in not paying a sum of money, and the sum stipulated is a sum greater than the sum which ought to have been paid. This, though one of the most ancient instances, is truly a corollary to the last test.

   (c) There is a presumption (but no more) that it is a penalty when "a single sum is made payable by way of compensation, on the occurrence of one or more or all of several events, some of which may occasion serious and others but trifling damage".

   (d) It is no obstacle to the sum stipulated being a genuine pre-estimate of damage that the consequences of the breach are such as to make precise pre-estimation almost an impossibility. On the contrary, that is just the situation when it is probable the pre-estimated damage was the true bargain between the parties.'

Lord Dunedin then went on to consider the facts:

'Turning now to the facts of the case, it is evident that the damage apprehended by the appellants owing to the breaking of the agreement was an indirect and not a direct damage. So long as they got their price from the respondents for each article sold, it could not matter to them directly what the respondents did with it. Indirectly it did. Accordingly, the agreement is headed "Price Maintenance Agreement", and the way in which the appellants would be damaged if prices were cut is clearly explained in evidence ... and no successful attempt is made to controvert that evidence. But though the damage as a whole from such a practice would be certain, yet damage from any one sale would be impossible to forecast.

It is just, therefore, one of those cases where it seems quite reasonable for parties to contract that they should estimate that damage at a certain figure, and provided that figure is not extravagant there would seem no reason to suspect that it is not truly a bargain to assess damages, but rather a penalty to be held in terrorem.'

The unanimous opinion of the four Law Lords was that the stipulated sum was by way of liquidated damages and the award of the trial master in favour of Dunlop was upheld.

The *Philips Hong Kong* case (1993)

Because of its importance the case of *Philips Hong Kong Ltd* v. *Attorney General of Hong Kong* (1993) is considered here at some length.

Philips entered into a contract with the Hong Kong Government to design, supply, install and commission a computerised supervisory system for the approach roads and twin tunnels for a new route in the New Territories. The contract was one of seven contracts, six of which contained a flow chart setting out the programme for the progress of the work and also flow charts for the work for five of the other contracts. The flow charts identified Key Dates at which Philips' work interfaced with the programmes of other contracts. Clause 27 of the Philips contract imposed an obligation on Philips to meet its Key Dates and Clause 29 provided that if the Key Dates were not met liquidated damages were payable. Additional liquidated damages were also payable if the whole of the work was not completed within a specified time.

Philips sought, and obtained from the court of first instance, declarations that clause 29 was unenforceable because it was penal and because it was uncertain. The Court of Appeal of Hong Kong allowed an appeal that Clause 29 had no application because there was no statement in the contract which permitted the takeover of sections before the issue of the certificate of taking over of the whole of the works. Philips appealed to the Privy Council.

The Privy Council rejected the appeal and held, amongst other things:

(1) The liquidated damages provision in the contract between the parties was a genuine pre-estimate and enforceable, since:
(2) The purpose of being able to agree beforehand the damage recoverable for a breach of contract was to the advantage of the parties since they should be able to estimate with a reasonable degree of certainty the extent of their liability and the risks which they run. This is particularly true of building and engineering contracts.
(3) The court should not adopt an approach to provisions as to liquidated damages which could defeat their purpose.
(4) To identify that a provision is penal on an objective assessment it will not normally be enough to identify situations where the application of the provision could result in a larger sum being recovered by the injured party than his actual loss.

(5) The test for determining whether a provision for the deduction of liquidated and ascertained damages is a penalty is whether or not it is a genuine pre-estimate of what the loss is likely to be.

(6) The fact that the issue has to be determined objectively, judged at the date the contract was made, does not mean that what actually happens subsequently is irrelevant. It can provide valuable evidence as to what could reasonably be expected to be the loss at the time the contract was made.

(7) In the case of a governmental body the nature of the loss is especially difficult to evaluate. The Government reasonably adopted a formula which reflected the loss of returns on the capital involved at a daily rate, to which were added figures for supervisory staff costs, the daily actual cost of making any alternative provision and a sum for fluctuations.

(8) The Government would not by this approach receive double compensation (i.e. for the delay in not meeting a Key Date and when the contract completion date was not met) as different losses arose from each.

(9) The provision for a minimum payment was not penal as the Government would inevitably have to incur expense of a standing nature irrespective of the scale of work outstanding.

(10) Although the provisions could have been drafted with greater clarity they were not so uncertain as to be unenforceable.

Lord Woolf set out the basis of the Philips attack on the liquidated damages clause as follows:

> 'At this stage Mr Nicholas Dennys QC does not suggest on behalf of Philips that the sum claimed by the government by way of liquidated damages is in fact exorbitant in view of the very substantial delay which in fact occurred in the execution of this contract by Philips. Instead he bases his argument on what could have happened in a number of different hypothetical situations. He suggests that if one or more of those situations had happened, the sum which would then be payable by way of liquidated damages would be wholly out of proportion to any loss which the Government was likely to suffer in that situation and that this is sufficient to establish that the provisions are penal in effect. If Philips' approach is correct this would be unsatisfactory. It would mean that it would be extremely difficult to devise any provision for the payment of liquidated damages in the case of a contract of this sort which would not be open to attack as being penal. As is the case with most commercial contracts, there is always going to be a variety of different situations in which damage can occur and even though long and detailed provisions are contained in a contract it will often be virtually impossible to anticipate accurately and provide for all the possible scenarios. Whatever the degree of care exercised by the draftsman it will still be almost inevitable that an ingenious argument can be developed for saying that in a particular hypothetical situation a substantially higher sum will be recovered

than would be recoverable if the plaintiff was required to prove his actual loss in that situation. Such a result would undermine the whole purpose of parties to a contract being able to agree beforehand what damages are to be recoverable in the event of a breach of contract. This would not be in the interest of either of the parties to the contract since it is to their advantage that they should be able to know with a reasonable degree of certainty the extent of their liability and the risks which they run as a result of entering into the contract. This is particularly true in the case of building and engineering contracts. In the case of those contracts provision for liquidated damages should enable the employer to know the extent to which he is protected in the event of the contractor failing to perform his obligations.

As for the contractor, by agreeing to a provision for liquidated damages, he is seeking to remove the uncertainty as to the extent of his liability under the contract if he is unable to comply with his contractual obligations.'

Lord Woolf then went on to consider the question:

'Is it sufficient for a contractor to identify hypothetical situations where the effect of the application of the clause may be to produce a sum payable to the employer substantially in excess of the damage which the employer is likely to suffer in order to defeat the intended effect of a clause freely entered into by the parties providing for the payment of liquidated damages?'

After reviewing the approach of the courts to liquidated damages clauses, Lord Woolf concluded that it was not sufficient. He said:

'Except possibly in the case of situations where one of the parties to the contract is able to dominate the other as to the choice of the terms of a contract, it will normally be insufficient to establish that a provision is objectionably penal to identify situations where the application of the provision could result in a larger sum being recovered by the injured party than his actual loss. Even in such situations so long as the sum payable in the event of non-compliance with the contract is not extravagant, having regard to the range of losses that it could reasonably be anticipated it would have to cover at the time the contract was made, it can still be a genuine pre-estimate of the loss that would be suffered and so a perfectly valid liquidated damage provision. The use in argument of unlikely illustrations should therefore not assist a party to defeat a provision as to liquidated damages.'

Regarding the argument which had been upheld in the case of *Arnhold & Co. Ltd* v. *Attorney General of Hong Kong* (1989) that a proportioning down clause with a stop figure was a penalty, Lord Woolf said:

'The second point arises due to the presence of the minimum payment provision. The argument is based on the judgment of Sears J in *Arnhold* v. *Attorney General of Hong Kong* (1989) 47 BLR 129 which Mayo J followed in this case. There can conceivably be circumstances where it is so obvious, before completion of the works as a whole, that the actual loss which will be sustained will be less than a specified minimum figure that to include that minimum figure in a provision for the payment of liquidated damages on a reducing sliding scale will have the effect of transforming an otherwise perfectly proper liquidated damages provision into a penalty, in so far as it prevents the liquidated damages from being reduced below that figure. However this is certainly not such a case and, so far as it is possible to ascertain the facts from the report [in the *Construction Law Journal*] which is available, nor was *Arnhold*.

To conclude otherwise involves making the error of assuming that, because in some hypothetical situation the loss suffered will be less than the sum quantified in accordance with the liquidated damage provision, that provision must be a penalty, at least in the situation in which the minimum payment restriction operates. It illustrates the danger which is inherent in arguments based on hypothetical situations where it is said that the loss might be less than the sum specified as payable as liquidated damages. Arguments of this nature should not be allowed to divert attention from the correct test as to what is a penalty provision – namely is it a genuine pre-estimate of what the loss is likely to be? – to the different question, namely are there possible circumstances where a lesser loss would be suffered? Here the minimum payment provision amounted to about 28% of the daily rate of liquidated damages payable for non-completion of the whole works by Philips. The government point out that if there is delay in completion it will continue inevitably to incur expenses of a standing nature irrespective of the scale of the work outstanding and that those expenses will continue until the work is completed. This being a reasonable assumption and there being no ground for suggesting that the minimum payment limitation was set at the wrong percentage, its presence does not create a penalty.'

And on the argument that the liquidated damages clause was void for uncertainty, Lord Woolf said:

'Finally it is contended that the manner in which the liquidated damages provisions are expressed in the contract results in such uncertainty as to the manner in which they were intended to operate that they are unenforceable. This contention is also misconceived. The effect of the provisions could have been drafted with greater clarity, but their meaning can be ascertained and therefore relied on by the government.'

Comparison – *Dunlop Tyre / Philips Hong Kong*

Comparison of the rulings in *Dunlop Tyre* and in *Philips Hong Kong* reveals the key point to be that in both cases the stated test for determining whether a provision is for liquidated damages or is a penalty is whether or not the stipulated sum is a genuine pre-estimate of loss. Additionally both rulings say that the question is to be determined objectively and judged as at the time the contract was made.

Where *Philips Hong Kong* goes further than *Dunlop Tyre* is that it emphasises that the court should not adopt an approach to liquidated damages which would defeat their purpose. It also makes the points that evidence as to what happened can assist in judgments on pre-estimates; that formulae methods of assessment may be permissible; and that insignificant or hypothetical drafting problems do not render liquidated damages provisions unenforceable.

A further point of note is that *Philips Hong Kong* apparently moves away from the 'in terrorem' aspect of distinguishing between liquidated damages and penalties.

Post *Philips Hong Kong* developments

The *Philips Hong Kong* case was subject to a great deal of scrutiny in legal circles and elsewhere as to whether the law had significantly changed in its approach to penalty clauses. Some courts and commentators seem to have concluded that there was such a change and that the underlying 'genuine pre-estimate of loss' derived from the *Dunlop Tyre* case had, by *Philips Hong Kong* and some Commonwealth cases, been superseded. Others were more cautious.

The following extracts from English High Court and Court of Appeal judgments illustrate the thinking of the courts:

Lordsvale Finance v. *Bank of Zambia* (1996)

The judgment of Mr Justice Colman in the case of *Lordsvale Finance Plc v. Bank of Zambia* (1996), although given in the High Court on an application for summary judgment contains rulings which have clearly influenced decisions subsequently given by the Court of Appeal.

The case related to rates of interest payable on defaults of complex financial instruments. One issue was whether the interest provisions should be classed as penalties. The judge started by stating that the issue was of far-reaching importance in the English law of banking. He explained this as follows:

'London is one of the greatest centres of international banking in the world. Here and in New York most of the world's international syndicated loans are set up. Such loans almost invariably provide for enhanced rates of default interest to apply. It would be highly regrettable if the English courts were to refuse to give effect to such prevalent provisions while the courts of New York are prepared to enforce them. In the absence of compelling reasons of principle or binding authority to the contrary there can be no doubt that the courts of this country should adopt in international trade law that approach to the problem which is consistent with that which operates in that nation which is the other major participant in the trade in question. For there to be disparity between the law applicable in London and New York on this point would be of great disservice to international banking.'

After examining world-wide case law, some old, some modern, the judge concluded:

'In my judgment, weak as the English authorities are, there is every reason in principle, for adopting the course which they suggest and for confining protection of the creditor by means of designation of default interest provisions as penalties to retrospectively-operating provisions. If the increased rate of interest applies only from the date of default or thereafter there is no justification for striking down as a penalty a term providing for a modest increase in the rate. I say nothing about exceptionally large increases. In such cases it may be possible to deduce that the dominant function is in terrorem the borrower. But nobody could seriously suggest that a 1 per cent rate increase could be such. It is in my judgment consistent only with an increase in the consideration for the loan by reason of the increased credit risk represented by a borrower in default.

For these reasons I conclude that clause 10.03A contains nothing in the nature of a penalty and that the default interest provision must be fully enforced.'

Discussion

Interesting as the above extracts are in themselves, the part of the judgment which has caught the attention of a number of judges and Lord and Lady Justices in later cases is the part which reads:

'The speeches in *Dunlop Pneumatic Tyre Co.* v. *New Garage & Motor Co.* show that whether a provision is to be treated as a penalty is a matter of construction to be resolved by asking whether at the time the contract was entered into the predominant contractual function of the provision was to deter a party from breaking the contract or to compensate the innocent party for breach. That the contractual function is deterrent rather than compensatory can be deduced by comparing the amount that would be payable on breach with the loss that might be sustained if breach occurred.'

For discussion on the question of whether or not this passage suggests that old tests for considering pre-estimates of loss should be replaced by a compensatory test based on comparison of stipulated damages with damages which can be proven, see the quoted extracts from the *Leisureplay* case below.

Indian Airlines v. GIA International (2002)

In *Indian Airlines Ltd* v. *GIA International Ltd* (2002) one of the claims the High Court had to consider in an application for summary judgment was a claim by Indian Airlines for US$5,550,000 as liquidated damages arising from breach of an airline leasing agreement. The relevant clause of the agreement read:

> 'If leasing of the aircraft pursuant to this agreement does not, other than due to any default by lessee, commence on or by the expected delivery date, the lessor shall, promptly on demand, pay to the lessee as liquidated damages the amount of $ 8,500 (the delay payment) for each day following expected delivery date until either (1) the aircraft is delivered to lessee in accordance with the provisions hereof and the lease period commences, or (2) the lessee exercises its option under clause 2.8 to terminate its obligation to lease the aircraft.'

GIA failed to deliver any of the promised aircraft and Indian Airlines terminated the lease agreement and sought to apply the provision for liquidated damages. GIA opposed the claim on grounds that the provision was a penalty clause in that the stipulated sum was not a genuine pre-estimate of loss and was oppressive. It was conceded by Indian Airlines that the stipulated sum was not a calculated pre-estimate of loss but it argued that it was a reasonable amount in the circumstances.

Mr Justice Tomlinson, after referring to the *Dunlop Tyre, Philips Hong Kong* and *Clydebank Engineering* cases, said this:

> '71. The Philips case in the Privy Council marks, in my judgment, something of a sea change in the approach of the courts to penalty clauses. I note, for example, that Lord Woolf, giving the advice of the Judicial Committee, cited, with approval, the view of Dixon J. in the Supreme Court of Canada in *Elsy* v. *JG Collins Insurance Agencies* [1978] 83 DLR at 15, where he said:
> "It is now evident that the power to strike down a penalty clause is a blatant interference with freedom of contract and is designed for the sole purpose of providing relief against oppression for the party having to pay the stipulated sum. It has no place where there is no oppression."
>
> 72. That passage has also received the approval of the High Court of Australia in *Isander Finance Corporation* v. *Plesnick* [1989] ALJ 238. Furthermore, in a powerful judgment delivered by Mason J. and Wilson J. in the High Court of Australia in *AMEV-UDC Finance Ltd* v. *Austin* (1986) 162 CLR 170, those learned Judges said this:
> "But equity and the common law have long maintained a supervisory jurisdiction not to re-write contracts imprudently made but to relieve against provisions which are so unconscionable or oppressive that their nature is penal rather than compensatory. The test to be

applied in drawing that distinction is one of degree, and will depend on a number of circumstances including (1) the degree of disproportion between the stipulated sum and the loss likely to be suffered by the plaintiff, a factor relevant to the oppressiveness of the terms of the defendant, and (2) the nature of the relationship between the contracting parties, a factor relevant to the unconscionability of the plaintiff's conduct in seeking to enforce the term. The courts should not, however, be too ready to find the requisite degree of disproportion lest they impinge on the parties' freedom to settle for themselves the rights and liabilities following a breach of contract. The doctrine of penalties answers in situations of the present kind an important aspect of the criticism often levelled against unqualified freedom of contracts, namely the possible inequality of bargaining power. In this way the courts strike a balance between the competing interests of freedom of contract and the protection of weak contracting parties – see generally Atiyah, The Rise and Fall of Freedom of Contract (1979)."

73. I find in the reference to the requisite degree of disproportion a distinct echo of the manner in which Lord Dunedin had put the matter in the *Dunlop Pneumatic Tyre* case, where he suggested that a provision would be held to be a penalty if the sum stipulated for is extravagant and unconscionable in amount in comparison with the greatest loss that could conceivably be proved to have followed from the breach.'

Later in his judgment the judge said:

'79. As Mr Coleman submitted, the Court when considering whether or not a clause should be struck down as an unenforceable penalty must stand back and look at the situation as it would have appeared to the parties at the time, and must apply the test which has been established by long authority, which is essentially a test of unconscionability. Notwithstanding what is said by Mr. Basu as to his anxiety to obtain the business on behalf of GIA, I simply cannot regard the situation which is revealed as being one of inequality of bargaining power of the sort which is referred to by Mason J. and Wilson J. in their judgment to which I have referred.

80. In my judgment, this is a case very similar to the Clyde Bank Engineering case, in which it was simply impossible for a precise pre-estimate to be made of the consequences of delay in delivery of the aircraft, which delay, I entirely accept, would have been anticipated not to be a delay running over many months, but to be a delay measured in days and weeks rather than in months and years. If an airline such as Indian Airlines had planned its schedules and its expansion on the expectation that it would have received five new aircraft within a period of four or five weeks, it would be obvious that the disruption to its schedules and the inability to earn profits

in consequence of delay in delivery would be considerable but incapable of precise calculation.'

He concluded, having found that the stipulated sum was neither extravagant nor unconscionable in comparison with the greatest loss that could conceivably be proved to have followed the breach:

> '83. In my judgment, to strike down this agreement made between two substantial parties operating in the international aviation market would indeed be a blatant interference with freedom of contract, and, bearing that in mind that the doctrine is intended to provide relief against oppression, I cannot believe that there is any room for the application of the doctrine here, where there may have been hard bargaining but, in my judgment, nothing which comes even close to oppression.
> 84. For all those reasons, therefore, which I fear I have expressed at undue length, I have concluded that the Defendants have no realistic prospect of establishing at trial that the liquidated damages clause, clause 2.7, is an unenforceable penalty, and I therefore conclude that the Claimant should be given summary judgment for liquidated damages in the sum of $ 5,550,000.'

Jeancharm v. Barnet Football Club (2003)

The case of *Jeancharm Limited* v. *Barnet Football Club Limited* (2003) concerned the supply of football kit and the like. The case is worthy of note for the manner in which the Court of Appeal rejected the proposition that the law had moved on from *Dunlop Tyre*. Mr Justice Jacob, after reviewing the early law on penalty clauses, said this:

> '10. Most recently, penalty clauses were considered by the Privy Council in *Philips Hong Kong Ltd* v. *The Attorney General of Hong Kong* [1993] 61 BLR 41. In that case the Privy Council considered decisions from Australia and Canada. But, as I read the decision, it did not depart from the law as laid down by Lord Dunedin in *Dunlop*.
> 11. Mr Kay suggested that following *Philips* and the Australian decision (referred to) the law had moved on from what was stated by Lord Dunedin to the extent that it had virtually abandoned it. In particular, he suggested that one should look at the contract as a whole, look at the risks being undertaken by both sides and ask whether the clause was an appropriate clause, having regard to the risk undertaken by the opposite party. Here, for instance, he said that his clients were at very considerable risk if they were in late delivery, having regard to the 20 pence per garment per day clause, and that should be balanced against the interest for late payments.
> 12. I can find nothing in the *Philips* case that suggests a departure of that gigantic nature from the law as laid down by Lord Dunedin. Lord Dunedin indicated that the question of whether or not a clause was

a penalty clause depended upon whether it could be regarded as a genuine pre-estimate of the damage caused if there was a breach. Mr Kay's formulation abandons that entirely.

13. If one goes to *Philips*, a passage from the joint decision of Mason and Wilson JJ in the Australian case, *AMEV-UDC Finance Ltd* v. *Austin* (1986) 162 CLR 170, is quoted by Lord Woolf in the advice to Her Majesty and reads as follows:

"But equity and the common law have long maintained a supervisory jurisdiction, not to rewrite contracts imprudently made, but to relieve against provisions which are so unconscionable or oppressive that their nature is penal rather than compensatory. The test to be applied in drawing that distinction is one of degree and will depend on a number of circumstances, including (1) the degree of disproportion between the stipulated sum and the loss likely to be suffered by the plaintiff, a factor relevant to the oppressiveness of the term to the defendant, and (2) the nature of the relationship between the contracting parties, a factor relevant to the unconscionability of the plaintiff's conduct in seeking to enforce the term. The courts should not, however, be too ready to find the requisite degree of disproportion lest they impinge on the parties' freedom to settle for themselves the rights and liabilities following a breach of contract. The doctrine of penalties answers, in situations of the present kind, an important aspect of the criticism often levelled against unqualified freedom of contract, namely the possible inequality of bargaining power. In this way the courts strike a balance between the competing interests of freedom of contract and protection of weak contracting parties: see generally Atiya, The Rise and Fall of Freedom of Contract (1979), especially Chapter 22."

14. Mr Kay particularly relies upon item (2) identified by Mason and Wilson JJ:

". . . the nature of the relationship between the contracting parties, a factor relative to the unconscionability of the plaintiff's conduct."

and elevates that to the leading principle for deciding whether or not a clause is or is not a penalty. But, immediately following that passage, Lord Woolf said this:

"It should not be assumed that, in this passage of their judgment, Mason and Wilson JJ were setting out some broader discretionary approach than that indicated as being appropriate by Lord Dunedin. On the contrary, earlier in their judgment they had noted that the '*Dunlop* approach' had been eroded by recent decisions and they stated that there was much to be said for the view that the courts should return to that approach."

15. It boils down to this, that since *Dunlop* the courts have continued to apply the rule in *Dunlop* but have held that one should be careful before deciding whether or not a clause is a penalty when the parties are of equal bargaining power. There was no abandon-

ment of the rule that the clause must be a genuine pre-estimate of damage.'

Lord Justice Keene, agreeing, went on to say:

'20. ... It is quite clear from the authorities that the concept of a penalty clause is not confined to situations where one party had a dominant bargaining power over the other, although it may, of course, often apply in such a situation: see the decision of the Privy Council in *Philips v Hong Kong Ltd* where the opinion was delivered by Lord Woolf, in particular the passage at the foot of page 58:
"~~Except possibly in the case of situations~~ where one of the parties to the contract is able to dominate the other as to the choice of the terms of a contract, it will normally be insufficient to establish that a provision is objectionably penal to identify situations where the application of the provision could result in a larger sum being recovered by the injured party than his *actual* loss."' *(emphasis added)*

His Lordship went on to explain that one can have a range of losses that could have reasonably been contemplated when the contract was made.

'21. It is perfectly clear from that passage that the Privy Council there was recognising that the situation where one party is dominant is not exhaustive of those contracts where a penalty may be identified. Indeed, were it otherwise, the concept would have little relevance in most commercial contracts, as Mr Kay himself recognises. That case also rightly rejected the proposition that there is some broader discretionary approach to be applied: see Lord Woolf at page 58 in the first paragraph.

22. The test, in my judgment, remains one of ascertaining whether the provision is a genuine pre-estimate of loss or is a penalty for non-performance of the contractual obligation, as was established in *Dunlop* and as *Philips*, more recently, has endorsed. The first type of provision is essentially compensatory in nature. The second is there to deter the party in question from breaking the contract by providing for a punitive level of payment. If one applies that test to the present case, one is bound to conclude that on its face an interest rate of 260% per annum would seem to be penal in nature. The evidence below did not establish that it was a genuine pre-estimate of loss, nor did the judge make any such finding.

23. In those circumstances, I cannot see how this clause can be seen as having anything other than a deterrent function. I conclude that it is unenforceable as a penalty.'

Lord Justice Peter Gibson, also agreeing, concluded his comments as follows:

'27. The principles that are relevant, in my judgment, for distinguishing a penalty provision, with which the courts will interfere from a valid contractual provision for a payment or payments in the event of default by a party are these:

(1) the court looks at the substance of the matter, rather than the form of words, to determine what was the real intention of the parties;

(2) the essence of a penalty is a required payment in terrorem of the party in default, as distinct from being a genuine pre-estimate of loss resulting from the default;

(3) the question whether a provision for payment on default is a penalty is a question of construction of the contract, and that is assessed at the time of the contract and not at the time of the breach;

(4) if the required payment is extravagant and unconscionable in amount in comparison with the greatest loss that could conceivably be established as the consequence of a default, it is a penalty.

28. It follows, therefore, that the court is concerned to construe the contract to see whether the intention of the parties at the time of the contract was to deter the paying party from falling into default. If it can be seen that the provision is a genuine pre-estimate of the loss that will result from the default, then there can be no penalty.

29. In my judgment, it is plain from the magnitude of the interest rate that in this case there was no genuine pre-estimate of loss. It may be that, as Mr Kay argued, the parties thought that if the purchaser was going to be allowed a damages clause for late delivery of the goods to be ordered, then the vendor should be given a very large rate of interest in the event of late payment. But, in my judgment, the authorities show that one concentrates on the relevant clause said to be a penalty, and, on the application of the test so clearly laid down in *Dunlop*, in my judgment it is plain that in this case the interest clause far exceeds anything that could be said to be a genuine pre-estimate of actual loss and amounts to a penalty.'

Alfred McAlpine v. Tilebox (2005)

The case of *Alfred McAlpine Capital Projects Ltd* v. *Tilebox Ltd* (2005) concerned substantial sums allegedly due as liquidated damages for late completion of an office development project. Mr Justice Jackson, after reviewing the authorities, made these observations:

'48. Let me now stand back from the authorities and make four general observations, which are pertinent to the issues in the present case.

1. There seem to be two strands in the authorities. In some cases judges consider whether there is an unconscionable or extravagant disproportion between the damages stipulated in the contract and the true amount of damages likely to be suffered. In other cases the courts consider whether the level of damages

stipulated was reasonable. Mr Darling submits, and I accept, that these two strands can be reconciled. In my view, a pre-estimate of damages does not have to be right in order to be reasonable. There must be a substantial discrepancy between the level of damages stipulated in the contract and the level of damages which is likely to be suffered before it can be said that the agreed pre-estimate is unreasonable.

2. Although many authorities use or echo the phrase "genuine pre-estimate", the test does not turn upon the genuineness or honesty of the party or parties who made the pre-estimate. The test is primarily an objective one, even though the court has some regard to the thought processes of the parties at the time of contracting.

3. Because the rule about penalties is an anomaly within the law of contract, the courts are predisposed, where possible, to uphold contractual terms which fix the level of damages for breach. This predisposition is even stronger in the case of commercial contracts freely entered into between parties of comparable bargaining power.

4. Looking at the bundle of authorities provided in this case, I note only four cases where the relevant clause has been struck down as a penalty. These are *Commissioner of Public Works v Hills* [1906] AC 368, *Bridge v Campbell Discount Co. Ltd* [1962] AC 600, *Workers Trust and Merchant Bank Ltd v Dojap Investments Ltd* [1993] AC 573, and *Ariston SRL v Charly Records* (Court of Appeal 13th March 1990). In each of these four cases there was, in fact, a very wide gulf between (a) the level of damages likely to be suffered, and (b) the level of damages stipulated in the contract.'

Mr Justice Jackson went on to conclude that the provisions were not unenforceable as a penalty.

Murray v. Leisureplay (2005)

The case of *Murray v. Leisureplay Plc* (2005) concerned termination of an employment contract which provided for liquidated damages equal to 12 months' salary and benefits in the event of termination without 12 months' notice. However, having terminated the contract at short notice the employer argued that the liquidated damages clause was unenforceable as a penalty because it failed to take into account the employee's duty to mitigate his losses by finding other employment and therefore it gave the employee more than he would receive in an ordinary claim for damages for breach. The judge at first instance agreed on the basis that an enforceable liquidated damages clause would have to make significant allowance for other income. On appeal that decision was reversed.

The principal judgment in the Court of Appeal was given by Lady Justice Arden. Although eventually coming to the conclusion that the clause was not a penalty, Lady Justice Arden said, amongst other things:

'39. In essence, this court [the Court of Appeal] held in the *Cine* case, that in determining whether provisions were a penalty the court had at the outset of its enquiry to look at the aggregate amount that would be payable on breach under the terms of the agreement, and compare that with what would have been payable if UIP had had to bring its claim under the common law. In other words the alleged genuine pre-estimate of loss in clause 17 had to relate to the overall net balance of losses payable on termination less the credits to which Cine would have been entitled at common law.'

and

'42. What, to my judgment, is striking about the statement of the law in the *Cine* case and its application is the way in which the court sought objectively to rationalise its conclusions as to whether the provisions of the agreement constituted a penalty. The court's reasoning turns on a comparison between the overall amount payable under the agreement in the event of a breach with the overall amount that would have been payable if a claim for damages for breach of contract had been brought at common law. The court proceeded on the basis that, if such a comparison discloses a discrepancy, which can be shown not to be a genuine pre-estimate of damage or to be unjustified, the agreement provides for a penalty.'

and

'54. With the benefit of the citation of authority given above, in my judgment, the following (with the explanation given below) constitutes a practical step by step guide as to the questions which the court should ask in a case like this:
i)   To what breaches of contract does the contractual damages provision apply?
ii)   What amount is payable on breach under that clause in the parties' agreement?
iii)   What amount would be payable if a claim for damages for breach of contract was brought under common law?
iv)   What were the parties' reasons for agreeing for the relevant clause?
v)   Has the party who seeks to establish that the clause is a penalty shown that the amount payable under the clause was imposed *in terrorem*, or that it does not constitute a genuine pre-estimate of loss for the purposes of the *Dunlop* case, and, if he has shown the latter, is there some other reason which justifies the discrepancy between i) and ii) above?'

Lord Justice Buxton, agreeing that the appeal should be allowed, arrived there by a different route. He said:

'109. I respectfully agree with my Lady in her paragraph 47, citing the observations of Mance LJ in the *Cine* case, that the language of

stipulations *in terrorem* sounds unusual in modern ears; and particularly when applied to a contract such as the present, where a
company well able to look after itself employed to play a leading
and entrepreneurial role in its affairs a Chief Executive who, as his
evidence cited by my Lady demonstrates, was motivated by a desire
to protect his own interests.

110. That insight requires a recasting in more modern terms of the classic
test set out by Lord Dunedin in *Dunlop* [1915] AC at p 86:

"The essence of a penalty is a payment of money stipulated as in
terrorem of the offending party; the essence of liquidated damages
is a genuine covenanted pre-estimate of damage."

That recasting is to be found in the judgment of Colman J in *Lordsvale
Finance plc v. Bank of Zambia* [1996] QB 752 at 762G, a passage cited with
approval by Mance LJ in paragraph 13 of his judgment in the *Cine* case
[2003] EWCA Civ 1699:

"whether a provision is to be treated as a penalty is a matter of
construction to be resolved by asking whether at the time the contract was entered into the predominant contractual function of the
provision was to deter a party from breaking the contract or to
compensate the innocent party for the breach. That the contractual
function is deterrent rather than compensatory can be deduced by
comparing the amount that would be payable on breach with the
loss that might be sustained if the breach occurred."

111. It is important to note that the two alternatives, a deterrent penalty;
or a genuine pre-estimate of loss; are indeed alternatives, with no
middle ground between them. Accordingly, if the court cannot say
with some confidence that the clause is indeed intended as a deterrent, it appears to be forced back upon finding it to be a genuine
pre-estimate of loss. That choice illuminates the meaning of the
latter phrase. "Genuine" in this context does not mean "honest";
and much less, as the argument before us at one stage suggested,
that the sum stipulated must be in fact an accurate statement of the
loss. Rather, the expression merely underlines the requirement that
the clause should be compensatory rather than deterrent.'

Lord Justice Buxton, explaining his disagreement with Lady Justice Arden's
approach then said:

'113. First, Colman J said no more than that the comparison was a guide
to the assessment of a provision as deterrent rather than compensatory. That also, in my view, is as far as this court went in the *Cine*
case itself. That was a summary judgment case, involving no more
than the identification of a triable issue: I would draw attention in
that connexion to the observations of Thomas LJ in his paragraph
[50] and of Peter Gibson LJ in his paragraph [54]. The approach that
should be applied at trial would be in more general terms than that
suggested by my Lady in her paragraph 42, that always requires a
comparison between the liquidated and the common law damages

to see if the comparison discloses a discrepancy; and then requires that discrepancy to be justified as a genuine pre-estimate of damages, or by some other form of justification.

114. I venture to disagree with that approach because it introduces a rigid and inflexible element into what should be a broad and general question. It is also inconsistent with warnings by judges of high authority that, at least in connexion with commercial contracts, great caution should be exercised before striking down a clause as penal; and with the tests that they have postulated to that end. My Lady has cited in her paragraph 66 the observations of Diplock LJ in *Robophone v. Blank* [1966] 1 WLR 1428 at p 1447. I would add the well-known passage of Lord Woolf in *Philips Hong Kong v. A-G of Hong Kong* (1993) 61 BLR 49 at pp 58–59:

"Except possibly in the case of situations where one of the parties to the contract is able to dominate the other as to the choice of the terms of a contract, it will normally be insufficient to establish that a provision is objectionably penal to identify situations where the application of the provision could result in a larger sum being recovered by the injured party than his actual loss. Even in such situations so long as the sum payable in the event of non-compliance with the contract is not extravagant, having regard to the range of losses that it could reasonably be anticipated it would have to cover at the time the contract was made, it can still be a genuine pre-estimate of the loss that would be suffered and so a perfectly valid liquidated damages provision."

And exclusive concentration on the factual difference between the liquidated and the contractual damages overlooks a principal test formulated by Lord Dunedin to identify a penalty, [1915] AC at p 87, that

"It will be held to be a penalty if the sum stipulated for is extravagant and unconscionable in amount in comparison with the greatest loss that could conceivably be proved to have followed from the breach."'

Lord Justice Clarke, preferring the broader approach of Lord Justice Buxton to that of Lady Justice Arden said:

'106. The essential reasons which have led me to the conclusion that clause 17.1 is not a penalty are these:
   i) Given the general principle that *pacta sunt servanda*, the courts should be cautious before holding that a clause in a contract of this kind is a penalty.
   ii) The modern approach to Lord Dunedin's test in *Dunlop Pneumatic Tyre v. New Garage and Motor Company Ltd* [1915] AC 67 at 86 is to be found in *Lordsvale Finance plc v. Bank of Zambia* [1996] QB 752 per Colman J at page 762G and *Cine Bes Filmcilik Ve Yapim Click v. United International Pictures* [2003] EWCA Civ 1699.

iii) It is perhaps no longer entirely appropriate to ask whether a payment on breach was stipulated *in terrorem* of the offending party but, as Colman J put it in the *Lordsvale* case at page 762G (in a passage quoted by both Arden and Buxton LJJ):

"whether a provision is to be treated as a penalty is a matter of construction to be resolved by asking whether at the time the contract was entered into the predominant contractual function of the provision was to deter a party from breaking the contract or to compensate the innocent party for breach."

iv) Colman J continued:

"That the contractual function is deterrent rather than compensatory can be deduced by comparing the amount that would be payable on breach with the loss that might be sustained if the breach occurred."

I do not read Colman J as saying there that, if that comparison discloses a discrepancy, it follows that the clause is a penalty. It seems to me that the comparison is relevant but no more than a guide to the answer to the question whether the clause is penal: see e.g. *Philips Hong Kong* v. *A-G of Hong Kong* (1993) 61 BLR 49 per Lord Woolf at 58–9.

v) In paragraph 15 of his judgment in the *Cine* case (set out by Arden LJ at paragraph 39) Mance LJ quoted a further passage from the judgment of Colman J in the *Lordsvale* case (at pages 763g–764a) where he said that a particular clause might be commercially justifiable, provided that its dominant purpose was not to deter the other party from breach.

vi) As I see it, each case depends upon its circumstances and, in considering those circumstances, the court should have in mind the warnings to which Arden and Buxton LJJ have adverted. They include the importance to the parties both of knowing what will be the financial consequences to them of a breach of contract (*Robophone Facilities* v. *Blank* [1966] 1 WLR 1428 per Diplock LJ at 1447) and of avoiding disputes (*Kemble* v. *Farren* (1829) 6 Bing 141 per Tindal CJ at 148). They also include the statements to the effect that a clause will only be held to be a penalty if the sum payable on breach is extravagant or unconscionable: see eg the *Philips Hong Kong* case per Lord Woolf at page 59 and *Dunlop* per Lord Dunedin at page 87.'

Comment

In short, to the extent that it may have appeared from some judgments that the broad tests of genuine pre-estimates of loss derived from *Dunlop Tyre* and *Philips Hong Kong* had been superseded by narrower comparison tests between stipulated amounts and common law damages, the majority ruling by the Court of Appeal in *Leisureplay* makes clear that the broad tests still apply.

Cine v. United International Pictures (2003)

The *Cine* judgment referred to in *Leisureplay* was given by the Court of Appeal in 2003. Its full name is *Cine Bes Filmcilik Ve Yapimcilik & Anr* v. *United International Pictures & Ors.*

One question for the court was whether there was a triable issue on whether certain clauses of a license agreement between the parties were unenforceable as penalty clauses. It was argued by Cine that application of the clauses would over-compensate UIP for any losses suffered. It was said that the stipulated amounts failed to take into account various benefits which UIP could realise on termination for breach by Cine.

In holding that there was a triable issue Lord Justice Mance referring to the judgment of Mr Justice Colman in *Lordsvale Finance Plc* v. *Bank of Zambia* (1996) said:

> '13. Although the phrase *in terrorem* has appeared in many cases since *Dunlop*, there is force in Lord Radcliffe's comment in *Campbell Discount Co. Ltd* v. *Bridge* [1962] AC 600, 622, that
>
> "I do not find that that description adds anything to the idea conveyed by the word 'penalty' itself, and it obscures the fact that penalties may quite easily be undertaken by parties who are not in the least terrorised by the prospect of having to pay them . . ."
>
> A more accessible paraphrase of the concept of penalty is that adopted by Colman J in *Lordsvale Finance Plc* v. *Bank of Zambia* [1996] QB 752, 762G, when he said that *Dunlop Pneumatic Tyre* showed that:
>
> "whether a provision is to be treated as a penalty is a matter of construction to be resolved by asking whether at the time the contract was entered into the predominant contractual function of the provision was to deter a party from breaking the contract or to compensate the innocent party for breach. That the contractual function is deterrent rather than compensatory can be deduced by comparing the amount that would be payable on breach with the loss that might be sustained if breach occurred."'

Lord Justice Thomas, examining the balance of account point, said:

> '46. The unusual feature of the dispute in relation to these clauses arises because in addition to making the payment of various specified sums due on breach, including the AB amount, clause 17 also expressly provided for the termination of all the rights to the films for which payment had already been made or was due. UIP therefore were in a position to exploit the balance of the licence period of such films for their own benefit, even though they had been paid in respect of that period by Cine 5. UIP were not, however, under the terms of the

clause required to bring into account against what was to be paid to them (including the AB amount) any benefit they received from that exploitation.

47. If damages were to be calculated in the ordinary way for the loss UIP had suffered from breach, it was accepted by UIP that any benefits that UIP obtained through exploiting the balance of the period by licensing the films to others would have to be brought into account; such benefits would be ones arising in the ordinary course of business as a consequence of the termination. However the contract provided no mechanism for such benefits to be brought into account under clause 17.

48. The short issue was whether UIP had for the purposes of summary judgment application under Part 24 shown that there was no realistic prospect of Cine 5 establishing that clauses 16 and 17 were penal, even though the clauses failed to provide a mechanism for bringing the benefits of exploitation into account against sums that were payable, other than those already accrued due.

49. It is clear from the authorities to which reference has already been made by Mance LJ that a clause may be penal if it cannot be justified as being a genuine pre-estimate of the loss which the innocent party will incur by reason of the breach. A decision on whether it is penal involves a careful examination of the circumstances which is not possible on a Part 24 application. There is, in my view, sufficient evidence to suggest that it would have been envisaged that on breach, the termination of the rights to the films for which payment had been made would confer benefits upon UIP; however, the terms of the clauses make it clear that these were not to be brought into account against the sums payable on termination including the AB amount.

50. In those circumstances I cannot reach a final conclusion that the clauses were not penal; they might not be, but there are two issues. First, in my view the question must be investigated at trial as to why what appear to be benefits which would accrue to UIP were not to be brought into account, if the clauses were to operate as a genuine pre-estimate of the loss to be suffered by the innocent party on breach. A genuine pre-estimate would ordinarily imply consideration being given to bringing into account the material and significant matters that went to the ascertainment of the actual loss suffered by the innocent party. Because there was a failure to include within the clauses a provision for bringing into account a benefit to the innocent party that would at first sight have been obvious to the parties at the time the contract was made, there is, in my view, a real issue as to whether the clauses were intended as a genuine pre-estimate of the loss. There might, of course, be reasons why the clauses were not penal, even though the benefits were not to be brought into account under their terms, but no conclusion can be reached, in my view, without the investigation appropriate to a trial.

51. Second, in my view clause 17 cannot be severed in the way suggested by UIP so that the AB amount remained payable, even if the remainder of the clause was penal. The clause was concerned with the sums to be paid on termination; apart from the provision which confirmed the obligation to pay the sums already accrued due, it should be looked at as a whole as to what is to be paid on termination. If the apparent benefit, arising from the termination of the film rights, which was to be derived from the opportunity to exploit the balance of the licence period of films for which Cine 5 had paid, had to be brought into account, it had to be brought into account against the sums which became payable on termination including the AB amount. That was because the AB amount was but one component of the whole of the clause the validity of which depended on whether it provided for a genuine pre-estimate of the loss.'

Comment

Given the complexities of the *Cine* case perhaps the most that can be taken from it is that there may be circumstances where a balance of account needs to be considered in making a genuine pre-estimate of loss. But, as explained in *Leisureplay*, it is not authority for the proposition that a genuine pre-estimate of loss requires comparison between liquidated and common law damages, nor should it be taken as authority that a balance of account is always necessary.

Summary review

1915

The House of Lords, in the *Dunlop Tyre* case, states that the essence of a penalty is payment of money 'in terrorem' and that the essence of liquidated damages is a genuine pre-estimate of damage. It provides various tests for distinguishing penalties from liquidated damages including the rule that a sum which is extravagant and unconscionable in comparison with the greatest loss which could be proved is a penalty. It confirms that it is no obstacle to a sum being a genuine pre-estimate of loss even if precise pre-estimation is almost impossible.

1915 to 1993

Different approaches by the courts to the application of the *Dunlop Tyre* rules to gradually emerge, with some judgments putting emphasis on pre-estimate of loss considerations and others emphasising freedom of contract and the upholding of commercial agreements.

1993–

The Privy Council ruling in the *Philips Hong Kong* case reasserts application of the *Dunlop Tyre* tests (in particular the genuine pre-estimate of loss test) but states that the courts should not adopt an approach to liquidated damages which would defeat their purpose.

1993 to 2002

Application of the purposeful approach to liquidated damages approved in the *Philips Hong Kong* case leads to some courts moving away from 'in terrorem' and genuine pre-estimate of loss tests towards compensatory / comparison tests. The 1996 High court judgment in the *Lordsvale* case recasts the rule in the *Dunlop Tyre* case and in the 2002 *Indian Airlines* case the High Court judge says that the *Philips* case marks 'something of a sea change in the approach of the courts to penalty clauses'.

2003

The Court of Appeal in the *Jeancharm* case re-emphasises application of the pre-estimate of loss rule and says there is nothing in the Philips case 'to suggest a departure of gigantic nature from the law as laid down by Lord Dunedin' (in the *Dunlop Tyre* case).

2005

Lord Justices Buxton and Clarke in the Court of Appeal ruling in the *Leisureplay* case both say that the modern approach to the *Dunlop Tyre* 'in terrorem' test is the recasting of the test as stated in the *Lordsvale* case:

> 'whether a provision is to be treated as a penalty is a matter of construction to be resolved by asking whether at the time the contract was entered into the predominant contractual function of the provision was to deter a party from breaking the contract or to compensate the innocent party for breach.'

They differ, however, with Lady Justice Arden on how Mr Justice Colman's view, 'That the contractual function is deterrent rather than compensatory can be deduced by comparing the amount that would be payable on breach with the loss that might be sustained if the breach occurred' should operate – preferring to see this as a guide to what constitutes a genuine pre-estimate of loss rather than a rule.

4.4 Pre-estimates of damage

As noted in many places above, the essence of liquidated damages is that they are a genuine pre-estimate of loss. In commercial projects this usually presents no difficulty; the concept of loss is easy to understand and the calculations can be based on figures which can be readily substantiated. In non-commercial projects, for public sector works and the like, the logic is not as straightforward and the argument is often heard that liquidated damages cannot be applied for late completion of a road contract or school building contract because the employer has suffered no loss.

Such an argument is wrong in fact and wrong in law. In fact, because the employer will usually have suffered a loss if only in extra supervision costs or financing charges and, in law, because the difficulty of precise calculation has long been recognised by the courts and provided that a genuine attempt is made at pre-estimating loss, such loss will be accepted as liquidated damages. Since loss need not be proved it then matters not what actual loss, if any, has been suffered.

Non-commercial loss

The problem of non-commercial loss came to be considered by Lord Halsbury in the case of *Clydebank Engineering & Shipbuilding Co. Ltd* v. *Yzquierdo y Castaneda* (1905) where a contract for the building of four warships provided that 'the penalty for later delivery shall be at the rate of £500 per week for each vessel'. The ships were delivered late but it was held that the sum of £500 per week was liquidated damages and not a penalty. The contractors argued that the £500 per week could not be a genuine pre-estimate of loss since there was no loss as 'a warship does not earn money', but Lord Halsbury, in refuting the argument and holding the sum stated to be liquidated damages said:

> 'It is a strange and somewhat bold assertion to say that, in the case of a commercial ship, the damages could easily be ascertained, but that the same principle could not be applied to a warship as it earned nothing. The deprivation of a nation of its warship might mean very serious damage, although it might not be very easy to ascertain the amount. But is that a reason for saying they were to have no damages at all? It seems to me hopeless to advance such a contention. It is only necessary to state the assertion to show how absurd it is.
>
> I should have thought that the fact that a warship is a warship, her very existence as a warship capable of use for such and such a time would prove the fact of damage if the party was deprived of it, although the actual amount to be earned by it, and in that sense to be obtained by the payment of the price for it, might not be very easily ascertained . . .'

In the case of *Multiplex Constructions Pty Ltd* v. *Abgarus Pty Ltd* (1992) the judge had this to say on non-commercial losses:

'Thirdly, if the arguments addressed by the builder are correct in relation to works of a public nature, such as dams or major road works, where traditionally such public works do not yield a cash flow, or any cost of capital incurred in the works, is, for instance in the case of a dam related to a water supply, to be recouped over a defined period of time at a defined interest rate, delay in completion of construction would simply defer commencement of that recoupment period such that it could be said, on one view, that delay caused the proprietor no loss. Conceptually I do not think it is correct to say that public works, because they may not yield a cash flow, cannot result in damages to the state or public authority if delay in construction occurs. Whilst the example may be peripheral to the one being here considered, it demonstrates that, at least in some instances, an appropriate measure of liquidated damages is the cost of capital tied up for the period of delay. I regard it as an inadequate answer, in the case of a public work, to say that if the work were delayed say six months, no damage is suffered, and no liquidated damages could be validly agreed, because there was no delay in receipt of cash flow, and there was mere deferment of a planned recoupment of capital and interest costs over time.'

Commercial losses

For commercial projects, the obvious heads of loss arising from late completion of a building project are:

(i)   loss of rent or delayed profit on sale;
(ii) additional financing charges;
(iii) additional supervision, administration costs.

Other costs which can easily arise are:

(i)   rent of alternative premises;
(ii) additional professional fees;
(iii) extra payments under variation of price clauses.

All of the above would appear to fall without difficulty into the first rule of *Hadley* v. *Baxendale* (1854) on remoteness of damage as arising naturally from the breach. They are therefore commonly included in pre-estimates of damage.
   A third category of costs covers items which could be argued to fall within the second rule of *Hadley* v. *Baxendale* as special damages which to be recoverable need to be within the contemplation of the parties at the time they made the contract. Such items could include:

(i)   additional costs of follow-on works;
(ii) loss of trading profit;
(iii) business disruption costs.

However, some caution on how a pre-estimate of damage is made for a commercial project needs to be exercised having regard to the following observations made by the judge in the *Multiplex* v. *Abgarus* case:

'In a large modern commercial development, as a result of the uncertain-
ties relating to the timing of any sale or lease, the quantum of any sale
price or rental, the extent to which a large modern development comprising multiple tenancies for varying uses can be let, and the uncertainty
regarding final terms and conditions of all or any such leases – all judged
or considered at the date of the construction contract some years earlier
– it cannot be said, in my view, that at the date of contract mere knowledge of the intended use of such a building results in it being able to be
said that the delayed performance by a contractor in achieving practical
completion results in delayed receipt of rentals or sale price (neither in
concept nor in specific quantum) being damages flowing from such a
breach of contract as being "such as may fairly and reasonably be considered either arising naturally, i.e., according to the usual course of things,
from such breach of contract itself, or such as may be reasonably be supposed to have been in the contemplation of both parties, at the time they
made the contract, as the probable result of the breach": *Hadley* v. *Baxendale* (at 354; 151).

   Nor do I think that, without more, knowledge of the proposed use of
such development satisfies the second rule in *Hadley* v. *Baxendale*. It will
be a question for determination in each case whether special circumstances relating to prospective loss were sufficiently drawn to attention
to satisfy that rule.'

In the *Multiplex* v. *Abgarus* case the judge made clear his preference for
the holding costs of accumulated expenditure as the pre-estimate of loss.
He said:

'The parties to the construction contract do, however, know at the date
of contract that delay in achieving practical completion will necessarily
result in additional holding costs. Such damages in my view fall within
the first rule in *Hadley* v. *Baxendale*.'

For further comment on such costs see the section below on holding
costs.

Application of rules of *Hadley* v. *Baxendale*

Simply stated, the rules of *Hadley* v. *Baxendale* are commonly expressed as:

'Such losses as may fairly and reasonably be considered as either arising:
(1st rule) "naturally", i.e. according to the usual course of things, or (2nd
rule) "such as may reasonably be supposed to be in the contemplation of
both parties at the time they made the contract, as the probable result of
breach of it".'

These rules provide the basis of the assessment of damages at common
law.

   There is legal authority for the proposition that the rules have some application to calculations of pre-estimates of loss for the purposes of liquidated

damages clauses. Thus, Lord Justice Diplock in the case of *Robophone Facilities Ltd* v. *Blank* (1996) said:

> 'Thus it may seem at first sight that the stipulated sum is extravagantly greater than any loss which is liable to result from the breach in the ordinary course of things, i.e., the damages recoverable under the so-called "first rule" in *Hadley* v. *Baxendale* (1854) 9 Exch 341. This would give rise to the prima facie inference that the stipulated sum was a penalty. But the plaintiff may be able to show that owing to special circumstances outside "the ordinary course of things" a breach in those special circumstances would be liable to cause him a greater loss of which the stipulated sum does represent a genuine estimate.'

Similarly the judge in the *Multiplex* v. *Abgarus* case was explicit in bringing the rules of *Hadley* v. *Baxendale* into calculations for liquidated damages. He said:

> 'It is important, in my view, to recognise the stages in a development project and the place which the construction contract occupies in that project. That is because agreement between the proprietor and the builder in the building contract regarding liquidated damages payable for tardy performance need not encompass all damages which in truth the proprietor may, although not necessarily will, suffer from such late performance. An agreement for damages limited to a segment of possible total damage may itself indicate an acceptance by the parties that other aspects of loss might occur but were treated by the parties, implicitly or explicitly, as not being losses which would arise in "the ordinary course of things" as contemplated by the first rule in *Hadley* v. *Baxendale* (1854) 9 Ex 341; 156 ER 145, and may, by design or default, not have been brought to sufficient attention of the builder by the proprietor so as to satisfy the second rule in *Hadley* v. *Baxendale*.'

Remoteness of damage

The question therefore is not so much, do liquidated damages automatically satisfy the test of 'within the contemplation of the parties at the time they made the contract' by virtue of their very stipulation? – that point seems difficult to oppose – but rather, is it permissible to include within the stipulated sums items which would not satisfy the tests of remoteness in *Hadley* v. *Baxendale* if challenged as general damages? In other words, are liquidated damages to be a genuine pre-estimate of forecast loss or a genuine pre-estimate of legally recoverable loss within the rules of remoteness? The correct answer it is thought lies in the second view, but the uncertainty can be avoided by revealing pre-contract the composition of the sums which make up the liquidated damages if there are items of doubtful recovery.

This may create problems in some instances. There are situations where for reasons of commercial confidentiality an employer may not wish to disclose the breakdown of the losses he will suffer from late completion.

However, if the employer could legitimately include confidential losses in liquidated damages, that would result in the employer being able to recover a higher level of damages by liquidating them than would be achievable if they remained unliquidated. It might well be asked – what is wrong with that if the pre-estimate of loss is genuine and the parties have agreed the amount of liquidated damages? The answer, if the approach in the *Multiplex v. Abgarus* case is to be taken as correct, is that the rules of *Hadley v. Baxendale* should apply to both liquidated and unliquidated damages. However, the answer remains troublesome as can be seen from the following extracts from the 1975 Law Commission Working Paper.

'(ii) The "loss" which is to be estimated

42. Penalty clauses are particularly valuable in cases in which it is likely to be difficult for the plaintiff's loss to be addressed. However, in cases in which a fairly accurate assessment of loss would be practicable, are the parties to the contract restricted to making a genuine pre-estimate of the damages which the court would award in an action if there were no penalty clause or can they agree that the loss which will be suffered on a breach of contract should be calculated on some other basis? Could they, to adopt the words of Asquith LJ in *Victoria Laundry (Windsor) Ltd* v. *Newman Industries Ltd.*, provide a "complete indemnity of all loss de facto resulting from a particular breach, however improbable, however unpredictable"? In cases of breach of contract the aggrieved party who sues for damages is "only entitled to recover such part of the loss actually resulting as was at the time of the contract reasonably foreseeable as liable to result from the breach". Can the parties stipulate for and thus, in effect, make foreseeable, damages on a scale more extensive than the court could otherwise award? There would seem to be no reason why the parties in *Hadley v. Baxendale* could not have contracted for liquidated damages assessed on the footing that the mill would continue to be at a standstill, or those in the *Victoria Laundry* case for the loss of profit on the lucrative government contract. Such a clause would, perhaps, have been doing no more than expressly invoking the so-called second rule in *Hadley v. Baxendale*. It may be, however, that the parties should be able to go even further and provide not merely for loss which is foreseeable in the light of their knowledge at the time of entering into the contract but also for loss directly resulting from the breach even if a court would, in a case in which there was no express provision for liquidated damages, regard such loss as unforeseeable or as irrecoverable for some other reason, for example, because of failure to mitigate the loss suffered or by reason of the incidence of taxation.

43. Diplock LJ discussed this aspect of penalty clauses in *Robophone Facilities Ltd* v. *Blank*:

"The onus of showing that [a stipulation for payment of a sum in the event of breach of contract] is a 'penalty clause' lies upon the party who is sued upon it. The terms of the clause may themselves be sufficient to give rise to the inference that it is not a genuine estimate of damage likely to be suffered but is a penalty. Terms which give rise to such an inference are discussed in Lord Dunedin's speech in *Dunlop Pneumatic Tyre Co.* v. *New Garage and Motor Co. Ltd.* but it is an inference only and may be rebutted. Thus it may seem at first sight that the stipulated sum is extravagantly greater than any loss which is liable to result from the breach in the ordinary course of things, i.e., the damages recoverable under the so-called 'first rule' in *Hadley* v. *Baxendale.* This would give rise to the prima facie inference that the stipulated sum was a penalty. But the plaintiff may be able to show that owing to special circumstances outside 'the ordinary course of things' a breach in those special circumstances would be liable to cause him a greater loss of which the stipulated sum does represent a genuine estimate. In the absence of any special clause in the contract, this enhanced loss due to the existence of such special circumstances would not be recoverable at common law from the defendant as damages for the breach under the so-called 'second rule' in *Hadley* v. *Baxendale* unless knowledge of the special circumstances had been brought home to the defendant at the time of the contract in such a way as to give rise to the inference that the defendant impliedly undertook to bear any special loss referrable to a breach in those special circumstances: see Asquith LJ's explanation of *British Columbia Sawmills Co.* v. *Nettleship* contained in *Victoria Laundry (Windsor) Ltd* v. *Newman Industries Ltd.*

The basis of the defendant's liability for the enhanced loss under the 'second rule' in *Hadley* v. *Baxendale* is his implied undertaking to the plaintiff to bear it . . . But such an undertaking need not be left to implication; it can be express . . . And so if at the time of the contract the plaintiff informs the defendant that his loss in the event of a particular breach is likely to be £x by describing this sum as liquidated damages in the terms of his offer to contract, and the defendant expressly undertakes to pay £x to the plaintiff in the event of such breach, the clause which contains the stipulation is not a 'penalty clause' unless £x is not a genuine and reasonable estimate by the plaintiff of the loss which he will in fact be likely to sustain. Such a clause is in my view enforceable whether or not the defendant knows what are the special circumstances which make the loss likely to be £x rather than some lesser sum which it would be likely to be in the ordinary course of things;"

44. The extent to which the parties to a contract should be free to go beyond invoking the "special circumstances" rule, and apply to the measurement of the loss caused by breach a yardstick which the court would not use, is a question on which we should value views. Our provisional view is that the proper yardstick by reference to which it should be determined whether the stipulated sum is a genuine pre-estimate is the damages which a court would award. If a party wishes

to ensure that he can recover compensation for a loss in excess of recoverable damage he should do so by an express provision, not by a penalty clause; so, too, if he wishes to intimate the existence of "special circumstances" to the other party he should not simply rely on a provision for a high stipulated sum.'

Summary

Overall the position seems to be that pre-estimates of loss for amounts included in contracts as liquidated damages can be made having regard to matters arising naturally and to special circumstances but, in the case of the latter, express provision should be made unless it can be assumed that the special circumstances are within the contemplation of both parties.

Public sector losses

For public sector projects there have traditionally been three main headings in pre-estimate of loss calculations:

(i)   notional interest on capital employed;
(ii) additional supervision/administration costs;
(iii) additional accommodation costs.

One approach, used by some central government departments and based on guidelines issued by the Treasury Solicitor in the 1960s, is to take the pretender estimate of contract cost, divide by 365, and apply 15% of this as the daily rate for liquidated damages. This equates to £2876 per week on a £1 million contract. The figure of 15% comprises 12.5% notional interest on the capital employed plus 2.5% supervision.
   Another formula, recommended for use in local government suggests three main headings:

(i)   interest on capital expended;
(ii) administrative costs;
(iii) additional accommodation costs.

In this formula, as with others in the public sector, it is reasonably assumed that 80% of the capital cost of the project will have been incurred at the point of delay, and then assuming an interest rate of 12%, the capitalised interest is:

$$\frac{80\% \times 12\%}{52} = 0.185\% \text{ of the capital cost/week}$$

If administration costs are taken as 2.75% of the capital cost per year, this adds a further 0.052% per week, making a total for items (i) and (ii) in the formula of 0.237% per week, or £2370 per £million/week.
   The two figures above, £2876 and £2370 per million/week, may well underestimate true loss but to this extent they are on the safe side and there

is no known case where they have been successfully challenged. Indeed, some authorities round up the figures to £3000/week per £million and can probably do so with safety.

Whether or not formulae are used, some record should always be kept of the calculations used to produce the liquidated damages sums and many authorities and private practices do so on specially designed tabulation forms.

In the case of *J F Finnegan Ltd* v. *Community Housing Association Ltd* (1993) it was a condition of the housing grant financing the development by Community Housing that the ultimate figure for liquidated damages in the building contract included a sum relating to the capital cost. The formula used produced the liquidated damages per week by taking:

$$\frac{80\% \times \text{estimated total scheme cost} \times \text{Housing Corporation Lending Rate}}{52}$$

In seeking a declaration that the liquidated damages clause was void the contractor argued, amongst other things, that:

(1) the formula was defective in that it, inter alia, fixed an arbitrary figure of 85%, making no allowance for any interim payments of HAG, nor for any delay that was likely to be incurred in obtaining and installing appropriate tenants, nor for any rent arrears that were likely to be incurred or any tax paid thereon;
(2) the formula was ultra vires the terms of section 29 of the Housing Act 1974 and ought not to have been imposed upon the contractor nor used by the housing association.

The judge rejected these arguments and held that the use of the formula was justified and it was not ultra vires. He said:

'(a) I conclude on the evidence I have heard that the figure of £2,500 per week was a genuine attempt by the parties and/or the defendants to estimate in advance the loss which the defendants were likely to suffer should the plaintiffs, in breach of contract, fail to complete the contract works;
(b) I reject the plaintiffs' assertion that the figure of £2,500 per week was extravagant and unconscionable;
(c) I find that the formula used was justified at the time the parties entered into the contract;
(d) since I have concluded that the figure of £2,500 per week was not a penalty, and/or the use of the formula was justified at the time the parties entered into the contract, I find that the formula was not ultra vires section 29 of the Housing Act 1974;
(e) I am satisfied that the plaintiffs' assertion that the defendants' loss was at £750.00 per week is (a) incorrect and (b) based upon an investigation of detail which in any event is one which the court should not consider. I am by no means satisfied that the defendants' loss which may be proved in the future is in fact less than £2,500;

it could even be more. This is, however, for a future tribunal to consider.'

His concluding comments were as follows:

'Finally, I should say that I have considered the general question of how can it be said that the defendants have made a genuine estimate of their damages when they are required by a third party to put into their calculation a formula over which the defendants had no control? The reality is that the defendants rely upon third parties for their sources of funding. It is not in my judgment unreasonable for a third party to protect their position by requiring the defendants to include in their contracts with others a clause of the nature which has been the subject of this action. The plaintiffs' position is safeguarded if the court then proceeds to examine the "imposed clause" as between the plaintiffs and the defendants and to consider the question of penalty accordingly.'

Employer's losses only

Where formulae are used some thought should be given to whether any broad assumptions made are applicable. There may be a problem where part of the capital cost of a public sector project is provided by way of interest free funds. In such a case it would not be appropriate for the employer to estimate his loss on the full capital cost. The loss stipulated as liquidated damages must be the employer's loss and not that of someone else. It is certainly not correct to include in liquidated damages for a roadworks project, estimated costs of delay to road users at large since whatever costs such road users might individually incur, or the economy in general might suffer, the costs do not fall directly on the employer and they cannot be brought within the rules on remoteness of damage.

The converse of this is that changes to funding of public sector projects or like projects may involve the employer in more damages than had previously been the case. Thus, at one time the Housing Corporation recommended to housing associations that 30% of the total capital cost of projects was used in liquidated damages calculations as the remaining 70% was interest free. This was changed to 55% as grants altered, and then, as housing associations moved towards private sector financing, it was left to individual housing associations to assess their own position on each project.

Losses paid by others

The fact that some public sector projects are funded by grants, or the employer may only be acting as agent for some other body does not interfere in the process. The situation is analogous to insurance.

In *Design 5* v. *Keniston Housing Association Ltd* (1986) the court was asked to decide in relation to a claim for general damages whether payment of a

Housing Association Grant for the full cost of a scheme prevented the housing association from suing its architect over increased expenditure. The court held that it did not.

Similarly in *Jones* v. *Stroud District Council* (1986), the court held that it was not concerned whether the cost of repair to a house had been met by the plaintiff or by some other person.

Note also the decision in the *Finnegan* case mentioned above where the judge rejected the argument that liquidated damages should not apply because the housing association was funded by a grant. He said:

> 'In any event I do not see how the financial arrangements entered into by the defendants with a third party can affect the position between the plaintiffs and the defendants given the fact that interest was being charged on the monies advanced.'

Average interest rates

Calculations for liquidated sums which rely on interest rates will not be invalidated by use of average rates. It is probably more appropriate to use average rates than currently prevailing rates which might be unusually high or low. Thus in the period 1981–1991 rates varied from 16% to 7%, but averaged 11.61% over the full ten years and 12.01% over the last five years.

In the *Multiplex* v. *Abgarus* case the liquidated damages clause provided for interest to be paid 'at a rate per annum equal to the maximum rate of interest then charged by Trading Banks on overdraft accounts over $100,000'. The contractor challenged this on the basis that it did not purport to reflect any actual rate charged to the employer and it was a 'worst case scenario' which could not be said to be a genuine pre-estimate of damage.

The judge, however, upheld the use of the maximum interest rate with this reasoning:

> 'It is, I think, common knowledge that most trading banks charge interest to major customers on accounts in excess of $100,000 on a base rate plus a percentage. The base rates normally do not differ significantly. The added percentage depends upon the bank's view of the customer, and perhaps other factors. In my view specifying a maximum rate charged to such a significant borrower by a trading bank does not prescribe a rate or loss that is "out of all proportion" to the damage "likely" to be suffered as a result of breach where such rate is for the purpose of determining holding charges.'

Holding costs of accumulated expenditure

In the great majority of construction contracts liquidated damages are specified as an amount per day or per week derived from calculations made by the employer or otherwise agreed between the parties. However, in principle there is no reason why a formula should not be specified with the figures of the amount payable left to be precisely determined at the time of the contractor's default.

Such a formula applied in the *Multiplex v. Abgarus* case. It was set out as follows:

'16. 10.14-Liquidated and Ascertained Damages.
If the Builder shall fail to bring the Works to Practical Completion by the Date for Practical Completion then:

10.14.01 The Architect may give notice in writing to the Builder and to the Proprietor not later than 20 days after the date on which the Works actually reached or are deemed to have reached Practical Completion that in his opinion the Works ought reasonably to have been brought to Practical Completion at some earlier date to be stated in that notice, not being earlier than the Date for Practical Completion.

10.14.02 If such notice is given then the Builder shall pay or allow to the Proprietor as liquidated and ascertained damages:
A Interest at a rate per annum equal to the maximum rate of interest then charged by Trading Banks on overdraft accounts over $100,000 calculated on daily balances of the total of the items listed hereunder for the period commencing on the date so specified by the Architect during which the Works shall remain or have remained not brought to Practical Completion:
(i) $30 million being the value of the Site at the date of this agreement.
(ii)   Payments made by the Proprietor under any contract relating to the execution of the Works.
(iii)   Preliminary expenses incurred by the Proprietor.
(iv)   Rates and taxes and other statutory charges assessed against or incurred by the Proprietor in connection with the Site or the Works.
(v)   Reasonable costs and expenses incurred by the Proprietor in enforcing or attempting to enforce any contract relating to the execution of the Works.
(vi)   Reasonable costs and expenses incurred by the Proprietor in insuring the Works.
(vii)   Fees paid to architects, surveyors, engineers, consultants, project managers, and other experts engaged in the execution of the Works.
(viii) Salaries paid to the building clerks of works and mechanical clerks of works.
(ix)   All other costs and expenses incurred by the Proprietor which were reasonably necessary to the execution of the Works.
(x)   Interest at a rate per annum equal to the maximum rate of interest then charged by major Trading Banks on

overdraft accounts over $100,000 calculated on daily balances on the amounts referred in items (i) to (ix) above inclusive from the respective dates upon which any such amounts were expended by the Proprietor. Such interest shall be capitalised on 31 December in each year prior to the Date for Practical Completion of the Works.

B All rates statutory charges and other reasonable outgoings in respect of the Works and the Site assessed against or incurred by the Proprietor in respect of the period commencing at the date so stated by the Architect and finishing when the Works reach Practical Completion.'

The judge in *Multiplex* v. *Abgarus* reviewed the authorities at length before concluding:

'I am satisfied that a clause which specifies as liquidated damages in respect of a major city building the accumulated costs of the proprietor to date of contractual practical completion and determines the holding charges of those costs for any period of delay occasioned by the builder constitutes a valid, enforceable liquidated damages clause. If it neglects to encompass additional damages which the proprietor may suffer, that does not derogate from its integrity as a valid liquidated damages clause.'

Benefits to the employer

One of the challenges made by the contractor in the *Multiplex* v. *Abgarus* case to the liquidated damages clause was that as the formula gave no credit for any benefit which the employer might derive from being able to commence fitting-out during the period of delay to practical completion the liquidated damages clause failed as a genuine pre-estimate of damage and was thus a penalty.

The judge rejected the contractor's contention as follows:

'In my view there are two matters which sufficiently dispose of this contention. First, any such "benefit" if it be capable of any form of financial assessment at the date of contract, which I doubt because of the multitude of varying circumstances which might arise in relation to leasing to which I have referred earlier, falls within that area of discretion which the law allows to the parties in agreeing upon a quantum of damage. The very circumstance that the liquidated damages are a "pre-estimate" involves that it will not be precise. The parties may elect not to refer to matters which they regard of little importance. In my view this circumstance falls within that tolerance referred to in the passages in *Esanda Finance Corporation Ltd* v. *Plessnig* (at 141–142) and in *AMEV-UDC* (at 190).

Secondly, the contract contemplated staged performance. Stage 1 was to be progressed to a level of completion excluding certain works to the ceilings, air conditioning and chilled water supplies so as to enable the proprietor to engage either the builder or others as fit-out subcontractors to commence fit-out twenty-four weeks prior to the date of practical completion (cl 15.05). By cl 15.07 the builder agreed to the proprietor having access "to each floor of the works as it becomes sufficiently complete to allow tenancy fit out to proceed". This was without prejudice to the remaining clauses of the agreement dealing with completion of stage 1 and date for practical completion. Tenancy fit-out was dealt with by cl 7.02A.

Clause 9.10 dealt with occupation before practical completion. By cl 9.10.01, if the builder failed to achieve practical completion by the contractual date, the proprietor could give a notice of intention to occupy portions of the building. By cl 9.10.02, any occupation by the proprietor under such a notice, or under the fit-out provisions referred to in ell 7.02, 7.02A, 15.05 or cl 15.07 did not result in practical completion being achieved. Occupation or use otherwise resulted in the works being deemed to be practically completed (cl 9.10.03).

Any occupation of the premises prior to achieving practical completion for the purpose of fit-out was thus a contemplated contractual right and was not a 'benefit' conferred upon the proprietor by delayed practical completion such as to cast doubt upon the integrity of the liquidated damages clause.'

Mitigation

The question of whether a party's duty to mitigate its losses should be considered in the drafting of liquidated damages provisions or in the calculation of stipulated sums, was considered in the *Leisureplay* case mentioned in Section 4.3 above.

In that case the judge at first instance held that the liquidated damages provision in an employment agreement was a penalty clause because it took no account of the employee's duty to mitigate his losses. The Court of Appeal reached a different conclusion. Lord Buxton said:

'115. Neither the literal wording of that test nor the spirit of it applies here. Mr Murray's terms were generous, but they were not unconscionable. As to the absence of any requirement of mitigation in clause 17.1, to which as we have seen the judge attached determinative importance, two comments have to be made. First, it must have been difficult to say with confidence at the time of entering into the contract what might happen to Mr Murray were he to be dismissed: provisions protecting an employee in the case of wrongful termination may take the form that they do because such an event can damage his future employability, at least in the short term. Second, in order to meet this criticism a pre-estimate of damages clause would have to be drafted to encompass not only the fact of mitigation in terms of income from other sources but also the duty to seek such mitigation Such a clause would directly invite disputes about the reasonableness of Mr Murray's behaviour after termination, of the kind that clauses stipulating the amount of compensation are precisely designed to avoid. As Tindal CJ put it in *Kemble* v. *Farren* (1829) 6 Bing 141 at p 148, a dictum approved in the *Dunlop* case, even where damages accruing from a breach can be accurately ascertained, a liquidated damages clause "saves the expense and difficulty of bringing witnesses to that point". And a clause that made reference to the duty to mitigate would also inevitably postpone payment under the clause well beyond the termination date: again, something that the inclusion of such a clause in the contract must have been intended to avoid. This last consideration strongly reinforces the general impression created by this case, that the traditional learning as to penalty clauses is very unlikely to fit into the dynamics of an employment contract, at least when the penalty is said to be imposed on the employer.'

It is doubtful that Lord Buxton's comments can be taken as indicating a general rule that mitigation is never a factor to be taken into account but they certainly indicate that there is no general rule that it should be taken into account.

4.5 *Particular aspects of penalty clauses*

This section covers cases which deal with unusual aspects of penalty clauses or which are otherwise of interest for historic or particular reasons.

Penalties other than for stated sums

Most penalty clauses, whether they be expressly described as such or described as liquidated damages, concern sums stated in money terms as amounts due on breach. However, from time to time cases reach the courts where relief is sought from contractual provisions on grounds that they are penalties albeit that the provisions to be examined contain no stipulated sums.

Thus in *Jobson* v. *Johnson* (1988) a provision relating to transfer of property was held to be unenforceable as a penalty. More recently in *City Inn Limited* v. *Shepherd Construction Limited* (2001, 2003 and 2007), the courts of Scotland had to consider whether a contractual provision requiring notice to be given as a condition precedent to rights of claim for loss and expense and extensions of time was itself a penalty clause by virtue of its consequences.

The *City Inn* case, of which more is said later in this book on other matters, concerned disputes relating to variations, extensions of time and liquidated damages. It started in the Outer House, Court of Session, in 2001 following an adjudication, proceeded to the Inner House for debate on certain legal matters in 2003, and returned to the Outer House for proof in 2007. One of the defences raised by the contractor to a claim for liquidated damages for late completion was that there was no relationship between failure to comply with notice provisions and the amount which became due if the provisions were breached. Both Houses decided against the contractor on the penalty point, the reasoning of the Inner House being that the notice provisions gave the contractor an option and it was a matter for the contractor whether or not it exercised that option. The court said:

> '(b) Whether clause 13.8.5 imposes a penalty
> 27. On the view that we have taken on the preceding question, this question does not arise. We should say, however, that if we were to treat the defenders' failure to operate clause 13.8 as a breach of contract, we cannot see how that would result in the payment of a penalty in the legal sense of that expression (e.g. *Dunlop Pneumatic Tyre Co. Ltd v. New Garage and Motor Co. Ltd, supra,* Lord Dunedin at p 87). The sum complained of is not payable at that stage, and may never be payable. What is, on this assumption, a breach of contract merely gives rise to the possibility, the likelihood of which will depend on the circumstances of the case, that liquidated damages will become due at a later date. If that liability should in due course arise, it will not arise as a consequence of the assumed breach of contract under clause 13.8, but as a consequence of the contractor's breach of clause 23 consisting in his failure, for whatever reason, to complete the contract works on or before the completion date. We agree with the reasoning of the Lord Ordinary on this point.'

Limitation of liability

The general rule is that a liquidated damages clause which is not a penalty acts as a limitation of liability in respect of damages for breach (see Section 3.4 above). That raises questions as to whether the claiming party can seek to avoid limitation of damages by alleging the liquidated damages provisions to be penalty clauses.

In the old shipping case of *Wall v. Rederiaktiebolaget Ruggude* (1915) the charter party contained the clause 'Penalty for non-performance of this agreement provides damages, not exceeding estimated amount of freight'. The shipowners, who had breached the agreement, contended that the clause provided limitation of liability against a claim for general damages. The court held that the clause provided a penalty not a limitation of liability and that it did not prevent the party complaining of non-performance from recovering actual damages exceeding the amount of freight.

An unusual aspect of limitation of liability was considered by the Court of Appeal in the case of *Bath and North East Somerset District Council v. Mowlem Plc* (2004). The Council had obtained an injunction against Mowlem restraining it from denying access to a replacement contractor. Mowlem argued that the Council had an adequate remedy for the problems on the contract by way of liquidated damages for late completion. The Court of Appeal disagreed and upheld the injunction.

Drafting matters

Chapter 6 of this book deals with the legal construction of liquidated damages clauses. However, for convenience, some notable cases where such clauses have been considered by the courts are listed here:

Bramhall & Ogden v. Sheffield City Council (1983)

Liquidated damages of reasonable amount expressed at a rate per dwelling per week were held to be penalties because a proportioning down clause in the contract, which was obviously not intended to apply, was not deleted.

Stanor Electric Ltd v. R Mansell Ltd (1987)

The main contractor, Mansell, sought to deduct liquidated damages for late completion by their electrical sub-contractor, Stanor, of work on two houses where liquidated damages were staged at £5000 per week. Judge Fox-Andrews held that, as a matter of construction of the particular clause in the contract where work was to be done on two houses, the clause was self evidently a penalty.

Arnhold & Co. Ltd v. Attorney General of Hong Kong (1989)

This was a contract for electrical and mechanical work which was more than one year late in completion. Liquidated damages were expressed as a range of sums and were held to be void for uncertainty. They were also held to be void as penalties since:

(i) the maximum figure was recoverable for delay to the whole of the works or any portion thereof and could not be a genuine pre-estimate;
(ii) a provision for proportionally reducing damages as the works were occupied had a stop figure which it was held could not be a genuine pre-estimate of loss once more than 85% of the works were occupied.

Comment

All the above cases preceded the Privy Council ruling in the *Hong Kong Philips* case that the courts should not adopt an approach to liquidated damages so as to defeat their purpose and it is questionable whether all would stand today. Recent cases indicate that the courts are now more tolerant of drafting discrepancies:

Impresa Castelli Spa v. *Cola Holdings Ltd* (2002)

The case arose from late completion of a large hotel development in London. The original rate of £10,000 per day for liquidated damages was reduced to £5000 per day as part of an agreement allowing the developer partial occupation. Notwithstanding that this created certain difficulties in the legal construction of the liquidated provisions the court held the reduced rate to be a genuine pre-estimate of loss and enforceable.

North Sea Ventilation Ltd v. *Consafe Engineering (UK) Ltd* (2004)

The judge dismissed various arguments derived from the drafting of the contract and having quoted from the *Hong Kong Philips* ruling ('Striking down a penalty clause is a blatant interference with freedom of contract, and can only be justified where there is oppression') went on to find there was no oppression and that:

> 'The provision of graduated sums increasing in proportion to the seriousness of the breach is characteristic of a liquidated damages clause which is commonplace in commercial contracts.'

CFW Architects v. *Cowlin Construction Ltd* (2006)

This case followed a series of adjudications between the architects and the contractor on a design and build housing renovation project. One of the issues was whether the amounts claimed by the contractor for liabilities it had incurred as liquidated damages for late completion should be dismissed on the basis that the liquidated damages were penalties. The arguments on this related to the manner in which the damages became payable on a house by house calculation. The judge, in finding that there was no penalty, said:

> '. . . the liquidated damages clause, although potentially harsh on Cowlin, was nonetheless enforceable. The relevant test, enunciated by Jackson J which I accept correctly states the applicable test binding on judges at first instance, is as follows:
>> "In my view, a pre-estimate of damages does not have to be right in order to be reasonable. There must be a substantial discrepancy between the level of damages stipulated in the contract and the level of damages which is likely to be suffered before it can be said that the agreed pre-estimate is unreasonable."'

Steria Ltd v. Sigma Wireless Communications Ltd (2007)

Amongst the many issues considered by the court in this case (see Chapter 5 for some of the details) there were arguments as to whether there were inconsistencies in the drafting of the liquidated damages clauses such as to render them unenforceable. The court considered them to be sufficiently coherent to be enforceable.

Braes of Doune Wind Farm (Scotland) Ltd v. Alfred McAlpine Business Services Ltd (2008)

The claimant sought leave to appeal an arbitrator's award that provisions in an engineering procurement and construction contract were unenforceable. The arbitrator had found:

> '... the provisions of Clause 8.7 are not capable of generating with certainty liquidated damages flowing from an identified breach by the [Contractor]. Accordingly, in accordance with established authority, Clause 8.7 should not be enforced.'

The judge who had the task of deciding whether the arbitrator was obviously wrong (applying Section 69(3) of the Arbitration Act 1996) concluded that he was not. The judge said:

> 'I have formed the view, perhaps contrary to my initial impressions, that the Arbitrator was not obviously wrong. Although my own analysis would have been different and I might disagree with part of the Arbitrator's reasoning, I consider that his decision was ultimately right. The most convincing argument advanced by Mr Bartlett QC for the Contractor was that the liquidated damages clause could well impose a liquidated damages liability on the Contractor in respect of delays to individual wind turbines caused by the Wind Turbine Contractor.'

He concluded:

> 'E. Because it was clearly intended that the Contractor was not as such to be responsible for the defaults of the Wind Turbine Contractor or at least those which good co-ordination by the Contractor would have avoided, the parties nonetheless agreed a liquidated damages clause which would impose such damages upon the contactor in certain foreseeable circumstances.
> F. In those circumstances, there is in law a penalty which English Law will not enforce.'

Some caution may need to be exercised in taking too much from the last quoted sentence since it is not entirely clear from the judgment whether the circumstances referred to by the judge were faulty drafting, inequitable risk allocation or possible prevention or whether the arguments put to the judge were hypothetical or were based on fact.

Loss suffered by a third party

Clark Contracts Ltd v. *The Burrell Co. (Construction Management) Ltd* (2002)

This was a Scottish case in which the employer sought liquidated damages from the contractor for late completion of the redevelopment of a block of flats. The twist in the case, as the judge put it, was that the employer was not the proprietor of the flats – they belonged to The Burrell Co. (Developments) Ltd.

The judge, having examined differences between Scottish law and English law on losses suffered by third parties, found the liquidated damages to be a penalty and not enforceable. He explained this as follows:

> 'The defenders did not own the flats. The defenders did not therefore incur lost credit interest nor did they incur debit interest. There are no adequate averments to set up a contractual relationship whereby the defenders were obliged to make payment of any such losses to Developments. All that the defenders say is that there was an "understanding" between the defenders and Developments that the defenders "would seek recovery from the building contractors (the pursuers) of those losses and the sums recovered would be payable to Developments as a debt by the defenders". There was no averment that there was a binding contract between the defenders and Developments which the latter could enforce. Accordingly, the defenders have sustained no loss, and could have sustained no loss, by virtue of any failure to complete the works by the revised completion date. The provision in the contract is thus not a reasonable pre-estimate of the damages which the defenders, as opposed to Developments, might incur in the event of there being a delay in the completion of the building contract works. It thus follows that what is sought from the pursuers in the counterclaim by way of liquidated and ascertained damages is a penalty and not recoverable.'

For the purposes of this chapter that is all that need be said about the case. However, see discussion of the English case of *McAlpine Construction Ltd* v. *Panatown Ltd* (2000) in Chapter 3 on third party losses.

Settlement agreements as penalty clauses

CMC Group Plc v. *Michael Zhang* (2006)

The parties in this case became embroiled in disputes about financial transactions undertaken by Mr Zhang. They were eventually settled on terms which included the following:

> 'For the avoidance of doubt, you hereby agree that any breach of this settlement and agreement will render you liable to us for the sum of

US$40,000 together with a claim for reimbursement of our legal costs against you in addition to a claim for damages in relation to loss of business. Such a claim could be considerable.'

CMC alleged that Mr Zhang breached the agreement and claimed damages including the US$40,000 stipulated sum. It also claimed, and obtained, injunctions restraining Mr Zhang's conduct. The matter which eventually reached the Court of Appeal was whether the stipulated sum was a penalty.

The Court of Appeal unanimously held it to be a penalty with Sir Charles Mansell, saying:

'Without reference to authority, and just on a reading of the letter, it would appear to my eyes that the provision for the payment of US$40,000 was a penalty. It had been introduced as a deterrent to Mr Zhang and as an inducement not to break any of the terms of that agreement, which it is quite unnecessary for me to read again.'

and

'It is quite impossible, I would say, to read the provisions to which I have just referred as being other than a penalty within the terms identified by Lord Dunedin or Colman J. This was included as a deterrent. That it was so is reinforced by the further observation in the letter that there could be an additional claim for damages, and I quote, "Such a claim could be considerable". It is, in my view, quite impossible to read the letter as containing other than a penalty clause.'

Take-or-pay clauses

M & J Polymers Ltd v. *Imerys Minerals Ltd* (2008)

A supply contract required the purchasers, Imerys, to pay for a minimum quantity of products even if they ordered less. It was held in the High Court that Polymers' case that its claim was for a debt and that the law of penalties did not arise was too simplistic. It was said that as a matter of principle the rule against penalties might apply but on the facts of the case the claim did not offend the rule.

4.6 *Evidential matters*

Burden of proof

It is well established that the burden of proving that a stipulated sum is a penalty and not liquidated damages rests on the party making the challenge.

See, for example:

- Lord Justice Diplock's comment in *Robophone Facilities Ltd* v. *Blank* (1966) "The onus of showing that [a stipulation for payment of a sum in the event of breach of contract] is a penalty clause lies upon the party who is sued upon it"
- Lady Justice Arden in the *Leisureplay* case: "The burden of showing that a clause for the payment of damages on breach is a penalty clause is on the party who seeks to escape liability under it."

The above rule should normally hold good in cases which reach the courts on appeal from arbitrator's decisions. However, an interesting point emerges from the *Braes of Doune* case mentioned above where the court was considering an application for leave to appeal an arbitrator's decision. The judge, referring to Section 69(3) of the Arbitration Act 1996, made the point that in approaching the question of leave to appeal he had to consider if the arbitrator was "obviously wrong" in reaching his decision. Viewed from this perspective there would seem to be some reversal of the burden of proof.

Factual evidence

One of the points considered at length in the *Multiplex* v. *Abgarus* (1992) case was whether evidence concerning the circumstances in which a liquidated damages clause (alleged by the contractor to be a penalty clause) came into existence was relevant to unconscionability, and therefore admissible.

The court after extensively reviewing the authorities held that it was. Commenting on the judgment in the case of *AMEV-UDC Finance* v. *Austin* (1986) the judge in *Multiplex* said:

> 'Their Honours distinguish as a basis for striking down a clause circumstances which may render it unconscionable to enforce the clause, as well as a clause which may be oppressive in consequence of its monetary impositions indicating that it is not of a compensatory nature. That must, in my view, render admissible evidence concerning the circumstances in which the clause came into existence, and the parties' understanding of its intent in order to rebut any attack upon the basis that the clause was agreed in circumstances and in a relationship between the parties rendering its enforcement unconscionable.'

And later, commenting on a statement in the judgment in *Esanda Finance Corporation Ltd* v. *Plissing* (1989) the judge in *Multiplex* said that:

> 'This clause is to be construed from the point of view of the parties at the time of entering into the transaction. The character of a clause as penal or compensatory is then to be perceived as a matter of degree depending on *all the circumstances, including the nature of the subject-matter of the agreement.*'

The judge in *Multiplex* went on to say:

> 'If one is to have regard to "all of the circumstances, including the nature of the subject matter of the agreement" in determining as a matter of

degree whether a clause is penal or compensatory, one would need to know of the relationships between the parties at the time of contract, the genesis of the clause, discussions concerning it, the bargaining position of the parties, whether they were each fully advised and whether, in all the circumstances, the party now claiming the ineffectiveness of the clause, at the time of contract appreciated the likely imposition under the clause in consequence of his breach yet nonetheless agreed to the clause presumably because the contract was perceived to be beneficial to him notwithstanding the existence of the liquidated damages clause.

There is, in my view, a qualitative difference of which the law is able to take account between a clause freely negotiated between major commercial organisations, in respect of a substantial contract, where the major commercial organisations have available and receive competent legal advice regarding the meaning, purpose and likely consequence of the clause, from a clause attacked as a penalty in a contract of adhesion between a major organisation and an individual or small company who has, in reality, no opportunity to negotiate the contract. That is not to say that the latter form of contract containing such a clause would be struck down: it is rather to recognise that, quite apart from whether a clause fails because it lacks a compensatory character, it may also fail as being penally imposed in circumstances rendering enforcement of the clause unconscionable. The degree of contractual freedom afforded to parties to determine a measure of damages departing from strict compensation will, in my view, be affected by those matters constituting aspects of the relationship between the parties, in particular in relation to the relevant clause, to which I have referred. That seems to me to be implicit in the passage in the judgment of Mason J and Wilson J in *AMEV-UDC* where their Honours said (at 193) "and (2) the nature of the relationship between the contracting parties, a factor relevant to the unconscionability of the plaintiff's conduct in seeking to enforce the term".

For those reasons in my view the material tendered which shows the relationship between the parties, the genesis of, negotiation concerning, understanding of, and advice received regarding the relevant clause is material and admissible.'

Opinion evidence

Questions of whether amounts stated as stipulated sums are reasonable pre-estimates or extravagant and unconscionable may in some circumstances obviously require expert opinion evidence.

4.7 *Bonus clauses*

It is not uncommon for commercial and construction contracts to include bonus clauses to encourage early completion. However, notwithstanding any sums which are payable for early completion, sums which are claimed for late completion still have to satisfy the tests for liquidated damages if they are to avoid being declared penalties. Some bonus clauses emphasise this point by including a statement to the effect that any sums deducted for late completion are liquidated damages and not penalties. On this basis the setting of reciprocal bonuses/damages cannot be arbitrary but must follow the rules for pre-estimation of liquidated damages.

The following material provides an example of a bonus clause. This particular clause is taken from a motorway improvement contract and based on the ICE 5th Edition.

Payment of charge for continued site occupation

47(1)(a) If the contractor fails to complete the whole of the Works or any section thereof within the time prescribed by Clause 43 or any extension thereof granted under Clause 44 the Contractor shall pay the Employer the sum stated in the Appendix to the Form of Tender under the heading 'Bonus for Early Completion / Charge for Continued Site Occupation' for every Working Day which shall elapse between the date on which the prescribed time or any extension thereof expired and the date of completion of the whole of the Works or the relevant Section thereof. The Employer may deduct the sums so due as payments from sums otherwise due to the Contractor under the Contract or any other Contract which the Employer or his Agents have with the Contractor.

Bonus for early completion

47(1)(b) If the Contractor completes the whole of the Works or any Section thereof within a shorter time than that prescribed by Clause 43 or any extension thereof granted under Clause 44 the Employer shall add to the sums otherwise due to the Contractor the sum stated in the Appendix to the Form of Tender under the heading 'Bonus for Early Completion / Charge for Continued Site Occupation' for every Working Day by which the date of completion of the whole of the Works or the relevant Section thereof precedes the due time (or extended time) for completion of the Works or the relevant Section thereof.

Addition to / deducted from final account settlement

47(2) Where following a review under Clause 44(3) and Clause 44(4) the Engineer has issued the Certificate of Completion of the work together with the accompanying certified statement of the overall extension of time (if any) to which the Engineer considers the Contractor to be entitled in respect of the whole of the Works or any

Section thereof the Employer shall add to or deduct from any payment due in settlement of the final account such sum which is equal to the sum stated in the Appendix to the Form of Tender for every Working Day by which the Contractor may complete the Works or the relevant Section thereof earlier or later as the case may be in accordance with the respective provisions of sub-clause (1)(a) and (1)(b) of this clause.

Nil effect on retention money calculation

47(3) In the calculation of the amount to be deducted for each Working Day that the completion of the Works or any Section thereof exceeds the due date for completion in accordance with Clause 43 such amount shall have no effect on the calculation of the retention money in accordance with the provisions of Clauses 60(2) and 60(4).

Reimbursement of charge for continued site occupation

47(4) If upon a subsequent or final review of the circumstances causing delay the Engineer shall grant an extension or further extension of time or if an arbitrator appointed under Clause 66 shall decide that the Engineer should have granted such an extension or further extension of time the Employer shall no longer be entitled to charges for continued site occupation in respect of the period of such extension of time. Any sums in respect of such period which may have been deducted from payments due to the Contractor or paid by him shall be reimbursable forthwith to the Contractor together with interest at the rate provided for in Clause 60(6) from the date on which such charges for continued site occupation were paid or deducted. In the event that such a review shows the Contractor was rightfully entitled to a bonus payment then it shall be calculated in accordance with Clause 47(1)(b) and paid forthwith with the addition of interest at the rate provided for in Clause 60(6).

Damages not a penalty

47(5) All sums deducted as payment by the Employer pursuant to this clause from sums otherwise due to the Contractor shall be paid as liquidated damages for delay and not as a penalty.

Comment

The above clause is included in forms with conventional provisions for extensions of time but there are no general rules that extension of the date for completion to allow payment of bonuses should take account of the same circumstances which would apply to extension of the date to avoid payment of damages. Each case depends on the wording of the contract.

Thus in *Ware v. Lyttleton Harbour Board* (1882) where contractors finished six weeks early and received a £600 bonus based on £100 / week, it was held that they could not claim a further bonus in respect of time taken on extra works which prevented them finishing earlier since the extension of time clause applied only to save them from liquidated damages.

4.8 Site occupation charges

As an alternative to bonus / damages clauses, some contracts require the contractor to pay a daily charge for occupation of the site.

Amongst the best known are the lane rental contracts used by the Highways Agency and other highway authorities for roadworks repairs and improvements. These contracts state a cost per day per lane and the contractor inserts in his tender the total sum required for lane rental according to his programme. The costs of lane rental are then deducted from payments due to the contractor with due allowance being made for extensions of time.

An example of lane rental provisions is as follows:

Appendix to form of tender

Lane rental charges 47(1)

(i)   The daily Lane Rental Charge shall be £3500 per lane occupied and £1000 per hardshoulder occupied subject to (ii) & (iii).
(ii) The daily charge for Contra Flow shall be £25,000 per day. Contra Flow is the closure of one complete carriageway and the operation of two way traffic on the other carriageway.
(iii) The charge for occupation of any lane or hardshoulder during the period between 2100 hours and 0600 hours the next day shall be £1000 per lane or hardshoulder per period. This charge shall not be additional to (i).

Daily lane rental charges clause 47

(1) The Contractor shall pay to the Employer Lane Rental Charges in such sums as are stated in the Appendix to the Form of Tender for each day or part of day that any lane or hardshoulder, or combination of lanes and / or hardshoulder (or parts thereof) is occupied for the purposes of

carrying out the Works from and including the Date of Commencement of the Works until and including the date certified in the Certificate of Completion of the Works and shall also pay to the Employer Lane Rental Charges in such sums as are stated in the Appendix to the Form of Tender for each day or part of day that any lane or hardshoulder, or combination of lanes and /or hardshoulder (or parts thereof) is occupied by the Contractor for work required under Clause 49. On any day when different traffic management systems are employed the charge shall be the highest of the individual charges. Payments shall be effected by means of deductions in accordance with Clause 60.

Provided that:

(a) if under Clause 49(3) the value of such work shall be ascertained and paid for as if it were additional work the Contractor shall not pay to the Employer a Lane Rental Charge in respect of the lane occupation relating thereto; and

(b) if under Clause 44 the Engineer shall grant an extension of lane occupation to the Contractor, the Contractor shall not pay to the Employer a Lane Rental Charge in respect of the period of such extension.

(2) The Bill of Quantities includes an item specifically provided for the purpose of indicating the total sum of the lane rental charges which at the time of tendering the Contractor expected he would incur for his occupation of the site during the execution of the Works including the Works required under Clause 49. All other items in the Bill of Quantities are deemed to be exclusive of lane rental charges.

Deduction of lane rental charges from other contracts

(3) The Employer may deduct and retain from any sum otherwise payable by the Employer to the Contractor under the Contract or any other Contract which the Employer or his Agents have with the Contractor any charges due under sub-clause (1) of this Clause.

Final Review

(4) If upon a subsequent or final review of the circumstances causing an extension of lane occupation the Engineer shall grant an extension or further extension of lane occupation or if an arbitrator appointed under Clause 66 shall decide that the Engineer should have granted such an extension or further extension of lane occupation, the Employer shall no longer be entitled to lane rental charges in respect of the period of such extension of occupation. Any sums in respect of such period which may have been recovered pursuant to this Clause shall be reimbursable forthwith to the Contractor together with interest at the rate provided for in Clause 60(6) from the date on which such lane rental charges were recovered from the Contractor.

Notification of changes in lane occupation

    (5) The Engineer shall inform the Contractor in writing within 24 hours after any change in lane occupation of the relevant new lane occupation charge applicable from the date of the change.

Comment

The advantage of site occupation charges over conventional liquidated damages charges is that account can be taken of matters beyond the employer's own costs. Thus lane rental charges on roadworks can include for notional traffic delay costs.

Chapter 5
Prevention

5.1 *Principle of prevention*

The principle of prevention is of general application in contracts and is to the effect that one party cannot impose a contractual obligation on the other party where he has impeded the other in the performance of that obligation. It is a long-standing principle going back as far as *Comyns' Digest of the Laws of England* (1822). In *Perini Pacific Ltd* v. *Greater Vancouver Sewerage and Drainage District* (1966), it was said 'Since the earliest times it has been clear that a party to a contract is exonerated from performance of a contract when that performance is prevented or rendered impossible by the wrongful act of the other party.'

The principle can be expressed in many ways as evident from the cases below, particularly the *SMK Cabinet* case, but note that some definitions focus on prevention arising from breach of contract and others focus on prevention arising from contractually legitimate acts (e.g. ordering of variations) whereas the principle itself covers both.

There is debate as to whether the principle is founded on a rule of law or a rule of construction which is of interest to the operation of conditions precedent and time-bars (considered later in this chapter) but that apart it is the effect of the principle which is important rather than its standing.

Effects of prevention

There is a distinction to consider between prevention as it affects the innocent party and prevention as it affects the guilty party. An act of prevention which is a breach of contract gives the innocent party the right to sue for damages or determine the contract depending on whether the breach is of a warranty or a condition. This is quite clearly a rule of law. However, the restrictions on the guilty party in limiting enforcement of contractual rights seem to arise more from implied terms than any firm rule of law.

Thus, Lord Blackburn in *Mackay* v. *Dick* (1881) said that:

'Where in a written contract it appears that both parties have agreed that something should be done which cannot effectively be done unless both concur in doing it, the construction of the contract is that each agrees to do all that is necessary to be done on his part for the carrying out of that thing though there may be no express words to that effect.'

Lord Justice Vaughan Williams in *Barque Quilpue Ltd* v. *Bryant* (1904) said:

> 'There is an implied contract by each party that he will not do anything to prevent the other party from performing a contract or to delay him in performing it. I agree that generally such a term is by law imported into every contract.'

However, whether or not the principle of prevention derives from a rule of law or from implied terms, there is no doubt of its general effect on an employer's right to recover liquidated damages for late completion. The principle proves perhaps the most effective and most used defence against liquidated damages.

Holme v. Guppy (1838)

The principle of prevention has long been applied to construction contracts. An early case is *Holme* v. *Guppy* (1838) where contractors failed to complete their work at a brewery within the stipulated time but avoided liquidated damages as part of the cause of late completion was delay by the employer in giving possession and delay by the employer's own workmen.

Dodd v. Churton (1897)

A later, much-quoted case is *Dodd* v. *Churton* (1897) where a two-week delay caused by the employer lost him the right to 25 weeks' liquidated damages. In *Dodd* v. *Churton* the contract provided for the whole of the works to be completed by 1 June 1892, under a penalty of £2 per week for every week that any part of the work remained unfinished after that date as liquidated damages. There was a provision that any alteration or addition in or to the works was not to vitiate the contract. There was apparently no provision for extending the time for completion if additional work was ordered. Additional works were ordered which involved a delay in the completion of the works beyond the specified date. The works were not completed until 5 December 1892. Evidence was given on the part of the defendant to the effect that a fortnight was a reasonable time for the additional work, and the defendant, allowing a fortnight's additional time for the completion of the works, claimed £2 per week in respect of the delay of 25 weeks. The county court judge held that by giving the order for additional works the defendant had waived the stipulation for penalties in respect of non-completion of the work by 1 June. His decision was upheld in the Court of Appeal. Lord Esher, Master of the Rolls said this:

> 'The principle is laid down in *Comyns' Digest,* Condition L(6) that, where one party to a contract is prevented from performing it by the act of the other, he is not liable in law for that default; and, accordingly, a well recognised rule has been established in cases of this kind beginning with *Holme* v. *Guppy* (1838), to the effect that, if the building owner has ordered

extra work beyond that specified by the original contract which has necessarily increased the time requisite for finishing the work, he is thereby disentitled to claim the penalties for non-completion provided for by the contract.'

Trollope & Colls Ltd v. NWMRHB (1973)

Lord Denning in the Court of Appeal hearing of *Trollope & Colls Ltd v. North West Metropolitan Regional Hospital Board* (1973) referred to *Dodd v. Churton* when commenting that the principle of prevention applies to legitimate conduct as well as to breach. He said:

'It is well settled that in building contracts – and in other contracts too – when there is a stipulation for work to be done in a limited time, if one party by his conduct – it may be quite legitimate conduct, such as ordering extra work – renders it impossible or impracticable for the other party to do his work within the stipulated time, then the one whose conduct caused the trouble can no longer insist upon strict adherence to the time stated. He cannot claim any penalties or liquidated damages for non-completion in that time.'

Lord Denning did go on to say that *Dodd v. Churton* established that:

'The time becomes at large. The work must be done within a reasonable time – that is, as a rule, the stipulated time plus a reasonable extension for the delay caused by his conduct.'

Whilst Lord Denning was corrected by Lords Pearson and Cross when the House of Lords came to hear the *Trollope & Colls* case for saying that *Dodd v. Churton* was authority for the proposition that time becomes at large as a result of prevention – that case is authority only for the proposition that liquidated damages cannot be deducted – he was not in error in the statement itself that time becomes at large as a result of prevention.

5.2 *Need for extension of time provisions*

In practical and financial terms for the parties it amounts to much the same thing – whether prevention has invalidated liquidated damages or whether prevention has put time at large. The only remedy left for the employer is to prove and sue for general damages for such late completion as can be established and the liability left on the contractor is to complete within a reasonable time or face general damages for failure.

Given the complexity of construction projects, large or small, the likelihood of extras or variations, the difficulties of co-ordination and the problems of the unforeseen, it is, unless some relief is available, almost impossible for the employer to avoid falling into the trap of prevention. That relief is provided by extension of time clauses.

In the early cases mentioned above, *Holme* v. *Guppy* and *Dodd* v. *Churton*, there were no provisions in the contracts to extend times for completion, and the employers in both cases were caught by strict provisions for completion which they had prevented being fulfilled. Consequently, the liquidated damages clauses failed. To avoid this legal predicament, extension of time provisions are included in construction contracts with the primary purpose of keeping liquidated damages clauses alive in the event of prevention.

There is nothing new about extension of time clauses – *Hudson* gives the case of *Legge* v. *Harlock* (1848) – but whether drafted as one-offs or included in standard forms of contract, they have suffered in the courts from strict interpretation. So notwithstanding the best efforts of contract draftsmen over the last two centuries, prevention remains a live obstacle to liquidated damages and extension of time clauses are effective only insofar as the courts hold them to operate or to apply.

Peak v. McKinney (1970)

The classic exposition of the difficulties facing employers was given by Lord Justice Salmon in *Peak Construction (Liverpool) Ltd* v. *McKinney Foundations Ltd* (1970) when he said:

'In my judgment, however, the plaintiffs are not entitled to anything at all under this head, because they were not liable to pay any liquidated damages for delay to the corporation. A clause giving the employer liquidated damages at so much a week or month which elapses between the date fixed for completion and the actual date of completion is usually coupled, as in the present case, with an extension of time clause. The liquidated damages clause contemplates a failure to complete on time due to the fault of the contractor. It is inserted by the employer for his own protection; for it enables him to recover a fixed sum as compensation for delay instead of facing the difficulty and expense of proving the actual damage which the delay may have caused him. If the failure to complete on time is due to the fault of both the employer and the contractor, in my view, the clause does not bite. I cannot see how, in the ordinary course, the employer can insist on compliance with a condition if it is partly his own fault that it cannot be fulfilled: *Wells* v. *Army & Navy Co-operative Society Ltd* (1902); *Amalgamated Building Contractors* v. *Waltham Holy Cross UDC* (1952); and *Holme* v. *Guppy* (1838). I consider that unless the contract expresses a contrary intention, the employer, in the circumstances postulated, is left to his ordinary remedy; that is to say, to recover such damages as he can prove flow from the contractors' breach.

No doubt if the extension of time clause provided for a postponement of the completion date on account of delay caused by some breach of fault on the part of the employer, the position would be different. This would mean that the parties had intended that the employer could recover liquidated damages notwithstanding that he was partly to blame for the failure to achieve the completion date. In such a case the architect would extend the date for completion, and the contractor would then be liable to pay liquidated damages for delay as from the extended completion date.

The liquidated damages and extension of time clauses in printed forms of contract must be construed strictly *contra proferentem*. If the employer wishes to recover liquidated damages for failure by the contractors to complete on time in spite of the fact that some of the delay is due to the employers' own fault or breach of contract, then the extension of time clause should provide, expressly or by necessary inference, for an extension on account of such a fault or breach on the part of the employer.

I am unable to spell any such provision out of clause 23 of the contract in the present case. In any event, it is clear that, even if clause 23 had provided for an extension of time on account of the delay caused by the contractor, the failure in this case of the architect to extend the time would be fatal to the claim for liquidated damages. There had clearly been some delay on the part of the corporation. Accordingly, as the architect has not made 'by writing under his hand such an extension of time', there is no date under the contract from which the defendants' liability to pay liquidated damages for delay could be measured. And therefore none can be recovered: see *Miller* v. *London County Council* (1934).'

Wells v. *Army & Navy* (1902)

In the case of *Wells* v. *Army & Navy Co-operative Society* (1902) referred to in the *Peak* case, there was a one-year delay in completing a one-year contract. It was held that the phrase 'other causes beyond the contractor's control' in the extension of time clause did not cover breaches by the employer in giving late possession and late information. Lord Justice Vaughan Williams, in making an interesting observation on time as a benefit as well as an obligation, said this:

'. . . in my mind that limitation of time is clearly intended, not only as an obligation, but as a benefit to the builder . . . In my judgment where you have a time clause it is always implied in such clauses that the penalties are only to apply if the builder has, as far as the building owner is concerned and his conduct is concerned, that time accorded to him for the execution of the works which the contract contemplates he should have.'

Bilton v. *GLC* (1982)

Lord Fraser of Tullybelton in *Percy Bilton Ltd* v. *Greater London Council* (1982) also referred to *Wells* in this exposition on prevention in a case where a nominated sub-contractor had gone into liquidation and re-nomination had caused delay. He said:

'1. The general rule is that the main contractor is bound to complete the work by the date for completion stated in the contract. If he fails to do so, he will be liable for liquidated damages to the employer.
2. That is subject to the exception that the employer is not entitled to liquidated damages if by his acts or omissions he has prevented the main contractor from completing his work by the completion date – see for example *Holme* v. *Guppy* (1838) and *Wells* v. *Army & Navy Co-operative Society* (1902).
3. These general rules may be amended by the express terms of the contract.'

Rapid Building v. *Ealing Family Housing* (1984)

The authority of the *Peak* decision was cited in another case, *Rapid Building Group Ltd* v. *Ealing Family Housing Association Ltd* (1984), by Lord Justice Stephenson. He said:

'In my judgment that authority is binding upon us; it quite clearly supports the decision of the learned judge that no counterclaim for liquidated damages under clause 22 of this contract can succeed. Presumably if the employer is responsible for any delay which does not fall within the *de minimis* rule, it cannot be reasonable for him to have completed the works on the completion date. Whatever the reasoning underlying the decision of this court it binds us and justifies the judge's decision that the counterclaim for liquidated damages is no answer to the plaintiff's claim.'

In that case the contractor was some 43 weeks late in completing the works, of which approximately three weeks was due to late possession of the site which at the date for commencement was occupied by squatters. The architect purported to grant an extension of time for this delay but the court found that the extension of time clause made no provision for breach of contract in failing to give possession of the site. Accordingly, the employer was barred from counterclaiming liquidated damages although a counterclaim for unliquidated damages was permitted.

5.3 Defining an act of prevention

One point which clearly emerges from the preceding cases is that an act of prevention may vary from an omission on the part of the employer, a fault, or even the ordering of variations and extras which might be fully contemplated by the contract.

In an Australian case, *SMK Cabinets* v. *Hili Modern Electrics Pty* (1984), Mr Justice Brooking summarised the law as follows:

'A wide variety of expressions have been used to describe the act of prevention which will excuse performance. At times words are employed which suggest that any act or omission preventing performance will suffice: *Dodd* v. *Churton* (1897) where all three members of the Court speak of an act: *Bruce* v. *The Queen* (1866) where the Court refers simply to prevention: *Percy Bilton Ltd* v. *Greater London Council* (1982). *Hudson Building and Engineering Contracts* (10th ed.) p. 631 (acts or omissions) speaks of acts, whether authorised by or breaches of the contract but at p. 700 refers to wrongful acts. In *Perini* v. *Greater Vancouver Sewerage and Drainage District* (1966) Bull JA with whose judgment Lord JA agreed, spoke of a wrongful act. The expressions used by Salmon LJ and Phillimore LR in *Peak Construction (Liverpool) Ltd* v. *McKinney Foundations Ltd* (1970) are "fault" and "fault or breach of contract". Another phrase to be found is "act or default" *Amalgamated Building Contractors Ltd* v. *Waltham Holy Cross Urban District Council* (1952). Words used by Lord Denning ("his conduct – it may be quite legitimate conduct, such as ordering extra work") appear in a passage cited with approval in the leading speech in the House of Lords: *Trollope & Colls Ltd* v. *North West Metropolitan Regional Hospital Board* (1973). I interpolate the observation that any formulation must accommodate the case of the ordering of extras, whether or not in the exercise of a power conferred by the contract. In the well known case of *Roberts* v. *Bury Improvement Commissioners* (1870) two different statements appear.

Blackburn and Mellor JJ, at p. 526, say that no person can take advantage of the non-fulfilment of a condition the performance of which has been hindered by himself, while Kelly CB and Channell B at p. 329 would ask whether performance has been prevented by a wrongful act; both statements are cited by Lord Thankerton in delivering the principal speech in *Panamena Europa Navigacion (Compania Limitada)* v. *Frederick Leyland & Co. Ltd* (1947). It is worth noting the formulation of Davis J of the Supreme Court of Canada in *Ottawa Northern and Western Railway Co.* v. *Dominion Bridge Co* (1905).'

Building Law Reports summary

The editors of *Building Law Reports* when commenting on the Court of Appeal decision in *Percy Bilton* v. *GLC* (1982) expressed the matter in more general terms:

' "Act of Prevention" is not easy to define but historically it has come to mean "virtually any event not expressly contemplated by the Contract and not within the Contractor's sphere of responsibility" – See *Hudson's Building Contracts* 10th edn, page 624 where the subject is treated fully. From the cases illustrated it may be seen that it is generally first necessary to determine whether there has been a breach of contract on the part of the employer or some other positive act or omission thereby preventing the contractor from completing the contract work by the due date and secondly, whether the contract did not make any express provision for extending time in such circumstances.

The older cases were largely decided in relation to contract where little or no provision was made for extending the time for completion so as to keep alive the Contract Completion Date and thus preserve the right to liquidated damages. Contracts nowadays generally contain extensions of time clauses drafted so as to cover the eventualities likely to constitute "acts of prevention" and are in many cases meticulous in their definition of the risks and responsibilities assumed by each party.

It is submitted that in a modern contract such as the Standard Form of Building Contract the correct analysis of events which may delay completion should not be between "acts of prevention" and "other acts" but rather between matters for which the contractor in law assumes the risk and matters for which he does not assume the risk. Such an approach is based upon the proposition that by undertaking to complete the work within the time stated a contractor assumes the responsibility of surmounting all risks other than those constituting breaches of contract or fault by the employer. It is sometimes useful to consider this apportionment of risk in terms of the "fault" of one party or the other, although "fault" is an emotive word.'

5.4 Prevention after the completion date

The *SMK* case offered guidance on the difficult issue of what is the employer's position if his prevention occurs when the contractor is in culpable delay – that is after the time for completion has expired and the contractor has still not completed.

In *SMK* it was said that:

> 'the ordering of variations after the completion date which substantially delays completion will, unless the contract provides otherwise and in the absence of an appropriate extension of time clause, prevent the employer from recovering or retaining liquidated damages which might otherwise have accrued after the giving of the order, although the employer's right in respect of amounts already accrued would not be affected.'

The decision in *SMK* was commented on in the case of *Balfour Beatty Building Ltd* v. *Chestermount Properties Ltd* (1993).

Mr Justice Colman said:

> 'Finally [the contractor] advanced an argument that the employer may not recover liquidated damages for the period after the date of the variation instruction, although he remains entitled to liquidated damages up to that date. He relied in support on the New Zealand decision in *Baskett* v. *Bendigo Gold Dredging Company (Ltd)* (1902) 21 NZLR 166 and the decision of the Supreme Court of Victoria in *SMK Cabinets* v. *Hili Modern*

Electrics Pty Ltd [1984] VR 391. Both were cases where it was contended that the employer's acts of prevention had discharged the liquidated damages obligation where there was no contractual provision for extension of time in the event of such acts of prevention. In both cases it was held that where the act of prevention took place during a period of culpable delay the liquidated damages already accrued remained unaffected but that no such damages were recoverable in respect of the delay after the act of prevention.'

And later in his judgment he said:

'In view of my decision upholding the award on the first question, this issue does not arise for decision in relation to this award. The arbitrator expressed no view on the matter. The essential issue between the parties on this appeal was whether, if the employer's act of prevention in the course of the period of culpable delay discharged the contractor's obligation to complete by the completion date and to pay liquidated damages should he fail to do so, the contractor remained liable for such liquidated damages as had accrued up to the date of the act of prevention and for general damages for failure thereafter to complete within a reasonable time or whether the obligation to pay such liquidated damages as had accrued was also discharged. In other words, was the solution to this problem that was arrived at in New Zealand in *Baskett* v. *Bendigo Gold Dredging Company* and in Victoria in *SMK Cabinets* v. *Hili Modern Electrics Pty Ltd*, to which I have already referred?

Although that solution is on the face of it a very practical one, it raises conceptual difficulties which suggest that the correctness of these decisions may have to be further reviewed. It would not be appropriate for this court to conduct that exercise in a case such as this where on the proper construction of the contract in question the point does not arise. I therefore think it best to express no concluded view on this issue.'

Comment

It needs to be noted that the *SMK* case and the above-quoted extracts from *Chestermount* relate to situations where the contractual provisions are deficient in dealing with prevention after the completion date. The reason the judge in *Chestermount* declined to express any concluded view on *SMK* was that he found that JCT 1980 was not so deficient.

Balfour Beatty v. *Chestermount* (1993)

The key issue in the *Chestermount* case was the long-standing question of whether an extension of time granted in respect of relevant events occurring during a period of culpable delay should be awarded on a gross basis or a net basis. That is to say whether the extension should include for the

contractor's culpable delay prior to the variation or whether the extension
should allow only for the time required for the variation itself such that only
this extra time should be added on to the time previously fixed.

However, to get to that issue (referred to in the case as the second question) the judge had first to decide whether clause 25 of JCT 1980 conferred
jurisdiction on the architect to grant an extension (the first question). The
judge concluded that it did, saying:

> 'It was common ground that if the contract failed to provide for power
> to grant an extension of time on account of delays caused by an act of
> prevention, the effect of the act of prevention was to prevent the employer
> relying on the completion date / liquidated damages provisions in the
> contract. The obligation to complete the works was to be performed
> within a reasonable time, there could be no extensions on account of
> relevant events and the employer's only hope of compensation would be
> to recover unliquidated damages for delay: see *Peak Construction
> (Liverpool) Ltd* v. *McKinney Foundations Ltd* (1970) 1 BLR 111. The remark-
> able consequences of the application of this principle could therefore be
> that if, as in the present case, the contractor fell well behind the clock and
> overshot the completion date and was unlikely to achieve practical com-
> pletion until far into the future, if the architect then gave an instruction
> for the most trivial variation, representing perhaps only a day's extra
> work, the employer would thereby lose all right to liquidated damages
> for the entire period of culpable delay up to practical completion or, at
> best, on the respondent's submission, the employer's right to liquidated
> damages would be confined to the period up to the act of prevention. For
> the rest of the delay he would have to establish unliquidated damages.
> What might be a trivial variation instruction would on this argu-
> ment destroy the whole liquidated damages regime for all subsequent
> purposes.
>   So extreme a consequence for the future operation of the contract could
> hardly reflect the common intention, particularly having regard to the
> very specific distribution of risk provisions which are agreed to be appli-
> cable in respect of relevant events occurring before the completion date.
> It is certainly a construction which is most improbable in the absence of
> some other express provision supporting it.'

And in summarising his views later, the judge said:

> 'In conclusion therefore, on the first question, in my judgment the con-
> struction for which the [Contractor] contends involves legal and com-
> mercial results which are so inconsistent with other express provisions
> and with the contractual risk distribution regime applicable to pre-com-
> pletion date relevant events that, in the absence of express words compel-
> ling that construction, it cannot be right. In this respect the contract is not
> ambiguous or so unclear as to call for application of the *contra proferentem*
> rule or the resolution of nicely-balanced issues of construction in favour
> of the employers for whose benefit the liquidated damages regime is

introduced. The apparently anomalous consequence of the application of the arbitrator's construction that the architect could refix a completion date before the issue of the variation instruction is in my view entirely consistent with the basic purpose of the liquidated damages regime for reasons which I have already explained. Moreover the retrospective post-ponement of the completion date to a date before the event causing delay was an eventuality contemplated with equanimity by Lord Denning MR in *Amalgamated Building Contractors Ltd* v. *Waltham Holy Cross UDC* [1952] All ER 452 at 454.'

On the second question as to whether a net extension or gross extension was due, the judge said:

'the function of the completion date is to identify the end of the period of time commencing with the date of possession within which the con-tractor must complete the works, including subsequent variations, failing which he must pay liquidated damages. The means by which that period is adjusted is by advancing or postponing the completion date which can be done prospectively or retrospectively. If it is advanced by reason of an omission instruction the consequence may well be that the adjustment required by way of reduction of the time for completion is sufficiently substantial to justify re-fixing the completion date before the issue of the instruction. Similarly, in the case of a variation which increases the works, the fair and reasonable adjustment required to be made to the period for completion may involve movement of the completion date to a point of time which may fall before the issue of the variation instruction or after it, depending on the extent to which the variation works have delayed completion of the works as a whole. The completion date as adjusted retrospectively is thus not the date by which the contractor ought to have achieved or ought in future to achieve practical completion but the date which marks the end of the total number of working days starting from the date of possession within which the contractor ought fairly and rea-sonably to have completed the works.'

And, later in his judgment, the judge said:

'Accordingly I conclude on the second question that it would be wrong in principle to apply the "gross" method, and that the "net" method repre-sents the correct approach. I therefore uphold the award on this point.'

However, the judge did go on to add a word of caution on the question of whether all relevant events would necessarily have the same application after the completion date had passed. In particular, whether the contractor could claim an extension for neutral events which occurred during his cul-pable delay. On this, the judge said:

'Before leaving this issue it is right to add that the application of the "net" method to relevant events occurring within the period of a culpable delay may give rise to particular problems of causation. These were discussed at some length in the course of argument. In each case it is for the architect exercising his powers under clause 25.3.3 to decide whether an adjustment of the completion date is fair and reasonable having regard to the incidence of relevant events. Fundamental to this exercise is an assessment of whether the relevant event occurring during a period of culpable delay has caused delay to the completion of the works and, if so, how much delay. There may well be circumstances where a relevant event has an impact on the progress of the works during a period of culpable delay but where that event would have been wholly avoided had the contractor completed the works by the previously-fixed completion date. For example, a storm which floods the site during a period of culpable delay and interrupts progress of the works would have been avoided altogether if the contractor had not overrun the completion date. In such a case it is hard to see that it would be fair and reasonable to postpone the completion date to extend the contractor's time. Indeed, where the relevant event would not be an act of prevention it is hard to envisage any extension of time being fair and reasonable unless the contractor was able to establish that, even if he had not been in breach of overshooting the completion date, the particular relevant event would still have delayed the progress of the works at an earlier date. Such cases are not likely to be of common occurrence.'

Although the case was brought under a JCT 80 contract its outcome is of wider application and is of relevance to most standard forms of construction contracts. In short on the first question the judge held that the extension provisions of JCT 80 were wide enough to apply to any relevant events and the architect was empowered to grant an extension of time after the due date for completion had passed. In respect of the second question the judge held that the net method of calculation should apply.

For further comment on this case see Chapter 14.

5.5 *Effect of late variations on unliquidated damages*

The effect of variations issued after the completion date was also considered in the case of *McAlpine Humberoak Ltd* v. *McDermott International Inc.* (1992). In that case there was a counterclaim for unliquidated damages for late completion. The contractor argued that no damages were due for the period of delay preceding the order for extra works.

The Court of Appeal rejected the contractor's argument. Lord Justice Lloyd said:

'[The contractor] submits that, since the extra work is covered by the definition of "the Work" in clause 1 of the contract, and since the extra work was not ordered until 11 June, the date for completion of the contract cannot precede that date. Accordingly the defendant's claim for damages cannot run from 1 May.

We do not agree. Even if the defendants were in a position to claim liquidated damages (which they are not) we doubt if the argument

would prevail, at any rate so far as the period prior to 11 June is concerned: see *SMK Cabinets* v. *Hili Modern Electrics* [1984] VR 391 at 398;
Keating on Building Contracts 5th Edn page 231 fn 18 [now 6th Edn page
250]. Here, as we have said, the defendants are claiming unliquidated
damages. Obviously they cannot recover damages for any additional
delay caused by the extra work. But this was taken care of by the three
weeks allowed by Mr McLaughlan, and by the ten and a half weeks
which we are allowing ourselves. If a contractor is already a year late
through his culpable fault, it would be absurd that the employer should
lose his claim for unliquidated damages just because, at the last moment,
he orders an extra coat of paint. On the facts of this case and the conditions of this contract the ordering of extra work on 11 June did not
have that effect. The defendants were not deprived of their right to
damages.'

5.6 Prevention and time at large

In relation to time for completion prevention clearly has the potential to set
time at large. Whether or not it does so on any particular construction project
depends on whether the act of prevention is causative of delay to completion
and whether the provisions in the contract for extending time cover the
preventive act.

Mr Justice Jackson in the case of *Multiplex Constructions (UK) Ltd* v.
Honeywell (2007) usefully summarised the law as follows:

'56. From this review of authority I derive three propositions:
  (i)   Actions by the employer which are perfectly legitimate under a
        construction contract may still be characterised as prevention, if
        those actions cause delay beyond the contractual completion
        date.
  (ii) Acts of prevention by an employer do not set time at large if the
        contract provides for extension of time in respect of those
        events.
  (iii) In so far as the extension of time clause is ambiguous, it should
        be construed in favour of the contractor.
57. The third proposition must be treated with care. It seems to me that,
    in so far as an extension of time clause is ambiguous, the court should
    lean in favour of a construction which permits the contractor to
    recover appropriate extensions of time in respect of events causing
    delay. This approach also accords with the principle of construction
    set out by Lewison in "The Interpretation of Contracts" (3rd edition,
    2004). That principle reads as follows:
    "Where two constructions of an instrument are equally plausible,
    upon one of which the instrument is valid and upon the other of
    which it is invalid, the court should lean towards that construction
    which validates the instrument." '

5.7 *Conditions precedent and time-bars*

It is increasingly common for standard forms of construction contracts, and even more so for ad-hoc construction contracts, to include notice requirements which effectively amount to conditions precedent to obtaining extensions of time and to include timing requirements which effectively time-bar late notices.

Various questions can then arise. What is the position if the contractor is prevented by the employer's breach or some legitimate preventive act (e.g. ordering variations) from completing on time but is debarred from obtaining any extension of time by application of conditions precedent or time-bars?. Can the contractor rely on the prevention principle to avoid liquidated damages for delay or can the employer enforce liquidated damages – apparently benefiting thereby from his own preventive act? In short, does the prevention principle have primacy over the contractual provisions when it comes to examining whether time is at large.

Some of these questions were considered by Mr Justice Jackson in the *Multiplex* v. *Honeywell* case under the heading of 'The Gaymark Point':

Part 6. The Gaymark Point

'95. Honeywell contends that, even if compliance with clause 11 remained possible, nevertheless Honeywell's failure to comply with that clause was sufficient to put time at large. If it were otherwise, says Honeywell, then Multiplex would be able to recover damages for a period of delay which Multiplex had caused. The legal basis for this argument is the Australian decision in *Gaymark Investments Pty Ltd v Walter Construction Group Ltd* [1999] NTSC 143; (2005) 21 *Construction Law Journal* 71.

96. I must therefore begin by reviewing the *Gaymark* decision. In that case the employer claimed liquidated damages against the contractor for delay in constructing an hotel in Darwin. Clause 19.1 of the Special Conditions of Contract imposed conditions in respect of giving notice of delay. Clause 19.2 of the Special Conditions provided:
"The Contractor shall only be entitled to an extension of time for Practical Completion where . . . (b)(i) the contractor has complied strictly with the provisions of sub-clause SC19.1 and in particular has given the notices required by sub-clause SC19.1 strictly in the manner and within the times stipulated by that sub-clause."

97. The Arbitrator made the following findings:
(1) That the contractor was delayed in completing the work, including a delay of 77 days by causes for which the employer was responsible, but the contractor's application for an extension of time was barred because of its failure strictly to comply with the notification requirements for the extension of time clause.

(2) That the 77 days' delay constituted acts of prevention by the employer with the result that there was no date for practical completion and the contractor was then obliged to complete the work within a reasonable time (which the Arbitrator found that it in fact did) with the consequence being that Gaymark was prevented from recovering liquidated damages for delay.

98. The Supreme Court of the Northern Territory of Australia refused leave to appeal and upheld the Arbitrator's award. Bailey J said this at paragraphs 69–71 of his judgment:

"69. Acceptance of Gaymark's submissions would result in an entirely unmeritorious award of liquidated damages for delays of its own making (and this in addition to the avoidance of Concrete Constructions' delay costs because of that company's failure to comply with the notice provisions of SC19). The effect of re-drafting GC35 of the contract (to delete GC35.4 and substitute SC19) has been to remove the power of the superintendent to grant of allow extensions of time. SC19 makes provision for an extension of time for delays for which Gaymark directly or indirectly is responsible but the right to such an extension is dependent on strict compliance with SC19 (and in particular the notice provisions of SC19.1). In the absence of such strict compliance (and where Concrete Constructions has been actually delayed by an act, omission or breach for which Gaymark is responsible) there is no provision for an extension of time because GC35.4 which contains a provision which would allow for this (and is expressly referred to in GC35.2 and GC35.5) has been deleted.

70. In *Peak Construction (Liverpool) Limited* v *McKinney Foundations Limited* [1970] 1 BLR 111, Salmon LJ held:

'The liquidated damages and extension of time clauses and printed forms contract must be construed strictly contra proferentum. If the employer wishes to recover liquidated damages for failure by the contractors to complete on time in spite of the fact that some of the delay is due to the employer's own fault or breach of contract, then the extension of time clause should provide, expressly or by necessary inference, for an extension on account of such a fault or breach on the part of the employer'.

71. In the circumstances of the present case, I consider that this principle presents a formidable barrier to Gaymark's claim for liquidated damages based on delays of its own making. I agree with the arbitrator that the contract between the parties fails to provide for a situation where Gaymark caused actual delays to Concrete Construction's achieving practical completion by the due date coupled with a failure by Concrete Constructions to comply with the notice provisions of SC19.1. In such circumstances, I do not consider that there was any 'manifest error of

law on the face of the award' or any 'strong evidence' of any error of law in the arbitrator holding that the 'prevention principle' barred Gaymark's claim to liquidated damages."

99. In reaching this conclusion Bailey J took a different view from that expressed obiter by Cole J in *Turner Corporation Limited (Receiver and Manager Appointed) v Austotel Pty Limited* (2nd June 1994); 1997 13 BCL 378 at 12. In that earlier judgment Cole J had said:

"If the Builder, having a right to claim an extension of time fails to do so, it cannot claim that the act of prevention which would have entitled it to an extension of time for Practical Completion resulted in its inability to complete by that time. A party to a contract cannot rely upon preventing the contract of the other party where it failed to exercise a contractual right which would have negated the effect of that preventing conduct."

100. The correctness of the *Gaymark* decision has been a matter of some debate. The editors of *Keating on Building Contracts* (8th edition 2006) note that there is no English authority on the matter but incline to the view that *Gaymark* was correctly decided (see paragraph 9-025). The editor of Hudson on Building Contracts, the late Ian Duncan Wallace QC, argues that *Gaymark* was wrongly decided (see paragraph 10.026 of the first supplement to the 11th edition of Hudson). Professor Wallace (a formidable commentator on construction law, who is now sadly missed) also wrote a trenchant article on this subject. See "Prevention and Liquidated Damages: a Theory Too Far" (2002) 18 Building and Construction Law, 82. In that article Professor Wallace refers to the *Turner* case, which I have previously mentioned, and certain other authorities. He points out the useful practical purpose which contractual provisions requiring a contractor to give notice of delay serve. Professor Wallace argues that both the arbitrator and the judge came to the wrong conclusion in *Gaymark*. In Professor Wallace's view, *Gaymark* extends the prevention theory too far.

101. In *Peninsula Balmain Pty. Limited v Abigroup Contractors Pty. Limited* [2002] NSWCA 211, the New South Wales Court of Appeal declined to follow *Gaymark* and preferred the reasoning of Professor Wallace. Hodgson JA gave the leading judgment with which other members of the court agreed. At paragraph 78 Hodgson JA said this:

"I accept that, in the absence of the Superintendent's power to extend time, even if a claim had not been made within time, Abigroup would be precluded from the benefit of an extension of time and liable for liquidated damages, even if delay had been caused by variations required by Peninsula and thus within the so-called 'prevention principle'. I think this does follow from the two *Turner* cases and the article by Mr. Wallace referred to by Mr. Rudge."

102. A year after *Peninsula,* the Second Division of the Inner House of the Court of Session gave judgment in *City Inn Limited v Shepard Construction Limited* 2003 SLT 885. In that case the employer contended that the contractor was not entitled to any extension of time, because the Contractor had not complied with clause 13.8 of the contract in relation to notices. The court held that the contractor could not obtain an extension of time if it did not comply with that provision (see paragraph 23 of the Opinion of the court). It appears, however, that the Australian cases were not cited.

103. I am bound to say that I see considerable force in Professor Wallace's criticisms of *Gaymark.* I also see considerable force in the reasoning of the Australian courts in *Turner* and in *Peninsula* and in the reasoning of the Inner House in *City Inn.* Whatever may be the law of the Northern Territory of Australia, I have considerable doubt that *Gaymark* represents the law of England. Contractual terms requiring a contractor to give prompt notice of delay serve a valuable purpose; such notice enables matters to be investigated while they are still current. Furthermore, such notice sometimes gives the employer the opportunity to withdraw instructions when the financial consequences become apparent. If *Gaymark* is good law, then a contractor could disregard with impunity any provision making proper notice a condition precedent. At his option the contractor could set time at large.

104. Although I have considerable doubts that *Gaymark* represents the law of England, nevertheless that is not a question which I am required finally to decide. This is because *Gaymark* should readily be distinguished from the present case. In *Gaymark* non-compliance with the notice clause exposed the contractor to an automatic liability for liquidated damages (if the liquidated damages clause were upheld). In the present case, non-compliance with clause 11.1.3 has no such automatic consequences. Even if (contrary to Mr. Thomas' submissions) Honeywell forfeits any entitlement to extension of time, that does not automatically make Honeywell liable to pay damages for delay. Under clause 12 of the Sub-Contract Conditions, Multiplex can only recover in respect of loss or damage "caused by the failure of the Sub-Contractor". If in reality the relevant delay was caused by Multiplex, not Honeywell, then (whatever the position under clause 11) Multiplex cannot recover against Honeywell under clause 12.

105. Let me now draw the threads together. If the facts are that it was possible to comply with clause 11.1.3 but Honeywell simply failed to do so (whether or not deliberately), then those facts do not set time at large. Honeywell is not entitled to the relief which it seeks in respect of the *Gaymark* point.'

The *City Inn* reasoning referred to by Mr Justice Jackson in his judgment comes from the second judgment in the case of *City Inn Ltd* v. *Shepherd Construction Ltd* (the 2003 appeal ruling). Mention has already been made to this judgment in Chapter 4 above in relation to whether the condition precedent provisions in the construction contract acted as a penalty clause. The court doubted that they did. A more significant aspect of the case, and apparently that referred to by Mr Justice Jackson, was whether the conditions precedent deprived the contractor of the opportunity to obtain an extension of time – thereby imposing on the contractor liability to pay liquidated damages and thereby turning the conditions precedent into a penalty for breach. The court held that they did not. Its reasoning was as follows:

'a.   Whether failure by the contractor to take action under clause 13.8.1 constitutes a breach of contract

[22] This issue arises because, in order to have clause 13.8.5 construed as a penalty clause, the defenders have to argue that where they do not operate the procedures laid down in clause 13.8.1, they themselves commit a breach of contract. In holding that in that case the contractor commits a breach of contract, the Lord Ordinary considered that the breach consisted in the contractor's failure to form an opinion on the question raised by the instruction and defined in the clause and to intimate the relevant estimates based on that opinion. In the argument before us, however, counsel for the defenders argued that the breach consisted in the contractor's proceeding to carry out the work.

[23] In our opinion, the contractor would not commit a breach of contract in either respect. Clause 13.8 does not impose any obligation on the contractor when he receives an architect's instruction. If the contractor receives such an instruction, he has to consider its likely effects, and in particular its likely effect on the duration of the building period. He may, for reasons of his own, decide to accept the instruction without resistance and hope that he will be able to complete the work, as varied by the instruction, within the building period. But if he wishes an extension of time, he must comply with the conditions precedent that clause 13.8 provides for in these specific circumstances (cf. cl. 13.8.5). In particular, he must serve notice on the architect of his estimate of *inter alia* the probable cost. In the light of that the employer may take the opportunity to withdraw the instruction. But if the contractor fails to take the steps specified in clause 13.8.1, then unless the architect waives the requirements of the clause under clause 13.8.4, the contractor will not be entitled to an extension of time on account of that particular instruction.

[24] In short, clause 13.8 provides the contractor with an additional right, in the specific case to which it applies, that would not be available to him in the case of an instruction issued under the general provisions of clause 4. But clause 13.8 does not oblige the contractor to

invoke its protection. It merely provides the contractor with an option to take certain action if he seeks the protection of an extension of time in the circumstances in which the clause applies. The provisions of clause 13.8, in our view, are merely conditions precedent to his doing so.

[25] If the contractor fails to take action in accordance with clause 13.8.1, it is a false analysis to describe him as being in breach of contract. The ultimate effect of that failure may be that he is unable to complete the works, as varied by the instruction, by the completion date; although that cannot be assumed at that stage. But the breach of contract that that would constitute would not be a breach of clause 13.8. It would be a breach of clause 23 (*supra*), of which the ultimate cause may be the contractor's failure to take advantage of clause 13.8.

[26] For these reasons, we consider that the clause does not oblige the contractor to respond to the instruction in accordance with the clause and thereby to give the architect the opportunity to reconsider the instruction. If the architect issues an instruction that may have such consequences, the chance that the contractor will not contest it is just one of the risks that the architect takes. If therefore the contractor goes ahead with the work in accordance with the instruction, he does not commit any breach of contract. In this respect we disagree with the Lord Ordinary's reasoning.'

Comment

All three judgments in the *City Inn* case attracted considerable interest – particularly for the way the first and second judgments dealt with what might be described as prevention type arguments. In holding that the conditions precedent did not render the liquidated damages provisions unenforceable the Scottish courts seemed to be providing a lead in an area of law previously lacking legal authority (save for the Commonwealth cases).

However, although the reasoning in *City Inn* is certainly interesting, some caution may need to be exercised before assuming that the ruling is of general application. The construction contract contained ad-hoc provisions on variations and the giving of notices and the dispute arose from clauses requiring the contractor to provide estimates of extra cost and time before acting on any instruction or the like which, in the opinion of the contractor, constituted a variation. In such circumstances it is not difficult to see why the Scottish courts took the view that it was at the contractor's option whether or not it gave notice and/or acted on an instruction.

It is questionable whether the courts would have reached the same view if the contract had required the contractor to act on all instructions validly given by the contract administrator and questionable as to what relevance it has to the effects of conditions precedent when the act of prevention is employer's breach (e.g. failure to give timely possession of the site).

5.8 *Steria Ltd* v. *Sigma Wireless Communications Ltd* (2007)

The *Steria* case concerned delays in completion of a subcontract on a computer-assisted mobilisation and communications project. The sub-contract was based on a heavily revised version of the model form for electrical and mechanical works, MF/1. Sigma, the main contractor, sought to recover by way of set-off and/or counterclaim, liquidated or alternatively general damages for Steria's delay in completing its works.

The 92-page judgment of Judge Stephen Davies covers a wide range of interesting matters relating to extensions of time and liquidated damages – albeit that some parts of the judgment are clearly stated to be views rather than decisions. The principal matters relate to application of the prevention principle on which the judge firmly states, following Mr Justice Jackson in *Multiplex*, that the prevention principle does not mean that failure to comply with notice provisions puts time at large. Other matters concern:

- interpretation
- notices
- penalty clauses.

The prevention principle

Rejecting Steria's arguments on prevention and time at large the judge said:

'78. The first issue is whether or not the requirement in clause 6.1 for Steria to give written notice of the circumstances giving rise to the delay within a reasonable period is a condition precedent to its right to an extension of time. Whilst Steria contends principally that clause 6.1 is not a condition precedent, or if it is that: (a) it complied with that requirement; alternatively (b) Sigma has waived the requirement for compliance / is estopped from complaining of non-compliance; the further issue of law which arises if those arguments are unsuccessful is Steria's contention that if delay has been caused due to acts of prevention by Sigma (including for this purpose CAMP East and/or Mason) and if Steria has failed to give notice in compliance with clause 6.1, then the result is that the time for completion is set at large. This argument relies on what has become known as the "prevention principle".

AND

'The application of the 'prevention' principle
93. It is convenient at this stage also to address the prevention principle argument, even though again the point will only arise for decision if I find first that Steria had failed to give the requisite notices and second that Sigma had not waived, or was not estopped from relying on the absence of, the requisite notices.

94. I am extremely fortunate in that I have the benefit of the analysis of Jackson J. in the *Multiplex* case of the conflicting Australian authorities (*Turner*, *Gaymark* and *Peninsula*), the decision of the Court of Session in *City Inn v Shepard Construction* 2003 SLT 885, and the views expressed both by the editors of *Keating on Building Contracts* and by the late Professor Wallace QC. In summary, Jackson J. concluded in paragraph 103 that:

"I am bound to say that I see considerable force in Professor Wallace's criticisms of *Gaymark*. I also see considerable force in the reasoning of the Australian courts in *Turner* and in *Peninsula* and in the reasoning of the Inner House in City Inn. Whatever may be the law of the Northern Territory of Australia, I have considerable doubt that *Gaymark* represents the law of England. Contractual terms requiring a contractor to give prompt notice of delay serve a valuable purpose; such notice enables matters to be investigated while they are still current. Furthermore, such notice sometimes give the employer the opportunity to withdraw instructions when the financial consequences become apparent. If *Gaymark* is good law, then a contractor could disregard with impunity any provision making proper notice a condition precedent. At his option the contractor could set time at large."

95. Although on the facts of that case Jackson J. did not, due to the particular wording of the extension of time and liquidated damages clauses employed, need to express a final decision on the point, nonetheless I gratefully adopt his analysis and agree with his preliminary conclusion. Generally, one can see the commercial absurdity of an argument which would result in the contractor being better off by deliberately failing to comply with the notice condition than by complying with it. Furthermore, when applied to the facts of this case, particularly acute difficulties arise when considering how the application of the prevention principle should work in practice. Thus clause 6.1 permits an extension of time in 3 relevant circumstances, one of which is "any circumstance which entitles the contractor to an extension of time under the main contract". Clause 33.1 of MF/1 (the main contract term relating to extension of time), allows an extension of time in 4 specified circumstances, one of which is "any industrial dispute" and the other of which is "circumstances beyond the reasonable control of the contractor arising after acceptance of the works". Does the prevention principle apply to such circumstances? It is difficult to see why they should, since these circumstances cannot readily be characterised as acts of prevention by the employer. If not, however, is the effect that the notice procedure is a condition precedent in relation to delays caused by those events, but not to delays caused by other events? That would produce an inconsistent and undesirable result. Furthermore, in addition to conferring a right to an extension of time, clause 6.1 also confers a right on Steria to recover "all extra costs incurred in relation to [the delay] together with a reasonable

allowance for profit". Does the prevention principle mean that Steria could obtain these benefits even if it had not complied with the notice condition precedent? Again, that would appear to involve the contractor obtaining a benefit from his own breach, which is the converse of the prevention principle and hence might be said to be equally objectionable, but to construe the clause such that Steria was entitled to an extension of time even if it did not comply with the notice condition, but not to an extra payment, would again in my judgment be inconsistent and undesirable.

96. In its closing submissions [§134] Steria invites me to distinguish *Multiplex* on the grounds that it contained a clear and unambiguous notice condition precedent clause, unlike that found here. It does not seem to me that strictly speaking any question of distinguishing *Multiplex* arises since, as I have already noted, in that case Jackson J. did not need to reach an actual decision on the point. However I must confess that I cannot see that the particular form of the clause used in that case is relevant to the analysis of the authorities and the provisional conclusion reached. In any event I am, as I have already said, respectfully in agreement with Jackson J. Gratefully adopting therefore the reasons which he gives in that case, together with the further reasons set out above in relation to the particular clause in this case, I conclude that the prevention principle does not mean that failure to comply with the notice requirement of clause 6.1 puts the time for completion at large.

97. A separate but connected argument advanced by Steria was that one cause of delay, namely delay due to negotiations between the DFB and its employees' trades union in relation to the introduction of the new system 5, fell outside the scope of the relevant circumstances provided for by clause 6.1 and thus, in the event that delay was caused by this circumstance, would amount to prevention for which Sigma was responsible for which an extension of time was not available, with the result that time for completion was set at large. Although ingenious, in my judgment the argument fails because that cause of delay would either fall within the definition of "circumstances beyond the reasonable control of the contractor arising after acceptance of the works" (and thus within clause 33.1 MF/1), or a "breach by the contractor" (because the effect of delay caused by such ongoing negotiations would amount to breaches by Sigma of its positive obligations in Schedule 11 of the sub-contract).'

Interpretation

On interpretation of the contractual requirements for notice the judge said:

'90. Turning to the wording of the clause, in my judgment the phrase "provided that the sub-contractor shall have given within a reasonable period written notice to the contractor of the circumstances giving rise to the delay" is clear in its meaning. What the sub-contractor is required to do is give written notice within a reasonable period from when he is delayed, and the fact that there may be scope for argument in an individual case as to whether or not a notice was given within a reasonable period is not in itself any reason for arguing that it is unclear in its meaning and intent. In my opinion the real issue which is raised on the wording of this clause is whether those clear words by themselves suffice, or whether the clause also needs to include some express statement to the effect that unless written notice is given within a reasonable time the sub-contractor will not be entitled to an extension of time.

91. In my judgment a further express statement of that kind is not necessary. I consider that a notification requirement may, and in this case does, operate as a condition precedent even though it does not contain an express warning as to the consequence of non-compliance. It is true that in many cases (see for example the contract in the *Multiplex* case itself) careful drafters will include such an express statement, in order to put the matter beyond doubt. It does not however follow, in my opinion, that a clause – such as the one used here – which makes it clear in ordinary language that the right to an extension of time is conditional on notification being given should not be treated as a condition precedent. This is an individually negotiated sub-contract between two substantial and experienced companies, and I would be loathe to hold that a clearly worded requirement fails due to the absence of legal "boilerplate".'

Notices

Regarding Steria's case that it could rely on minutes of meetings or its pleadings the judge said:

'82. I also consider that the written notice must emanate from Steria. Thus for example an entry in a minute of a meeting prepared by Mason which recorded that there had been a delay by CAMP East in approving the FDS, and that as a result the sub-contract works had been delayed, would not in my judgment by itself amount to a valid notice under clause 6.1. The essence of the notification requirement in my judgment is that Sigma must know that Steria is contending that relevant circumstances have occurred and that they have led to delay in the sub-contract works.'

Penalty clause points

In rejecting various arguments by Steria that the provisions for liquidated damages amounted to penalty clauses, the judge said:

'98.   I must now deal with Steria's contention that clause 7.1 is a penalty clause.

99.   Steria has referred me to *Murray v Leisureplay* [2005] EWCA Civ 963, a case in which the Court of Appeal subjected the law on penalty clauses to a detailed scrutiny in the context of a clause requiring the defendant employer to pay the claimant employee a year's gross salary in the event of wrongful termination without notice. Steria has also referred me to *Alfred McAlpine Capital Projects v Tilebox* [2005] EWHC 281 (TCC), in which Jackson J. considered the law on penalty clauses in the context of a dispute arising under a construction contract. Sigma has referred me to the Tilebox case and also to the case of *Philips Hong Kong v AG for Hong Kong* (1993) 61 BLR 49, a decision of the Privy Council which was discussed in both cases.

100.   So far as I can discern there is no significant dispute between the parties as to the legal principles which I should apply. Thus Sigma contended that the question for me, in the light of Murray, was whether Steria could show that the liquidated damages provision in clause 7.1 was 'extravagant, unconscionable and not a genuine pre-estimate of loss' (see the judgment of Clarke L.J. at §106(vii)). Steria referred me to the same judgment at §106(iv) where Clarke L.J., referring to the decision of Colman J. in *Lordsdale Finance v Bank of Zambia* [1996] QB 752, concluded that the real question for the court was whether the contractual function of the clause was deterrent or compensatory, and that one guide to answering this question was to compare the amount which would be payable on breach with the loss that might be sustained if the breach occurred. In *Tilebox* Jackson J. considered that there must be a substantial discrepancy between the level of damages stipulated in the contract and the level of damage which is likely to be suffered before it can be said that the agreed pre-estimate is unreasonable. I must however also bear in mind that this is only a guide, and does not necessarily always provide the answer by itself, because – as was emphasised by Buxton L.J. in Murray at §111 and by Jackson J. in *Tilebox* at §48.3 – the question is a broad and general question, and that in commercial contracts the courts should exercise great caution before striking down a clause as penal.'

And later in the judgment, having examined the details of the liquidated damages clauses of the main contract and the sub-contract, the judge concluded:

'106.   In such circumstances, in my judgment: (i) there is no substantial discrepancy between the liquidated damages provisions of the sub-contract and the level of damages likely to be suffered by Sigma; (ii) on the facts of this case I am unable to conclude that the clause was – objectively considered as at the date the contract was entered into – intended to be deterrent rather than compensatory. Overall, this being a commercial contract entered into between two substantial and experienced companies with knowledge of the difficulties which can occur where after the event one party seeks to recover general damages from the other for delay, I am not prepared to strike down the clause as penal.'

Capping point

On the question of whether a stated cap on the amount of liquidated damages (in this case 10% of the subcontract price) acts as a limit on the amount of recoverable general damages when the liquidated damages provisions are held to be inoperable, the judge made these observations:

'114. Having upheld the liquidated damages provisions of clause 7.1 and Schedule 6, it is unnecessary for me to consider the further argument as to whether the cap in those provisions would also apply to cap the alternative claim for general damages. It is clear in my judgment from the concluding words of clause 7.1 that the entitlement to liquidated damages is Sigma's sole remedy for delay by Steria, so that it is not possible for Sigma to advance its claims for general damages as an alternative. If I had needed to decide the point, I would have inclined to the view that if the liquidated damages provision is held to be penal, then it prevents either party from relying on it, so that the cap also disappears.'

Chapter 6
Legal construction of liquidated damages clauses

6.1 Rules of construction

The object of the courts in construing written contracts is to discover the intentions of the parties. The courts will apply rules of construction to the express terms to resolve ambiguities or inconsistencies; they may add implied terms to a contract to provide business efficacy; but they will not make a contract for the parties or re-make a contract which the parties have made for themselves which turns out to have unexpected results. Lord Pearson in *Trollope & Colls Ltd* v. *North West Metropolitan Regional Hospital Board* (1973) put it this way:

> 'The basic principle is that the court does not make a contract for the parties. The court will not even improve the contract which the parties have made for themselves, however desirable the improvement might be. The court's function is to interpret and apply the contract which the parties have made for themselves. If the express terms are perfectly clear and free from ambiguity, there is no choice to be made between different possible meanings: the clear terms must be applied even if the court thinks some other term would have been more suitable. An unexpressed term can be implied if and only if the court finds that the parties must have intended that term to form part of their contract; it is not enough for the court to find that such a term would have been adopted by the parties as reasonable men if it had been suggested to them: it must have been a term that went without saying, a term necessary to give business efficacy to the contract, a term which, though tacit, formed part of the contract which the parties made for themselves.'

The rules of construction are briefly as follows:

Intention to be found from the contract itself

In construing written contracts, the courts will not go outside the written documents and substitute the presumed intentions of the parties. The intentions must be ascertained from the contract itself.

Ordinary/plain meaning

Words are to be given their ordinary or plain meaning as generally understood. This rule is departed from only to avoid absurdity or inconsistency. In trade or technical contracts, the customary meaning of words applies.

Valid meaning

Words, phrases or clauses capable of different meanings will be construed to make a contract valid rather than void or ineffective.

Contract to be read as a whole

The intention of the parties is to be derived by construing a contract as a whole and, as far as practicable, giving effect to each of its provisions. To do this, the court may first have to decide which documents form the contract and to discover what order of precedence the parties may, or may not, have given to the various documents.

Particulars prevail over standards

Unless the documents expressly provide otherwise, written words on a printed form will have greater effect than the printed words and the provisions in a written document will prevail over provisions in any incorporated documents. Thus, special or particular conditions will generally prevail over standard conditions.

'Expressio unius'

Express mention of a certain thing will exclude other things of a similar nature. Thus, a contract for the sale of a foundry plus two houses, together with the fixtures in the houses, was held to exclude the fixtures in the foundry.

'Ejusdem generis'

Where words of a particular class are followed by general words, the general words are taken to apply to things of the same class. Thus, in a charter contract it was held that the words 'any other cause' in the phrase 'war, disturbance or any other cause' did not include ice, but were restricted to events of the same kind as war and disturbance.

'Contra proferentem'

Where there is ambiguity in a document, the words are to be construed against the party who put forward the document.

Rules of construction / rules of law

Before considering the last rule in more detail, it is worth noting the distinction between rules of construction and rules of law. Rules of construction are applied to enable the courts to ascertain the intentions of the parties, as expressed, and to give effect to those intentions. Rules of law are applied to bring the parties within the framework of established legal rulings and these rules apply even though the parties may have expressed, and intended, something contrary.

The *West Bromwich* case (1997)

Modern authority on the interpretation of contracts is found in the House of Lords ruling in the case of *Investors Compensation Scheme* v. *West Bromwich Building Society* (1997) – a ruling concerning the meaning of a clause in a home income plan. Lord Hoffman set out five principles for construing contract documents:

'(1) Interpretation is the ascertainment of the meaning which the document would convey to a reasonable person having all the background knowledge which would reasonably have been available to the parties in the situation in which they were at the time of the contract.

(2) The background was famously referred to by Lord Wilberforce as the "matrix of fact", but this phrase is, if anything, an understated description of what the background may include. Subject to the requirement that it should have been reasonably available to the parties and to the exception to be mentioned next, it includes absolutely anything which would have affected the way in which the language of the document would have been understood by a reasonable man.

(3) The law excludes from the admissible background the previous negotiations of the parties and their declarations of subjective intent. They are admissible only in an action for rectification. The law makes this distinction for reasons of practical policy and, in this respect only, legal interpretation differs from the way we would interpret utterances in ordinary life. The boundaries of this exception are in some respects unclear. But this is not the occasion on which to explore them.

(4) The meaning which a document (or any other utterance) would convey to a reasonable man is not the same thing as the meaning of its words. The meaning of words is a matter of dictionaries and gram-

mars; the meaning of the document is what the parties using those words against the relevant background would reasonably have been understood to mean. The background may not merely enable the reasonable man to choose between the possible meanings of words which are ambiguous but even (as occasionally happens in ordinary life) to conclude that the parties must, for whatever reason, have used the wrong words or syntax. (See *Mannai Investments Co. Ltd. v. Eagle Star Life Assurance Co. Ltd.* [1997] 2 WLR 945.)

(5) The "rule" that words should be given their "natural and ordinary meaning" reflects the common sense proposition that we do not easily accept that people have made linguistic mistakes, particularly in formal documents. On the other hand, if one would nevertheless conclude from the background that something must have gone wrong with the language, the law does not require judges to attribute to the parties an intention which they plainly could not have had. Lord Diplock made this point more vigorously when he said in *The Antaios Compania Neviera S.A. v. Salen Rederierna A.B.* (1985) 1 AC 191, 201:
"...if detailed semantic and syntactical analysis of words in a commercial contract is going to lead to a conclusion that flouts business commonsense, it must be made to yield to business commonsense."'

The comments of Lord Hoffman which immediately preceded and followed the above-quoted statement of principles are worth noting – if only for their levity and their bluntness in reference to 'old intellectual baggage':

'In the Court of Appeal, Leggatt LJ said, on the authority of *Alice Through the Looking Glass*, that the judge's interpretation was "not an available meaning of the words". "Any claim (whether sounding in rescission for undue influence or otherwise)" could not mean "Any claim sounding in rescission (whether for undue influence or otherwise)" and that was that. He was unimpressed by the alleged commercial nonsense of the alternative construction.

My Lords, I will say at once that I prefer the approach of the learned judge. But I think I should preface my explanation of my reasons with some general remarks about the principles by which contractual documents are nowadays construed. I do not think that the fundamental change which has overtaken this branch of the law, particularly as a result of the speeches of Lord Wilberforce in *Prenn v. Simmonds* [1971] 1 WLR 1381, 1384–1386 and *Reardon Smith Line Ltd. v. Yngvar Hansen-Tangen* [1976] 1 WLR 989, is always sufficiently appreciated. The result has been, subject to one important exception, to assimilate the way in which such documents are interpreted by judges to the common sense principles by which any serious utterance would be interpreted in ordinary life. Almost all the old intellectual baggage of "legal" interpretation has been discarded. The principles may be summarised as follows.
[The five principles as quoted above]
If one applies these principles, it seems to me that the judge must be right and, as we are dealing with one badly drafted clause which is

happily no longer in use, there is little advantage in my repeating his reasons at greater length. The only remark of his which I would respect-fully question is when he said that he was "doing violence" to the natural meaning of the words. This is an over-energetic way to describe the process of interpretation. Many people, including politicians, celebrities and Mrs. Malaprop, mangle meanings and syntax but nevertheless com-municate tolerably clearly what they are using the words to mean. If anyone is doing violence to natural meanings, it is they rather than their listeners.'

Lord Hoffman went on to say:

'Finally, on this part of the case, I must make some comments upon the judgment of the Court of Appeal. Leggatt LJ said that his construction was "the natural and ordinary meaning of the words used". I do not think that the concept of natural and ordinary meaning is very helpful when, on any view, the words have not been used in a natural and ordinary way. In a case like this, the court is inevitably engaged in choosing between competing unnatural meanings. Secondly, Leggatt LJ said that the judge's construction was not an "available meaning" of the words. If this means that judges cannot, short of rectification, decide that the parties must have made mistakes of meaning or syntax, I respectfully think he was wrong. The proposition is not, I would suggest, borne out by his citation from *Alice Through the Looking Glass*. Alice and Humpty Dumpty were agreed that the word "glory" did not mean "a nice knock-down argument". Anyone with a dictionary could see that. Humpty Dumpty's point was that "a nice knock-down argument" was what *he* meant by using the word "glory". He very fairly acknowledged that Alice, as a reasonable young woman, could not have realised this until he told her, but once he had told her, or if, without being expressly told, she could have inferred it from the background, she would have had no difficulty in understand-ing what he meant.'

6.2 Contra proferentem *rule*

Of the various rules of construction, the *contra proferentem* rule has histori-cally proved the most significant in connection with cases on liquidated damages and extensions of time.

The rule itself comes from the Latin maxim: *'verba chartarum fortius accipiuntur contra proferentem'* – the words of written documents are con-structed more forcibly against the party offering them.

There is an abundance of cases going back to the early nineteenth century where the courts have shown a traditional hostility to liquidated damages clauses and have consistently construed contracts against their application. Modern authority for the strict application of the rule to liquidated damages and extension of clauses of construction contracts comes from the judgment of Lord Justice Salmon in *Peak* v. *McKinney* (1970) where he said:

'The liquidated damages and extension of time clauses in printed forms of contract must be construed strictly *contra proferentem.*'

Rapid Building v. *Ealing* (1984)

Many judgments in construction cases have referred to the rule in *Peak* – albeit not always with enthusiasm.

Thus Lord Justice Lloyd in the *Rapid Building* case mentioned in Chapter 3 and elsewhere in this book said:

'Like Phillimore J L in *Peak Construction (Liverpool)* v. *McKinney Foundations Ltd* (1970), I was somewhat startled to be told in the course of the argument that if any part of the delay was caused by the employer, no matter how slight, then the liquidated damages clause in the contract, clause 22, becomes inoperative.

I can well understand how that must necessarily be so in a case in which the delay is indivisible and there is a dispute as to the extent of the employer's responsibility for that delay. But where there are, as it were, two separate and distinct periods of delay with two separate causes, and where the dispute relates only to one of those two causes, then it would seem to me just and convenient that the employer should be able to claim liquidated damages in relation to the other period.

In the present case the relevant dispute relates to the delay, if any, caused by the presence of squatters. At the most, that could not account for more than the period from 23 June 1980 to 17 July 1980, a period of some 24 days. It ought to be possible for the employers to concede that there is a dispute as to that period, and then deduct the 24 days from the total delay from 22 September 1982 (when, according to the architect's certificate, the work ought to have been completed) and 23 July 1983, (that being the date of practical completion) and claim liquidated damages for the balance. But it was common ground before us that that is not a possible view of clause 22 of the contract in the light of the decision of the Court of Appeal in *Peak's* case, and therefore I say no more about it.'

Standard forms

It may seem a little odd that standard conditions of contract which are produced by bodies with representatives of employers' and contractors' organisations should fall within the *contra proferentem* rule. Moreover in 1963, in *Tersons Ltd* v. *Stevenage Development Corporation* (1963) some seven years before the *Peak* case, Lord Justice Pearson expressed what must surely be the common sense view of the situation, when he said:

'[Counsel for Tersons] has contended that the maxim *verba accipiuntur fortius contra proferentem* should be applied in this case in favour of the contractor against the Corporation on the ground that the General

Conditions were included in the invitation to tender sent by the Corporation to the contractor. In my view the maxim has little, if any, application in this case. The General Conditions are not a partisan document or an "imposed standard contract" as that phrase is sometimes used. It was not drawn up by one party in its own interest and imposed on the other party. It is a general form, evidently in common use, and prepared and revised jointly by several representative bodies including the Federation of Civil Engineering Contractors. It would naturally be incorporated in a contract of this kind, and should have the same meaning whether the one party or the other happens to have made the first mention of it in the negotiations.'

Application of the *contra proferentem* rule

It might well be asked, given that operation of the *contra proferentem* rule can lead to some unexpected results, why the courts apply the rule with such vigour? The explanation seems to be that there is, and perhaps always has been, a modification of the rule so that it is applied not only against the party who offered the conditions but also because of the very nature of certain conditions. So where the courts have a hostile or even cautionary approach to matters before them, whether these be liquidated damages, forfeiture or exemptions, they are not slow to use the *contra proferentem* rule to arrive at just solutions.

In *Monmouthshire County Council* v. *Costelloe & Kemple Ltd* (1965), a case which concerned the issue of an engineer's clause 66 decision under ICE Conditions of Contract, Lord Justice Harman revealed such thinking in these words:

'The other consideration which moves me is this. This is a process by which the defendants can be deprived of their general rights at law and therefore one must construe it with some strictness as having a forfeiting effect. It is not a penal clause, but it must be construed against the person putting it forward who is, after all, trying to shut out the ordinary citizen's right to go to the courts to have his grievances ventilated. Therefore, I think it would require very clear words and a very clear decision by the appointed person, namely the engineer, to shut the defendants out of their rights.'

Present thinking

Following the guidance in the *West Bromwich* case as to how contracts should generally be construed and the ruling in *Philips Hong Kong* that the court should not adopt an approach to liquidated damages clauses which would defeat their purpose it is doubtful if the *contra proferentem* rule will retain in the future the influence it had in the past. But for the time being it remains an important force in debate on the applicability of liquidated damages provisions.

6.3 Restrictions on implied terms

The law on the implication of terms was summarised by Lord Simon in *BP Refinery (Westernport) Property Limited* v. *Shire of Hastings* (1978) where he said:

> 'for a term to be implied, the following conditions (which may overlap) must be satisfied: (1) it must be reasonable and equitable; (2) it must be necessary to give business efficacy to the contract, so that no term will be implied if the contract is effective without it; (3) it must be so obvious that "it goes without saying"; (4) it must be capable of close expression; (5) it must not contradict any express term of the contract.'

The rules of construction are rarely used by the courts to imply terms which would permit extensions of time to be granted to keep alive provisions for liquidated damages. Nor are they used to imply terms to introduce or clarify liquidated damages provisions. In matters concerning extensions of time and liquidated damages, the courts stick firmly to the rule that implied terms are introduced only to give a contract business efficacy and since general damages can be sought when liquidated damages provisions fail there is no lack of business efficacy.

The effect of this is that liquidated damages provisions in contracts should be complete and consistent. If they are not, they may fail. They may be declared penalties, as for example in *Stanor Electric* v. *Mansell* (1988), when there was no provision for proportioning down damages, or they may be disallowed, as in *Peak* for prevention or as in *Bramall & Ogden* for inconsistency.

Exactly the same rules and principles apply to extension of time provisions because of their legal link with liquidated damages. However, the point is not always recognised, possibly because the contract draftsmen have more in mind the benefits to contractors than to employers in extension clauses, and so the lessons of *Holme* v. *Guppy* (1838), *Dodd* v. *Churton* (1897) and *Peak*, that the courts will not provide or improve extension of time clauses, continue to be re-learned in modern cases. Thus JCT 80 was found wanting in the *Rapid Building* case when the absence of a specific provision in the extension of time clause covering late possession of the site prevented the employer from recovering liquidated damages and was amended in the light of the case.

6.4 'Catch all' phrases

One device used by contract draftsmen to avoid the problem of deficiencies in extension of time clauses is to insert 'catch all' phrases of the type:

'all causes beyond the contractor's control'

or

'any other circumstances of any kind whatsoever'.

The courts have shown reluctance to accept such phrases as including unspecified breaches of contract by the employer. This may be because, as a matter of construction, applying the *ejusdem generis* rule, the phrases are taken to include only matters of the same class as preceding words, and not uncommonly these relate to neutral events such as weather or strikes and not to breach events. Alternatively, it may be that the courts apply the *contra proferentem* rule to stop the employer benefiting from lack of clarity in the documents.

In *Wells* v. *Army & Navy Co-operative Society Ltd* (1902) the contract was for completion within one year 'unless delayed by alterations, strikes, sub-contractors or other causes beyond the contractor's control'. The court held that the words 'other causes beyond the contractor's control' could not include for breaches of contract or other acts of the employer in failing to give possession of the site and drawings on time.

The ruling in *Wells* was followed in *Perini Pacific Ltd* v. *Greater Vancouver Sewerage* (1966) where the employer caused delay by delivering machinery in a defective condition such that it required repair. The extension of time clause provided for 'extras or delays occasioned by strikes, lock-outs, force majeure or other cause beyond the control of the contractor'. The court ruled that these words did not cover the employer's breach.

In *Fernbrook Trading Company Ltd* v. *Taggart* (1979) it was held that the words 'any special circumstances of any kind whatsoever' were not wide enough to empower the engineer to extend time for delays caused by the employer's breaches of contract in making late payments on interim certificates.

In the *Peak* case, Lord Justice Edmund Davies ruled that the extension of time provisions which read 'by reason of any enlargements or other additions to the works, or in consequence of any local combination of workmen or general strikes or lock-outs, force majeure or other unavoidable circumstances' did not extend to delay caused by the employer. He said: 'Delay due to the employer cannot be said to have been an unavoidable circumstance to anyone save the contractor.'

Surprisingly ICE Conditions of Contract continued to use the phrase 'other special circumstances of any kind whatsoever which may occur' without inclusion of a further provision for delay caused by the employer's breach until the advent of ICE 7th edition in 1999.

Beyond the contractor's control

Although the phrase 'beyond the contractor's control' has been given restricted application by the courts in relation to extensions of time for delays caused by the employer's breach, it was given an unexpectedly wide meaning by the House of Lords in *Scott Lithgow Ltd* v. *Secretary of State for Defence* (1989) on another matter. In that case it was ruled that a provision that the contractor should be paid for the effect of 'exceptional dislocation and delay arising during the construction of the vessel due to alterations, suspensions of work or any other cause beyond the contractor's control', extended to failure by suppliers or sub-contractors in breach of the contractual obligations to the contractor. Lord Keith said:

> 'In my opinion the first division reached the correct conclusion. The terms of clause 20A.3 do not reveal any genus capable of restricting the broad meaning of the words "any other cause beyond the contractor's control". Scott Lithgow did not fail in their contractual obligation to deliver submarines fitted with sound cables. They suffered exceptional dislocation and delay in doing so because BICC delivered faulty cables. Whether or not this was a matter within the control of Scott Lithgow is a question of fact. Prima facie it is not within the power of a contracting party to prevent quality breaches of contract on the part of a supplier or sub-contractor such as lead to delay.
>
> The contractor has no means in the ordinary case of supervising the manufacturing procedures of his supplier. He specifies his requirements but has no means of securing that they are met and the circumstance that he may have a claim against the supplier for breach of contract is irrelevant to the question whether delay consequent on the breach was due to a cause within his control. If the contractor failed to stipulate a time for delivery, consequent delay would be his own responsibility, but if he did so stipulate and delivery was late the position would be different. In this case the contract provided for a wide variety of equipment to be purchased by Scott Lithgow from many nominated suppliers or sub-contractors, including in some instances the Australian Navy. Failures by such suppliers or sub-contractors, in breach of their contractual obligations to Scott Lithgow are not matters which, according to the ordinary use of language can be regarded as within Scott Lithgow's control.'

The decision in the *Scott Lithgow* case has been the subject of much critical analysis. By applying a factual test to the question of control the decision appears at first sight to undermine the general rule that, in contract, the contractor is responsible for the performance of his sub-contractors. However, it is most unlikely that the decision has any impact at all on the general rule. The point at issue in the case was the determination of the contractor's entitlements not his responsibilities. And that determination rested on the express inclusion in the contract of the phrase 'beyond the contractor's control'. Consequently it is difficult to see how, in the ordinary run of things, a contractor can rely on the *Scott Lithgow* decision to escape responsibility for a defaulting sub-contractor.

Nevertheless, where a contractual entitlement is expressed in terms of 'beyond the Contractor's control' whether it be for money as in *Scott Lithgow* or for extra time as in many standard extension of time clauses then the correct approach, following *Scott Lithgow,* must be to apply a factual test. For further comment on this see Chapter 13.

Duty of care

With regard to the broader questions on the responsibility of main contractors for the performance of sub-contractors it may be worth adding here a few words on the important case of *D & F Estates Ltd* v. *Church Commissioners for England* (1988) which concerned, amongst other things, a contractor's liability in tort for defective work undertaken by a sub-contractor. The Court of Appeal held that Wates, the contractor, had discharged their duty of care by appointing a sub-contractor they believed to be reasonably competent, and if there was any duty to supervise the sub-contractor it was solely in contract. The House of Lords upheld the decision of the Court of Appeal and on the matter of a duty of care Lord Bridge likened the position of a contractor and his sub-contractor to that of employer and contractor. He said:

> 'If the mere fact of employing a contractor to undertake building work automatically involved the assumption by the employer of a duty of care to any person who may be injured by a dangerous defect caused by the negligence of the contractor, this would obviously lead to absurd results. If the fact of employing a contractor does not involve the assumption of any such duty by the employer, then one who has himself contracted to erect a building assumes no such liability when he employs an apparently competent independent sub-contractor to carry out part of the work for him.'

Later, Lord Bridge did go on to say:

> 'If in the course of supervision the main contractor in fact comes to know that the sub-contractor's work is being done in a defective and foreseeably dangerous way and if he condones that negligence on the part of the sub-contractor, he will no doubt make himself potentially liable for the consequences as a joint tortfeasor.'

However, it must be emphasised that the *D & F* case concerned negligence and its decision does not in any way relieve contractors of their contractual obligations to employers in respect of the work of sub-contractors.

6.5 Inconsistencies in drafting

In many of the cases mentioned earlier the courts have taken the view that inconsistencies or omissions in drafting have rendered liquidated damages provisions inoperable under the principles of prevention or penalties.

The case of *Bramall & Ogden Ltd* v. *Sheffield City Council* (1983) is particularly instructive, however, because the ruling against the employer's right to deduct liquidated damages was based not on a penalties argument, which was put forward for the contractor and apparently accepted, but on the broader argument that the inconsistencies themselves rendered the provisions inoperable. The case is important because Sheffield were following what many regarded as normal and proper practice of the time and few had seen the potential legal difficulties of construction.

Bramall & Ogden (1983)

The contract which was under JCT 63 was for the erection of 123 dwellings and in the appendix liquidated damages were stated to be at the rate of £20 per week for each uncompleted dwelling.

This was, and still is, a common enough way of expressing liquidated damages for multiple units and there was no doubt that the employer's intention was that they should be able to recover £20 per week for each dwelling remaining incomplete after the due date for completion of the contract. The problem for the employer was that they failed to complete the provisions for sectional completion contained in the contract and they failed to delete the provision for reducing the sum stipulated in the appendix in proportion to the value of work completed within time.

The following passages (i)–(iii) are some extracts from the judgment of Judge Hawser:

(i) '[Counsel for Bramall & Ogden] has submitted that there are errors of law on the face of the award in the arbitrator's reasons. His principal argument, though he also developed the others, was that the contract does not provide for sectional completion, as indeed it does not and as the arbitrator found it did not. The arbitrator's finding is in paragraph 8.02: "There was no provision for sectional completion in the Articles of Agreement." Therefore, according to its terms the respondents would have been entitled to deduct liquidated damages on all the dwellings and indeed on all the works as defined in condition 1(1) and as found by the arbitrator, up to the date of practical completion of the works, irrespective of whether they had taken possession of dwellings during the course of the work. The result, he said, would be that liquidated damages would turn into a penalty, since they could exceed substantially the actual loss sustained. The fact that the respondents chose not to exercise their rights and only to deduct the liquidated damages after the extension date of 4 May 1977 and until the time when they took possession of the last house on 29 November 1977 in respect of houses not taken into possession during that period does not and cannot affect the position in law. He submitted that they may well have operated the provisions in a reasonable manner but this cannot affect their validity. He said that the liquidated damages provisions here have to be construed strictly and contra proferentem – a proposition which seems to be borne out by the decision of the Court of Appeal in *Peak Construction (Liverpool) Ltd* v. *McKinney Foundations Ltd* (1969).'

(ii) 'The way in which the liquidated damages are dealt with is set out in the Appendix. This does not allow for the calculation to be made which is required by condition 16(e), and one cannot operate the Appendix and conditions 16(e) in the circumstances of this case. The inconsistency can only be reconciled if provision is made in the contract for sectional completion of those parts which are taken over and to which specific liquidated damages provisions are applied. In *M J Gleeson (Contractors) Ltd* v. *London Borough of Hillingdon* (1970), Mocatta J came to the conclusion that, as there were no sectional completion provisions in the contract before him, condition 12(1) of the RIBA conditions prevented the provisions of the contract bills from overriding the contract itself and that therefore sectional completion could not be relied upon to justify the deduction of liquidated damages.'

(iii) 'There is no doubt that partial possession was taken by the respondents from time to time of completed houses, and this indeed is how the liquidated damages were computed. Accordingly, it seems to me that the respondents operated under condition 16 and it cannot be said now that it is condition 22 and not condition 16 which applies. I do not think one can avoid the conclusion that condition 16(2) would apply to the present situation, and it does not seem to be consistent with the liquidated damages as set out in the Appendix.'

Ongoing problem

The decision in *Bramall & Ogden* v. *Sheffield* (1983) came as an unwelcome surprise to the many employers who found themselves in the same position as Sheffield and yet, even now, despite the wide publicity of the case, contracts are still being compiled with similar deficiencies. See, for example, the cases mentioned in Chapter 4 under the heading 'Drafting matters'. See also the judgment in *Avoncroft Construction Ltd* v. *Sharba Homes (CN) Ltd* (2008) where Judge Frances Kirkham said:

'4. The claimant's case is that there is no underlying entitlement to LADs on the part of the defendant. This is because the LADs clause fails according to the principle in *Bramall & Ogden* v *Sheffield City Council* (1983) 29 BLR 73.

5. There appears to be no doubt that the defendant took partial possession of some of the work. That is clearly stated by Mr Loftus, the defendant's solicitor, in his witness statement. Further, by its letter dated 15 February 2007, the defendant in effect acknowledges (by its calculation of what it claims to be entitled to by way of LADs) that it took partial possession of some elements of work. There is no

provision in the building contract for sectional completion. Accordingly, the principle in *Bramall & Ogden* applies.

6. The defendant's case is that it would be wrong to allow the claimant to rely on that principle because of what it contends are the unusual circumstances of this case. In his statement, Mr Loftus states that the claimant threatened to barricade a show home if the defendant refused to pay £20,000; the defendant refused; on the evening of 14 September 2007, the claimant proceeded to barricade a show home and prevent public access to it. There was no challenge to that evidence. Mr Maguire, for the defendant, submits that those matters should be tried, and that the grant of partial possession was frustrated by the claimants' actions. The defence to payment of LADs following Bramall & Ogden is a technical argument. The court should take into account the conduct of the claimant. To permit the claimant to rely on that defence, the claimant must be blameless.

7. I am not persuaded by that submission. It is a question of law whether, partial possession having been obtained, LADs are payable at all. The claimant's defence to a claim for LADs is not an equitable defence but one available pursuant to the contract.'

For another recent case on the legal construction of liquidated damages clauses see *Steria Ltd* v. *Sigma Wireless Communications Ltd* (2007) which is covered in some detail in Chapter 5.

Chapter 7
Effects of determination

7.1 The question of continuing responsibility

Liquidated damages provisions are patently made on the assumption that it is the contractor who will complete the works. Providing there is no defect in the provisions, then, in the eyes of the law, the contractor and employer have agreed and settled the rate of damages for late completion. But what is the position if the contractor does not complete the works either because the forfeiture or determination clause in the contract is exercised, or because either party alleges repudiation and terminates under common law? Do the provisions for liquidated damages still apply?

The question is far from straightforward. Where breach rather than bankruptcy is the cause of the default, it might seem that the party exercising determination, whether under the contract or at common law, is making a choice to forgo his right to liquidated damages. The point being, how can the employer enforce liquidated damages when by his action, albeit lawful, he has prevented the contractor from completing? But if the liquidated damages clause does fall on determination, would legal successors of the original contractor face claims for general damages for late completion whether or not they took on the burden of completing the works? Then there is this consideration, has the determination taken place before or after the due completion date? This is clearly a matter of some importance since some liquidated damages may have already been deducted. Finally there is the question, and this may be of decisive importance, what does the contract say?

Although determination is a common enough feature of the construction industry there are few legal authorities on the subject as it affects liquidated damages. This is probably because most determinations follow financial failure, usually of the contractor, and pursuit of damages is pointless. There are, however, a few pointers.

7.2 British Glanzstoff

The case of *British Glanzstoff Manufacturing Company Ltd* v. *General Accident Assurance Corporation Ltd* (1912) provides the firmest rule. That is, in the absence of express provisions in the contract, the contractor is not liable for any liquidated damages accruing after the date of determination. In *British*

Glanzstoff the contract provided that if the contractor suspended the works, the employer was entitled to engage another contractor to complete them. As a result of bankruptcy, the contractor did in effect suspend and the employer engaged another contractor who completed the works six weeks after the due date. The employer claimed liquidated damages for delay from the guarantor of the original contractor, but the House of Lords held that the liquidated damages clause applied only where the contractor himself completed and was ineffective where control was taken out of his hands.

The rule in *British Glanzstoff* has alternatively been stated to the effect that future obligations cannot arise when a contract has been terminated, except where the express terms of the contract so provide.

7.3 Contractual provisions

Most standard forms of construction contract include in their forfeiture provisions some reference to the financial liabilities of the parties in the event of forfeiture. Even then there can be problems. Indeed in some cases the forfeiture clauses themselves have been held to be unenforceable as penalties. Thus in *Ranger v. G W Railway* (1854) a clause that payments already made were to be taken as full satisfaction of all work completed and all outstanding moneys, tools and materials were to become the property of the employer was held to be a penalty and the entitlement of the employer was to do no more than recoup his actual losses.

Provisions in JCT forms

In none of the commonly used standard forms is the position totally clear on whether liquidated damages survive determination. Clause 8.12 of JCT 2005 provides that in the event of determination of the contractor's employment by the employer, the contractor shall pay 'the amount of any direct loss and / or damage caused to the employer by the determination'. This brings out the point that there are two sources of cost for the employer arising from determination – the cost of actually completing the works and the cost of late completion. Clause 8.12 may relate to both sources, in which case they are both clearly unliquidated, but there is a view that it may relate only to the costs of completing the works and that clause 2.32 on liquidated damages remains operative for the costs of late completion. This view relies on the argument that it is only the employment of the contractor which has been determined under the contract, not the contract itself, and clause 2.32 says that the contractor is responsible for liquidated damages up to the date of practical completion. It might actually suit a defaulting contractor or his legal successors to argue that this was the case rather than face general damages under the heading loss and expense in clause 8.12.

Provisions in ICE forms

Clause 65 of ICE 7th Edition requires the amount payable by either party on termination to include 'damages for delay'. It is not entirely clear whether these are liquidated or unliquidated. An argument can be mounted for liquidated damages on an earlier part of clause 65 which states that determination occurs 'without thereby avoiding the contract or releasing the contractor from any of his obligations or liabilities under the contract', and clause 47 supports the argument that damages are liquidated with its statement that the contractor is liable for liquidated damages until the whole of the works are completed.

7.4 Novations

Assignments of construction contracts are comparatively rare since it is a general rule of law that liabilities under a contract cannot be assigned. In any event, the normal intention of assignment in construction contracts is that one party, usually the contractor, should be relieved of his responsibilities and another contractor should take his place. In law, the tripartite agreement which this process involves is termed a novation. In effect, the contract between the first two parties is rescinded in consideration of a new contract being made between one of the parties and a third party. The terms of the two contracts will normally be the same but it is open to the parties to vary them by agreement if they so wish.

Novations commonly occur when one firm is taken over by another and in such circumstances the terms of the first contract are almost invariably carried through without change into the new contract. If a contracting firm is in delay when it is taken over, the new firm takes on the burden of any liquidated damages arising from such delay. When, however, the novation follows a bankruptcy or receivership, which itself has led to determination of the first contractor's employment, it is normal practice for the replacement contractor to seek some relief from liquidated damages by negotiating with the employer a revised date for completion which will, in whole or in part, cover the time lost by the determination itself and any earlier delay by the first contractor. It is a matter of commercial judgment for the employer how far he allows such relief but some pressure to be flexible may come from the need to mitigate loss in respect of any dealings with the receiver.

Except to the extent that he obtains revised terms the replacement contractor in a novation is fully responsible for obligations and liabilities of the first contractor and he takes on the full burden of liquidated damages for delay. Thus in *Re Yeadon Waterworks Company & Wright* (1895) a contract was terminated and one of the contractor's sureties took an assignment to complete the works but failed to finish on time. It was held that the words 'without thereby affecting in any other respects the liabilities of the said contractor' in the termination clause kept alive the employer's right to deduct liquidated damages from the replacement contractor.

7.5 Summary on liquidated damages

The points arising from the above can be summarised as follows:

(i)   damages arising from determination will in the normal course of things be unliquidated but the express terms of the contract may say otherwise;

(ii)   liquidated damages deducted prior to determination will not normally be affected by the determination;

(iii) where the contractor's employment is determined before the due date for completion, the contractor will not be liable for liquidated damages unless there is an express clause to that effect;

(iv) where the contractor's employment is determined after the due date for completion, the contractor will only be liable for liquidated damages up to the time of determination unless express provisions say otherwise;

(v)   a replacement contractor, brought in by novation, will assume full responsibility for liquidated damages, whether or not the delays occurred before or after he took control, unless the novated contract contains revised terms;

(vi) a replacement contractor brought in by the employer after determination will only assume such responsibilities for liquidated damages as his new contract stipulates.

7.6 Determination and limitation on liability

A question which arises from point (i) above that damages from determination will normally be unliquidated is whether a defaulting contractor can rely on the limiting effect of liquidated damages for late completion to avoid the full effects of unliquidated damages. The answer to that is probably that he cannot.

A similar question was addressed by the House of Lords in *Bovis Construction (Scotland) Ltd* v. *Whatlings Construction Ltd* (1995). In that case a package sub-contract was terminated on the basis that the sub-contractor was not proceeding diligently and in response to a claim for £2,741,000 for breach of contract the sub-contractor attempted to rely on a clause in the sub-contract limiting its liability in respect of time-related costs to £100,000.

The House of Lords rejected the sub-contractor's defence, holding:

'(i) A clause limiting liability should state clearly and unambiguously the scope of the limitation and will be construed with a degree of strictness albeit with the same extent as an exclusion or indemnity clause.

(ii) The exclusion clause relied upon was not framed to cover damages flowing from a repudiatory breach of contract leading to termination and hence non-performance of the contract.'

Lord Jauncey explained the position with these words:

'Time is relevant to the performance of a contract during its existence but once the contract is determined by a repudiatory breach of whatever nature time ceases to have relevance. Damages thereafter flow from the repudiation resulting in non-performance and the need to provide for substitute performance.'

Chapter 8
Problems with sectional completion

8.1 Discovering the parties' intentions

There are good reasons in construction contracts why the parties might wish to have completion in phases or in sections. The benefits to the employer are that he gets earlier occupation and use of the parts, rather than waiting for full completion; and the benefits to the contractor are that he is relieved of some of his contractual obligations – insurance of the works being an obvious example. There are also good reasons why, if the parties have thought it appropriate to liquidate the damages payable for late completion of the whole of the works, they might wish to see the same principle applied to phases and sections. However, whenever the parties depart from the straightforward rule that a contract has a single due date for completion and that liquidated damages should be payable from that date, or a properly extended later date, they run the risk of invalidating the liquidated damages provisions.

The terminology itself is not important but it certainly helps to reduce the risk if the parties are clear in their understanding of the phrases used and the contractual importance they wish to attach to them. A section usually means a part of the works separately identified in the appendix to the form of tender and to which is given a stipulated date for completion. A phase may be a requirement expressed elsewhere in the tender documents, which may or may not be incorporated into the contract, or it might be something arising out of the contractor's programme.

The difficulties arise either because the contract documents do not make clear what, if any, are the contractor's liabilities for failure to meet phased or sectional completion dates, or because on technical grounds the expressed provisions for liquidated damages fall foul of the *contra proferentem* or penalty rules.

8.2 Proportioning down clauses

The first point to consider is, to what extent do provisions which permit the employer to take possession of parts of the works with the consent of, or by application from the contractor, differ from provisions which stipulate that the contractor shall complete in phases or in sections?

The key factor is the consent element; the contractor has no obligation to complete early and the employer has no obligation to occupy early. Unless,

therefore, the contract provides a mechanism for the issue of partial completion certificates and a corresponding mechanism for proportioning down
liquidated damages, the contractor remains liable for full damages up to the
date of total completion – as in *BFI* v. *DCB* (1987) mentioned in Chapter 4,
where the employer avoided loss by taking partial possession but still
obtained full damages.

However, where a contract does make provision for the issue of partial
completion certificates, albeit on a consent basis, then it is essential to
have corresponding provisions for proportioning down liquidated damages.
The courts will not imply such a term if it is missing and the liquidated
damages clause will then fall for uncertainty or as a penalty. Thus, in *Stanor
Electric* v. *Mansell* (1988) when there was a single sum for damages for
two houses and one was completed late, the absence of any contractual
machinery for proportioning down the damages led to them being declared
penalties.

Standard forms

Most standard forms of construction contracts have proportioning down
clauses. That in JCT 2005 is to be found in clause 2.37 – Partial Possession
by the Employer – whilst that in ICE 7th Edition is to be found in the liquidated damages clause itself – clause 47. Wording is typically to the effect
that if any part of the works is certified as complete before completion of
the whole of the works, the stipulated sum for liquidated damages shall be
reduced proportionately to the value of the work completed compared with
the value of the work as a whole.

No doubt attempts could be made in unusual cases to show that this
method of scaling down liquidated damages resulted in technical penalties,
but generally proportioning down clauses serves the industry well and are
a strong incentive for contractors to finish and hand-over as much work as
they can prior to full completion.

In the few cases where proportioning down clauses have run into trouble
with the courts they have been done so on fairly narrow technical grounds.
The best known is *Bramall & Ogden* v. *Sheffield* (1983) where in a JCT 80
contract it was found that clause 16 allowing for damages to be proportioned
down was incompatible with damages expressed in the appendix at a rate
per week per dwelling. There is no objection to damages being expressed at
a rate per week per unit or similar but that in itself is intended to provide
a scaling down mechanism and any other scaling down mechanism in the
contract should then be deleted or amended.

For comment on the practice of inserting a stop figure or minimum
payment into proportioning down clauses see the cases of *Arnhold & Co. Ltd*
v. *Attorney General of Hong Kong* (1989) and *Philips Hong Kong Ltd* v. *Attorney
General of Hong Kong* (1993) discussed in Chapter 4.

8.3 Provisions for sectional completion

Where there are express requirements for sectional or phased completion, any corresponding requirements for liquidated damages need to be expressed with both certainty and consistency if they are to be effective.

In *M J Gleeson (Contractors) Ltd* v. *London Borough of Hillingdon* (1970) on a JCT 63 contract the bills of quantities set out detailed requirements for sectional completion with related liquidated damages. There was, however, a standard provision in the contract that nothing in the bills should override the Articles of Agreement or the Conditions of Contract. Mr Justice Mocatta held that the requirements in the bills could not be relied on to justify the deduction of liquidated damages. The employer had sought to deduct liquidated damages at the rate of £5 per dwelling per week as stated in the appendix from the specified sectional completed dates, but his entitlement was to deduct only from the final completion date.

Subsequent to the *Gleeson* case the Joint Contracts Tribunal issued in 1975 a sectional completion supplement for use with later JCT forms.

ICE forms approach the precedence of documents differently from JCT forms. Clause 7 of ICE 7th Edition empowers the engineer to explain and adjust ambiguities or discrepancies and says that 'the several documents forming the Contract are to be taken as mutually explanatory of one another'.

It should therefore be possible under ICE forms to set out requirements on sectional completion elsewhere than in the appendix providing the intention is made clear enough. But in fact, with ICE 7th Edition, this should never be necessary – the appendix deals with sectional completion in the simplest of ways. Times for completion and rates of damages are stated either for the whole of the works or for any number of sections; and liquidated damages for sections can run concurrently where circumstances so dictate.

8.4 Requirements not fully specified

Some of the problems discussed above are a warning against over specification of contractual requirements and the incompatibility which is likely to result. But when it comes to liquidated damages it is equally fatal to fall short of the requirements necessary to produce certainty. The principle enunciated by Lord Pearson in *Trollope & Colls* v. *North West Metropolitan Regional Hospital Board* (1973) that the courts will not make a contract for the parties nor will the courts improve a contract however desirable that improvement might be, will leave any deficiency exposed. The courts will not imply terms for sectional completion where none exist and the courts will not imply that liquidated damages are due where none are stated.

In *Bruno Zornow (Builders) Ltd* v. *Beechcroft Developments Ltd* (1989) a contract was negotiated for a housing development on the basis of a first tier tender which showed a detailed programme to complete in sixteen months and second stage agreements for completion of the work in two overlapping phases. The architect calculated liquidated damages of £40,000 based on the stipulated rate of £200 per week per block from the date shown on the original works programme and the contractor sued for the return of this amount. It was held by Judge Davies, after considering Lord Pearson's dicta in *Trollope & Colls* and the very complicated facts of the case:

'(i)   the contract did not incorporate documents which specified dates for sectional completion but only phased provisions for the transfer of possession;

(ii)   a claim for liquidated damages could only be made in respect of failure to meet specified completion dates and not failure to meet transfer of possession dates – which operated on a consent basis;

(iii) no term would be implied for any sectional dates for completion.'

A similar situation arose in *Turner* v. *Mathind* (1986) where there was a clear requirement in the bills for phased completion but the sectional completion supplement was not used and the appendix contained only a rate for liquidated damages for late completion of the whole of the works of £1000 per week. All attempts by the employer to justify deduction of liquidated damages for failure by the contractor to meet the phasing dates failed. It was not appropriate that the employer should either pro-rata the stipulated damages to the number of phases or apply the stipulated rate to each phase.

These cases illustrate that neither programmes nor phasing requirements linked to programmes will ever create, by themselves, any liability for liquidated damages.

Finally, reverting to *Trollope & Colls*, which was a case in which the time remaining for phase III of a hospital building contract after extensions granted on phases I and II was 16 months instead of the 30 months originally intended. The employers finding themselves unable to nominate subcontractors for phase III who could complete in the shorter time, argued for an implied term in the contract that an extension should be granted to phase III to accommodate the delays in phases I and II. The contractors opposed the granting of any such extension. They were, in the words of Lord Pearson:

'turning the situation to their own advantage, because, if the contract could not be carried out, a new arrangement would have to be made for the work to be done at the prices prevailing in or about 1971, which were considerably higher than the contract prices. The difference between the contract prices and the prices prevailing in or about 1971 is said to be in the region of one million pounds.'

The Court of Appeal found that the contract was clear and free from ambiguity in stating that the date for completion of phase III was 30 April 1972. Accordingly, and in any event, no term could be implied.

Chapter 9
Application to sub-contractors

9.1 Effect of 'stepping-down' provisions

To minimise financial risk it is regarded as good business in many commercial transactions to deal with sub-contractors and the like on essentially the same terms as in main contracts; the main contractor having, one would expect, the benefit of a margin in the figures. In construction, it is particularly commonplace for the terms of main contractors to be 'stepped down' into sub-contracts and accordingly many standard forms of sub-contract incorporate the provisions of corresponding forms of main contract. This works perfectly well in covering the majority of contractual obligations but if it is applied to liquidated damages, the effect is not to indemnify the contractor against loss caused by late completion of a sub-contractor, but is to restrict the contractor's recovery to the amount of liquidated damages.

Gleeson v. Taylor Woodrow (1989)

The case of *M J Gleeson plc* v. *Taylor Woodrow Construction Ltd* (1989) illustrates the problem. Taylor Woodrow, as management contractors for work at the Imperial War Museum, entered into a sub-contract with Gleeson. The management contract provided for liquidated damages at £400 per day and clause 32 of the sub-contract provided for liquidated damages at the same rate. Clause 11 (2) of the sub-contract also provided that if the sub-contractor failed to complete on time the sub-contractor should pay:

> '. . . a sum equivalent to any direct loss or damage or expense suffered or incurred by (the management contractor) and caused by the failure of the sub-contractor. Such loss or damage shall be deemed for the purpose of this condition to include for any loss or damage suffered or incurred by the authority for which the management contractor is or may be liable under the management contract or any loss or damage suffered or incurred by any other sub-contractor for which the management contractor is or may be liable under the relevant sub-contract.'

Gleeson finished late and they received from Taylor Woodrow a letter as follows:

> 'We formally give you notice of our intention under clause 41 to recover moneys due to ourselves caused by your failure to complete the works

on time and disruption caused to the following sub-contractors. The following sums of money are calculated in accordance with clause 11(2) for actual costs we have incurred or may be liable under the management contract.'

Then followed a summary of accounts showing deductions of £36,400 for liquidated damages, being £400 per day from 31 May 1987 to 31 August 1987, and £95,360 in respect of 'set-off' claims from ten other sub-contractors.

Gleeson applied for summary judgment under Order 14 in respect of the sum of £95,360 and were successful. Judge Davies found that Taylor Woodrow had no defence:

'On the evidence before me, therefore, TWL's course of action against Gleeson in respect of set-offs is for delay in completion. It follows that it is included in the set-off for liquidated damages, and to allow it to stand would result in what can be metaphorically described as a double deduction.'

Comment

There is a salutary lesson here for all main contractors for Taylor Woodrow were doing nothing more in their sub-contract than trying to pass on losses they could suffer from the sub-contractor's default and they were doing so in a routine way with comprehensively drafted provisions. But they had failed to recognise that liquidated damages for late completion are the whole of the sum payable and not just part of the sum. Insofar as a claim under clause 11(2) of the sub-contract was made for late completion, it had no effect as it duplicated clause 32.

However, had the claim under clause 11(2) been for disruption, the decision would almost certainly have been different.

9.2 Can there be a genuine pre-estimate of loss?

The *Gleeson* case reveals primarily the restricting effect of stating liquidated damages in sub-contracts but it also reveals the difficulty for a main contractor in making a genuine pre-estimate of loss caused by his sub-contractor's late completion.

That loss has three main elements:

(i)   contractor's own costs of delay;
(ii)   contractor's liability for liquidated damages;
(iii) claims arising from delay to other sub-contractors and suppliers.

At first sight, items (i) and (ii) appear straightforward. The contractor can calculate with some precision his own costs of delay – site costs, supervision, overheads and financing charges; the contractor also knows the rate of liquidated damages in the main contract. The problem appears to be only with item (iii) in that the effect of one sub-contractor's delay on the progress and costs of others is virtually impossible to pre-estimate.

This, however, is only part of the problem. Few sub-contractors require the whole of the main contract period for their work and it does not automatically follow that sub-contractor's delay leads to main contractor's delay. Even if it can be shown that a sub-contractor's work is a critical path activity there is main contractor's float time to consider. Then there is the problem of duplication – can the main contractor, if he has stipulated liquidated damages, recover from more than one sub-contractor for the same loss?

Not surprisingly in view of these difficulties in making a genuine pre-estimate of loss, and in view of the limiting effect of that pre-estimate if stipulated as liquidated damages, most of the better known standard forms of sub-contract used in the construction industry omit provisions for liquidated damages payable by the sub-contractor and do no more than draw attention to the rate of such damages in the main contract.

9.3 *Commercial considerations*

The commercial interests of main contractors and sub-contractors are in opposition when it comes to damages for late completion. Main contractors benefit, although they do not always see it that way, from the certainty and limitation that liquidated damages bring to their contracts with employers. However, in their dealings with sub-contractors they want to remove that limitation and to recover in full any losses they have suffered. Sub-contractors would like the benefits of certainty and limitation but without liquidated damages clauses in sub-contracts they face uncertainty and potentially ruinous damages.

The commonly used forms of domestic sub-contract favour main contractors. Thus the standard form of domestic sub-contract known as DOM 1, once probably the most used form in the building industry, states in clause 12, which is headed 'Failure of Sub-Contractor to complete on time . . .':

'the Sub-Contractor shall pay or allow to the Contractor a sum equivalent to any loss or damage suffered or incurred by the Contractor and caused by the failure of the Sub-Contractor as aforesaid.'

The Civil Engineering Contractors' Association form of sub-contract, generally known as the Blue Form and widely used with ICE Conditions of Contract, states in clause 3(4):

'The sub-contractor hereby acknowledges that any breach by him of the sub-contract may result in the contractor's committing breaches of and becoming liable in damages under the main contract and other contracts made by him in connection with the main works and may occasion further loss or expense to the contractor in connection with the main works and all such damages loss and expense are hereby agreed to be within the contemplation of the parties as being probable results of any such breach by the sub-contractor.'

The intention in this latter clause in bringing all loss within the contemplation of the parties is to ensure that the sub-contractor cannot defend a claim which includes the main contractor's liquidated damages liability or other sub-contractors' claims by reference to remoteness within the rules of *Hadley* v. *Baxendale* (1854). If they fail the first test, they have, by express terms, been included as special damages under the second rule, i.e. within the contemplation of the parties. This is why the sub-contract will normally state the level of liquidated damages in the main contract.

Limitation of liability

Where the sub-contractor has bargaining power which he can exert to improve the terms of the sub-contract in his favour, he may well insist on the inclusion of a conventional liquidated damages clause or of some other limitation provisions. For example, a clause might be included to limit the liability of the sub-contract in respect of all claims arising from late completion to a percentage of the sub-contract sum with particular reference, for the avoidance of doubt, that this percentage figure included for any liability of the main contractor for liquidated damages. See, for example, the case of *Pigott Foundations Ltd* v. *Shepherd Construction Ltd* (1993) discussed in Chapter 3.

9.4 Nominated sub-contracts

With domestic sub-contracts the main contractor remains fully responsible for the acts, neglects and defaults of his sub-contractors and the contractual chain of liability from employer to main contractor to sub-contractor is intact. The ruling by the House of Lords in *Scott Lithgow* v. *Secretary of State for Defence* (1989) (mentioned in Chapter 6) that failures by sub-contractors were beyond the contractor's control turned on the particular wording of the contract in respect of the contractor's rights of claim and the ruling does not establish any general rule of law that main contractors are not responsible for their sub-contractor's performance.

With nominated sub-contracts there has always been some reluctance to place the whole burden of responsibility on the main contractor, on the not unreasonable proposition that if the employer wants to impose his choice of sub-contractor he should bear some if not all of the responsibility for that sub-contractor's performance. So main contracts have in varying degrees and various ways given indemnities to main contractors in respect of loss and expense, or excused them in respect of breach caused or committed by nominated sub-contractors. But time after time, the application of the rules of law to such considerations has shown that any disturbance of the chain of contractual liability has serious and frequently unexpected consequences.

Bickerton **(1970)**

The case which first drew widespread attention to the defects of nominating sub-contracting was *North West Metropolitan Regional Hospital Board* v. *T A Bickerton & Sons Ltd* (1970). In that case a nominated heating sub-contractor went into liquidation before the works commenced. The main contractor undertook the work himself, but contended that the employer was bound to nominate a second sub-contractor and pay the amount of that price, whereas the employer maintained there was no duty to re-nominate or pay more than the first price. The House of Lords found that there was a duty to re-nominate. Eminent construction lawyers of the day agreed that by extension of reasoning employers would be obliged to re-nominate and stand the costs of any repudiation by a nominated sub-contractor. The consequence of this would be that a defaulting nominated sub-contractor would go scot-free since he could not be sued by the main contractor who, as a result of payment by the employer, would suffer no loss and he could not be sued by the employer since he had no contract with him.

Bilton **(1982)**

The alarm created by *Bickerton* may to some extent have been exaggerated and the House of Lords in *Percy Bilton Ltd* v. *Greater London Council* (1982) examined the doctrine and, refuting that *Bickerton* had established automatic liability of the employer in the event of nominated sub-contractor withdrawal, restated the general law that the main contractor takes the risk unless there is fault by the employer. Passages (i)–(iii) below are extracts from the judgment of Lord Fraser:

(i) 'This appeal is concerned with the legal consequences of failure by the main contractor to complete work under a building contract by the due date, where the delay has been partly caused by the withdrawal of a nominated sub-contractor at a time when withdrawal inevitably delays completion of the works. In particular, the question is whether such a withdrawal which causes delay in completion of the works by the main contractor, prevents the employer from relying upon a clause in his contract with the main contractor giving the employer the right to deduct liquidated damages for delay in completion.'

(ii) 'It is common ground between the parties that the delay which followed the dropping out of Lowdells should be divided into two parts – first, the part arising directly from the withdrawal, and secondly, that arising from the failure of the respondent to nominate a replacement with reasonable promptness. The respondent was clearly responsible for the second part, and there is no doubt that, if it had been the only delay, the appellant would have been entitled to a reasonable extension of time to allow for it in accordance with clause 23(f). The dispute centres on the consequence of the first period of the delay which, as parties are agreed, does not fall with any of the provisions of clause 23. The appellant contends that the loss directly caused by the withdrawal of the nominated sub-contractor must fall on the respondent, on the ground that it has a responsibility not only to nominate the original sub-contractor and any necessary replacement, but to maintain a sub-contractor in the field so long as work of the kind allotted to him needs to be done. This is said to flow from the decision of your Lordships' House in *Bickerton* . . . What was actually decided in that case was that, where the original nominated subcontractor has gone into liquidation and dropped out, the main contractor had neither the right nor the duty to do any of the subcontractor's work himself, and that it was the duty of the employer to make a new nomination. Consequently (so it was argued for the appellant), if the nominated sub-contractor withdraws at a time when his withdrawal must inevitably cause delay, the main contractor is disabled from performing his obligations for want of a sub-contractor whom only the employer can provide, and the main contractor is thus "impeded" from working . . . In these circumstances it was said that the contractual time limit ceases to apply, the time for completion becomes at large, and the employer cannot rely on the provisions for liquidated damages in clause 22.

   If the argument is correct, its effect would be to turn the employer's duty of nominating a sub-contractor, and if necessary a replacement, into a duty to ensure that the main contract is not impeded by want of a nominated sub-contractor. That would be virtually a warranty that a nominated sub-contractor would carry on work continuously, or at least that he would be available to do so.

   But I see nothing in clauses 22 or 23, or elsewhere in the conditions of contract, to impose such a high duty on the employer. Such a warranty would, in my opinion, place an unreasonable burden on the employer, particularly as he has no direct contractual relationship with a nominated sub-contractor, and no control over him. When the nominated sub-contractor withdrew, the duty of the employer, acting through his architect, was in my opinion limited to giving instructions for nomination of a replacement within a reasonable time after receiving a specific application in writing from the main contractor under clause 23(f). In this case, the employer failed to perform that duty. It did not give instructions within a reasonable time, and the second part of the delay occurred, with the result that the appellant became entitled to an extension of the time for completion to cover the second part. But they never became entitled to any extension to cover the first part of the delay.'

(iii)

'(1) The general rule is that the main contractor is bound to complete the work by the date for completion stated in the contract. If he fails to do so, he will be liable for liquidated damages to the employer.

(2) That is subject to the exception that the employer is not entitled to liquidated damages if by his acts or omissions he has prevented the main contractor from completing his work by completion date – see for example *Holme* v. *Guppy* (1838) and *Wells* v. *Army & Navy Co-operative Society* (1902).

(3) These general rules may be amended by the express terms of the contract.

(4) In this case, the express terms of clause 23 of the contract do affect the general rule. For example, where completion is delayed "(a) by force majeure, or (b) by reason of any exceptionally inclement weather" the architect is bound to make a fair and reasonable extension of time for completion of the work. Without that express provision, the main contractor would be left to take the risk of delay caused by force majeure or exceptionally inclement weather under the general rule.

(5) Withdrawal of a nominated sub-contractor is not caused by the fault of the employer, nor is it covered by any of the express provisions in clause 23. Paragraph (g) of clause 23 expressly applies to "delay" on the part of a nominated sub-contractor but such "delay" does not include complete withdrawal; (this was accepted in argument by counsel for the appellant, rightly in my opinion).

(6) Accordingly, withdrawal falls under the general rule and the main contractor takes the risk of any delay directly caused thereby.

(7) Delay by the employer in making the timeous nomination of a new sub-contractor is within the express terms of paragraph (f) of clause 23, and the main contractor, the appellant, was entitled to an extension of time to cover that delay. Such an extension has been given.'

Fairclough v. *Rhuddlan* (1985)

Both *Bickerton* and *Bilton* were cases concerning nominated sub-contractors who had withdrawn after going into liquidation and there was some doubt as to how far they applied to wider issues. That was tested by the Court of Appeal in *Fairclough Building Ltd* v. *Rhuddlan Borough Council* (1985) where a nominated sub-contractor, Gunite, repudiated its contract when eight weeks late and with extensive remedial work necessary. Clause 23 of the contract giving the contractor entitlement to extension for delay on the part of a nominated sub-contractor was amended by the addition of 'but such delay will only be considered for those reasons which the contractor could obtain an extension of time for under the contract'. The architect re-nominated but the re-nomination did not cover the remedial work. The issues before the Court of Appeal were stated as follows:

(i)   Did the time provided in the proposed sub-contract entitle the contractor to refuse the nomination?
(ii) Was the nomination invalid by reason of the fact that the proposed sub-contract did not cover remedial works?
(iii) Was the contractor entitled to an extension for the eight weeks' delay incurred by Gunite before they withdrew?
(iv) Was the employer entitled to charge the contractor with the full costs of remedial work or only obtain credit for the amount which it had already paid in respect of Gunite's work before their withdrawal?

It was held:

(i)   The architect's instruction nominating sub-contractors who would not complete within the time allowed under the main contract was invalid and the contractor was therefore entitled to refuse the nomination.
(ii) The instruction nominating a new sub-contractor was also invalid because the proposed sub-contract did not include remedial work.
(iii) The contractor was not entitled to an extension of time for the delay incurred by Gunite before they withdrew because clause 23(g) as amended only applied if the sub-contractor's delay was itself due to one or other of the causes of the delay specified in the other sub-clauses of clause 23.
(iv) There was no basis upon which the employer could charge the contractors with the full costs of the remedial work when the obligation to re-nominate included the obligation to include remedial work in the work to be done by the re-nominated sub-contractor, and the contractor was neither entitled nor obliged to do such work.

Standard forms

Standard forms of contract issued since *Bickerton* have endeavoured to preserve the chain of liability as far as practicable, or as far as compromise in the drafting committee will allow. Thus ICE conditions go most of the way to placing full responsibility for nominated sub-contractors on the main contractor with no extensions of time allowed for nominated sub-contractor delay, with a statement in clause 59 of the 7th Edition saying:

'Except as otherwise provided in Clause 58(3) the Contractor shall be responsible for the work carried out or goods materials or services supplied by a Nominated Sub-contractor employed by him as if he had himself carried out such work or supplied such goods, materials or services.'

Some other standard forms do not go as far as this and still permit the granting of extensions of time for delay on the part of nominated sub-contractors.

Note, however, that 'on the part of' is not the same as 'delay caused by'. In *Westminster Corporation* v. *J Jarvis & Sons Ltd* (1970) the contractor could not get an extension for delay caused when it was found, after the date of completion of nominated sub-contract piling work, that the piles were defective.

Effect of indemnities

Although *Bilton* and *Fairclough* may have clarified the law as far as main contractor and employer relationships are concerned, problems will continue to arise under sub-contracts whenever indemnities are given to the contractor. To this extent *Bickerton* lives on and the recovery of damages is impeded accordingly.

In *Mellowes PPG Ltd* v. *Snelling Construction Ltd* (1989) Mellowes were supplying windows under a nominated sub-contract to Snelling main contractor for new county council offices in Hampshire. The architect issued a certificate of delay against Mellowes which entitled Snelling to claim from Mellowes whatever losses they had suffered. In a set-off from payments due to Mellowes, Snelling included for liquidated damages. Mr Recorder Fernyhough had this to say:

> 'Of course that certificate of delay entitles Snelling to claim whatever losses they can prove they suffered from Mellowes, but of course those losses cannot include liquidated and ascertained damages because if the main contract works have been delayed on the part of Mellowes, Snelling were entitled to an extension of time to the same extent of that delay so they will not have suffered any liquidated damages from that cause.'

Chapter 10
Recovery of liquidated damages

10.1 When do liquidated damages become payable?

When a contract provides that liquidated damages are payable for late completion, it would appear to be self-evident that they become payable on late completion. But the matter is by no means as straightforward as this: firstly, because conditions of contract frequently make the issue of certificates on extensions of time or non-completion conditions precedent to the deduction of liquidated damages; secondly, because the employer may be required to give prior notice of his intention to deduct liquidated damages under the terms of the contract or under statute and thirdly, because there is wide scope for dispute on what constitutes 'lateness' and what constitutes 'completion'.

With a contract in simple form the employer might well take the view that once the due date for completion had passed he could deduct damages at the appropriate rate from any further sums due to the contractor. He would probably not consider it necessary, or commercially sensible, to wait until the works were completed until making his deduction. Construction contracts, however, are rarely simple so it is necessary in every case to analyse the wording of the particular conditions of contract to see what conditions precedent they impose and when they permit damages to be deducted. Added to which the requirements for withholding notices under the Housing Grants, Construction and Regeneration Act 1996 may need to be considered.

Deduction by a single sum

Uncertainty on these matters often exists in well used forms. For example, clause 24 of JCT 80, prior to amendment, stated that the contractor shall pay or allow to the employer: 'the whole or such part as may be specified in writing by the employer of a sum calculated at the rate stated in the Appendix . . . for the period between the completion date and the date of practical completion.' This could be taken as applying to a single once and for all figure becoming due only when its full extent was known – that is, after practical completion and when all extensions had been granted. If so, any deductions from interim certificates prior to practical completion would be invalid. Lawyers argued the point for ten years before the uncertainty was removed by amendment 9 to JCT 80 such that the clause read:

'Subject to the issue of any certificate under clause 24.1, the contractor shall as the employer may require in writing but not later than the date of the final certificate pay or allow to the employer liquidated and ascertained damages at the rate stated in the Appendix (or such lesser rate as may be specified in writing by the employer) for the period between the completion date and the date of practical completion and the employer may deduct the same from any moneys due or to become due to the contractor under this contract . . .'

This amendment, by removing the reference to 'of a sum', made it clear that liquidated damages became due after the issue of a certificate of non-completion and there was no need for the employer to wait until practical completion before making his deduction.

Most ICE contracts clearly envisage deductions prior to the final review since they make provision for reimbursement of sums previously deducted. Moreover the wording of recent sets of conditions such as ICE 6th Edition and ICE 7th Edition make the position even clearer in that liquidated damages become due: 'if the contractor fails to complete the whole of the works within the time so prescribed . . .' The employer may then: 'deduct and retain the amount of any liquidated damages becoming due . . . from any sums due to the contractor.'

Deductions before completion

Although there may be scope for argument in some cases on whether liquidated damages become due immediately after the due date for completion has passed, or at some later date when completion is achieved, there is no legal case for arguing the proposition that liquidated damages can be deducted in advance of the due date even if it is apparent that completion will not be achieved on time. There may well be good commercial grounds for arguing such a proposition, since the employer who goes on paying interim certificates in full may have nothing left other than the retention fund from which to make his deductions for damages. He would then, if this was insufficient, be put to the expense of suing for recovery.

In the eyes of the law, however, the position is clear. Damages follow breach of contract, they do not anticipate it, and where liquidated damages are stipulated for failure to complete by a specified date, they become due only when that failure has materialised and the date has passed.

The matter is so obvious that it has rarely troubled the courts but it was a feature, although not a contentious point, in the case of *Lubenham Fidelities & Investments Co. Ltd* v. *South Pembrokeshire District Council* (1986). There the architects had made a number of errors, including making deductions for liquidated damages on the face of interim certificates and making such deductions prior to the date for completion. In summarising the background to the case in his judgment, Lord Justice May said:

'In seeking to make these deductions it is now common ground that the architects erred and, subject to the question of causation, which pays a substantial part in this case, were negligent towards each of the other two parties . . . In so far as any delay was concerned, any liability on Lubenham's part to pay liquidated damages in respect of it could not have arisen at least until 3 September 1977, which was the date for completion set out in the usual appendix to these two contracts.'

The *Lubenham* case shows the danger of contract administrators making up their own rules through ignorance or by design. As a result of the architects' errors, the contractor purported to determine his own employment alleging breach of under-payment; the employer treated the contractor's action as repudiation and, rightly as the court so found, determined the contractor's employment. Both ended up suing each other and the architects. In rejecting a claim by the contractor that the architects had intended to interfere with the performance of the contracts, Lord Justice May commented:

'This was a straightforward case of negligence by professional men. In issuing the two certificates and in subsequently maintaining their stance as to the correctness of those two certificates, they were not intending to interfere with the performance of the contracts. On the contrary, albeit in a misguided manner, they were seeking to further the performance of those contracts. As the judge correctly said, they were doing their incompetent best.'

Delay on programme or in progress

In addition to the premature deduction point which arose in *Lubenham,* two other points are worth noting. Firstly, the error by the architects in assuming that failure by the contractor to proceed to programme was a breach of contract, and from this to deduce wrongly that such a breach gave liability for liquidated damages; and secondly, the point that the employer is obliged to pay only the amount shown on the face of a certificate – even if he knows or suspects that it is incorrect. The course for the contractor in such a case is to invoke arbitration, not to determine the contract.

As for failing to proceed to programme, or failing to proceed with due expedition, regularly and diligently, or whatever other such phrases a contract may contain, the employer's remedy for each breach (and failing to proceed to programme will rarely be a breach), will not lie in liquidated damages but in such alternative contractual provisions as exist.

In most standard forms these usually amount to determination, so in effect the employer has a choice of remedies to exercise on commercial judgment: to remove the dilatory contractor and sue for general damages; or to let the contractor proceed at his own pace and face liquidated damages.

10.2 *Meaning of completion*

Various phrases are used in construction contracts to define completion:

- completion
- practical completion
- substantial completion.

Building contracts have traditionally used the term 'practical completion' whereas civil engineering contracts have traditionally used the term 'substantial completion'.

The significance of completion in construction contracts, however expressed, is generally that it marks:

- the transfer of risks for care of the works from the contractor to the employer
- the commencement of the defects liability period
- the end of the employer's entitlement to damages for late completion
- the employer's entitlement to repossess the site.

Disputes on completion are commonplace. Contractors may want early completion to reduce liabilities for liquidated damages and insurances and perhaps to secure part payment of retention monies. Employers may want later completion to ensure that the works are better finished or because they wish to delay occupation.

Completion in entire contracts

In its precise legal sense 'completion' means strict fulfilment of obligations under the contract, and when used in the context of 'entire contracts' which attract the doctrine of substantial performance failure to complete produces the apparently harsh result that no payment is due. Thus in the case of *Cutter v. Powell* (1795), when the second mate on a ship bound to Liverpool from Jamaica died before the ship reached Liverpool, his widow was unsuccessful in a claim for a proportion of his lump sum wages of 30 guineas.

Fortunately for contractors, construction contracts rarely fall into the category of 'entire contracts'. Indeed, if they did the risks for contractors would be immense since an employer unwilling to pay anything would only have to point to a modest default or item of unfinished work to escape the obligation of making payment.

The courts take a practical view of construction contracts as illustrated by this extract from the judgment of Lord Justice Denning in the case of *Hoenig v. Isaacs* (1952) which concerned the decorating and fitting-out of a one-room flat:

> 'In determining this issue the first question is whether, on the true construction of the contract, entire performance was a condition precedent to payment. It was a lump sum contract, but that does not mean that the entire performance was a condition precedent to payment. When a contract provides for a specific sum to be paid on completion of specified work, the courts leap against a construction of the contract which would deprive the contractor of any payment at all simply because there are

some defects or omissions. The promise to complete the work is, therefore, construed as a term of contract, but not as a condition. It is not every
breach of that term which absolves the employer from his promise to pay
the price, but only a breach which goes to the root of the contract, such
as abandonment of the work when it is only half done. Unless the breach
does go to the root of the matter, the employer cannot resist payment of
the price. He must pay it and bring a cross-claim for the defects and omissions, or alternatively, set them up in diminution of the price. The measure
is the amount which the work is worth less by reason of the defects and
omissions, and is usually calculated by the cost of making them good.'

And in *Bolton* v. *Mahadeva* (1972) Lord Justice Cairns made this comment on
substantial performance:

'In considering whether there was a substantial performance I am of
the opinion that it is relevant to take into account both the nature of the
defects and the proportion between the cost of rectifying them and the
contract price. It would be wrong to say that the contractor is only entitled
to payment if the defects are so trifling as to be covered by the *de minimis*
rule.'

In cases concerned with substantial performance the issue is generally the
employer's payment obligation and little else. Under most standard forms
of construction contracts the issues are likely to be wider – liability for liquidated damages, release of retention, etc. Generally, therefore, substantial
performance is not relevant to the meaning of 'completion' as mentioned in
construction contracts or to determination of terms such as 'practical completion' and 'substantial completion' used in such contracts.

Practical completion

Practical completion is the phrase commonly used in building contracts
to define the point at which the works are fit to be taken over by the
employer.
   It is also used in the ICE Minor Works Conditions where it is stated:

'Practical completion of the whole of the Works shall occur when the
Works reach a state when notwithstanding any defect or outstanding
items therein they are taken or are fit to be taken into use or possession
by the Employer.'

In the case of *Emson Eastern Ltd* v. *EME Developments Ltd* (1991) the court
had to decide whether the issue of a certificate of practical completion under
a JCT 80 contract constituted 'completion of the works' as mentioned in the
determination clause of the contract. Judge Newey QC held that it did. He
said:

'In my opinion there is no room for "completion" as distinct from "practical completion". Because a building can seldom if ever be built precisely

as required by drawings and specification, the contract realistically refers to "practical completion", and not "completion" but they mean the same.'

From the cases of *H W Neville (Sunblest) Ltd* v. *William Press & Sons Ltd* (1981) and *Westminster Corporation* v. *J Jarvis & Sons Ltd* (1970) the following rules to determine practical completion have been developed:

- practical completion means the completion of all the construction work to be done
- the contract administrator may have discretion to certify practical completion where there are minor items of work to complete on a *de minimis* basis
- a certificate of practical completion cannot be issued if there are patent defects
- the works can be practically complete notwithstanding latent defects.

Both *Westminster* v. *Jarvis* and *Nevill* v. *Press* concerned latent damage. In the *Jarvis* case the question was, could the contractor get an extension of time for carrying out replacement of faulty piling undertaken by a nominated sub-contractor but not discovered until after that sub-contractor had been given a certificate of completion of his work? The House of Lords ruled that he could not. Viscount Dilhorne said:

'From these provisions there are, in my opinion, two conclusions to be drawn: first that the issue of the certificate of practical completion determines the date of completion, which may of course be before or after the date specified for that in the contract; and secondly, that the defects liability period is provided in order to enable defects not apparent at the date of practical completion to be remedied. If they had been then apparent, no such certificate would have been issued.

It follows that a practical completion certificate can be issued when, owing to latent defects, the works do not fulfil the contract requirements; and that under the contract, works can be completed despite the presence of such defects. Completion under the contract is not postponed until defects which became apparent only after the work had been finished have been remedied.'

In *Nevill* v. *Press* a problem arose from defective groundworks in a preliminary works contract. Judge Newey, considering whether the remedies in respect of defective work found after the issue of a certificate of practical completion under a JCT 63 contract were restricted to clause 15, made this comment:

'I think that the word "practically" in clause 15(1), gave the architect a discretion to certify that William Press had fulfilled its obligation under clause 21(1), where very minor *de minimis* work had not been carried out, but that if there were any patent defects in what William Press had done the architect could not have given a certificate of practical completion.'

From these rules, and from basic legal principles, it has to be taken that the discovery of latent defects after the issue of a certificate of completion does not re-activate the employer's right to liquidated damages, even if the contractor has to return to the site and the employer has to give up possession. The employer's remedy is in general damages.

Substantial completion

The phrase 'substantial completion' is probably a more flexible concept than 'practical completion' and the provisions for dealing with outstanding works in ICE forms suggest that it is not the *de minimis* principle which applies to such works but whatever is acceptable to the engineer.

It is worth noting that under ICE forms the initiative for the issue of a completion certificate comes from the contractor – he applies to the engineer when he considers one due. Under JCT forms, the initiative is supposedly to be taken by the architect who is to issue a certificate when in his opinion it is due. In practice the contractor will, of course, normally make his views known and make an application.

Effects of occupation

Questions sometimes arise as to whether the test for completion includes occupation by the employer. The answer seems to depend on the wording of the contract. In the JCT case of *BFI Group of Companies Ltd* v. *DCB Integration Systems Ltd* (1987) the employer, BFI, was able to recover liquidated damages notwithstanding that it took possession on the extended date for completion. The point was not even argued before the court and the case turned on whether BFI had suffered any loss.

However, the position under ICE forms is different since they generally make occupation or use by the employer of any substantial part of the works grounds for the issue of a certificate of substantial completion. ICE Minor Works form puts the matter beyond any doubt in stating:

> 'Practical completion of the whole of the works shall occur when the works reach a state when notwithstanding any defect or outstanding items therein they are taken or are fit to be taken into use or possession by the employer.'

In the case of *Skanska Construction (Regions) Ltd v. Anglo-Amsterdam Corporation Ltd* (2002) the matters considered by the court in an appeal against an arbitrator's award concerned the impact of a partial possession clause on practical possession. The clause was of the standard JCT type reading:

> '17.1 If at any time or times before Practical Completion of the Works the Employer wishes to take possession of any part or parts of the Works and the consent of the Contractor (which consent shall not be unreasonably withheld) has been obtained, then,

notwithstanding anything expressed or implied elsewhere in this
Contract, the Employer may take possession thereof.

17.1.1 For the purposes of clauses 16.2, 16.3 and 30.4.1.2 Practical Completion of the relevant part shall be deemed to have occurred and
the Defects Liability Period in respect of the relevant part shall be
deemed to have commenced on the relevant date . . .'

Before practical completion was achieved Skanska, the contractor, was obliged
to allow a fitting-out contractor, ICL, on to site. That raised issues as to
whether clause 17 was activated and, if so, whether clause 17 applied only
in respect of possession of parts of the works as opposed to possession of the
whole of the works. The judge, in reversing the award, found that the clause
was activated and that it could apply to the whole of the works saying:

'55. . . . The irresistible conclusion is that Skanska handed possession of
the whole of the Works to ICL on 12 February 1996 at the request
of Anglo-Amsterdam's agent. Thereby, Skanska gave up possession
of the works. However, Skanska was permitted back to the site by
ICL on the occasions it returned to site on the express condition
imposed by Anglo-Amsterdam that Skanska itself made adequate
security arrangements for such visits.

56. It follows that Skanska gave up possession of the whole of the Works
on 12 February 1996 and that, whilst out of possession, was granted
a sub-licence by ICL for relevant parts of the Works for the purpose
of finishing off work left incomplete or in a defective state on 12
February 1996. This sub-licence had been granted to Skanska by ICL
once ICL had been granted possession of the works by Skanska following Skanska's giving up possession to Anglo-Amsterdam on
Anglo-Amsterdam's instructions.'

and

'58. The order that should be made is that the award should be varied so
as to provide that Skanska is entitled to the repayment of the liquidated damages it has paid out. This is because deemed Practical
Completion under clause 17.1 of the whole of the Works occurred on
12 February 1996 on account of partial possession of the whole of the
Works being taken by Anglo-Amsterdam and, through it, ICL on that
date . . .'

See also the decision in the *Multiplex* case, discussed in Chapter 4, regarding
the effect of early occupation by the employer for fitting out purposes on
the making of a pre-estimate of loss. The court held the benefit to the
employer to be insignificant and capable of being disregarded.

Practical completion and substantial performance

In *Big Island Contracting (HK) Ltd* v. *Skink Ltd* (1990), a contractor was entitled
to 25% of the contract price on practical completion. The employer occupied

the building but refused to pay because there were defects. It was held, dismissing the contractor's claim:

(i) practical completion could not be distinguished from substantial performance: the question was whether the work contracted for was 'finished' or 'done' in the ordinary sense;
(ii) on the facts practical completion had not been achieved.

Commenting on the judgment, the editors of *Building Law Reports* say this at page 112, 52 BLR:

> 'It may be doubted whether cases concerned with substantial performance are relevant to the determination of practical completion as the term is used in the standard form contracts where the event is not directly related to a payment obligation (as in the contract considered by the Hong Kong court) but rather with consequences such as liability for liquidated damages, re-possession and the release of retention.
>
> The plaintiffs did not in this case rely upon the fact that the defendants had taken possession of the building as constituting a waiver of the condition precedent to the payment of the 25% instalment. They did, however, rely upon that fact as supporting an argument that practical completion was achieved and that fact was acknowledged when the defendants went into occupation. This argument seems to have had less impact than the plaintiffs expected.
>
> In principle re-taking possession is not an acknowledgement that the works are practically complete. In practice where the employer has gone into possession without reservation of rights it may be difficult to contend there has not been practical completion because there will usually be beneficial occupation. Where there is no express provision for partial or other possession to be taken by the employer before practical completion it would be prudent to clarify the position before occupation by the employer is resumed.'

10.3 *Certificates and conditions precedent*

Standard forms of contract often contain conditions precedent to the deduction of liquidated damages. These work principally to the benefit of the contractor in forewarning him of likely deductions from amounts due, but they are also of benefit to the employer in drawing attention to his entitlement to damages. As a mechanism for placing key facts on record they work to the benefit of both parties.

JCT 2005 has three stated conditions precedent:

(i)   the contractor shall fail to complete on time;
(ii) the architect shall issue a certificate to that effect;
(iii) the employer shall give written notice of his intention to deduct damages.

However, under recent versions of ICE contracts, ICE 7th Edition and NEC 3, damages are stated to be payable simply on failure to complete on time.

These are the conditions precedent to be found in the liquidated damages provisions themselves, but it may not be possible to operate those provisions without the architect or engineer having first given attention to any specified obligations placed on him, elsewhere in the contract, to consider the contractor's entitlement to extensions of time.

When such obligations exist, as they do in most standard forms, these also can operate as conditions precedent to the deduction of liquidated damages.

Failure to comply with conditions precedent will render the deduction of liquidated damages unlawful and the contractor will be able to sue for their return.

In *Token Construction Co. Ltd* v. *Charlton Estates Ltd* (1973) clause 16 of the contract read:

> 'If the contractor fails to complete the works by the date stated in Appendix C to this contract or within any extended time fixed under clause 2(e) of these conditions and the architect certifies in writing that in his opinion the same ought reasonably so to have been completed, the contractor shall pay or allow to the employer a sum calculated at the rate stated in Appendix C as liquidated and ascertained damages for the period during which the same works shall so remain or have remained incomplete and the employer may deduct such damages from any moneys otherwise payable to the contractor under this contract.'

Some two years after the works were completed the architect made an interim certificate in favour of the contractor for £16,374 but wrote at the same time to the employer saying that the works had been completed 24 weeks late, that he was considering an extension of time for 13 weeks and that under clause 16, the employer was entitled to deduct 24 weeks' damages. At £800 per week these came to more than the amount certified and the employer refused to pay on the certificate. It was held:

(i) There is no reason why liquidated damages for delay should not be deducted from an interim certificate if the contract expressly gave that right.
(ii) Under the contract the right to deduct is subject to the condition precedent that there shall be a valid certificate by the architect.
(iii) On the facts there was no valid extension of time under clause 2(e) and there was no certificate of delay under clause 16.
(iv) While no set form of certificate is provided, the document relied upon must be the physical expression of a certifying process.
(v) The architect cannot certify for delay until he has first adjudicated upon the contractor's applications for extension of time.
(vi) The burden of proof that there has been certification of delay rests upon the one who alleges it.

Two similar cases are considered in Chapter 12: *Miller* v. *London County Council* (1934); and *Amalgamated Building Contractors Ltd* v. *Waltham Holy Cross UDC* (1952).

Repeat certificates

On the matter of repeat certificates the case of *A Bell & Son (Paddington) Ltd* v. *CBF Residential Care and Housing Association* (1989) is of interest. The contractor was granted extensions of time and, when he failed to complete by the extended date, the architect issued a certificate of non-completion and the employer gave written notice of his intention to deduct liquidated damages.

The architect subsequently gave a further extension of time and the employer deducted damages from this later date to the eventual date of practical completion. The issues before the court were, under clause 24 does the architect have to issue fresh certificates of completion after every grant of extension in a delay period, and does the employer have to issue fresh written notices of intention to deduct? The court held that when certificates of non-completion or notices of intention to deduct have been superseded by extensions which fix later completion dates, then new certificates and extensions are required. Judge Newey said:

'Construing clause 24.1 strictly, and in accordance with its plain and ordinary meaning, it demands issue of a certificate when a contractor has not completed by "completion date". A "completion date" is one fixed by the architect under clause 25.3.2. I think that when a new completion date is fixed, if the contractor has not completed by it, a certificate to that effect must be issued and it is irrelevant whether a certificate had been issued in relation to an earlier, now superseded, completion date. I think that this construction accords with the setting of the contract: contractors and employers using it need above all certainty and the issue of a fresh certificate will provide it.

Construing clause 24.2.1 in a similar manner to clause 24.1, since the giving of notice is made subject to the issue of a certificate of non-completion, if the certificate is superseded, then logically the notice should fall with it. Here the setting of the contract may point in the opposite direction, for once an employer has informed a contractor of his intention to recover liquidated damages he is unlikely to change his mind. However, I think that once again certainty is the greatest need and that if a new completion date is fixed any notice given by the employer before it, is at an end.'

Amendment 9 of JCT 80 dealt with the issues raised in *Bell* by clarifying the need for fresh certificates of non-completion and by eliminating the need for fresh notices of intention to deduct. In a case subsequent to *Bell* on the same form of contract, *Jarvis Brent Ltd* v. *Rowlinson Construction Ltd* (1990), the employer before deducting liquidated damages sent to the contractor, 'for information only', a copy of a letter from his quantity surveyor showing the calculation of the damages. The contractor some time before the issue of the final certificate, which in this form limits the time for deductions, became aware that he could challenge the deductions but he waited until after the issue of the final certificate before commencing proceedings for their recovery. Judge Fox-Andrews held, in a judgment which cannot be reconciled with *Bell*:

(i) the employer's letter was an adequate request in writing as it got the message across to the contractor;
(ii) in any event:
    (a) the employer's written request was not a condition precedent to his right to deduct liquidated damages;
    (b) the contractor by his conduct had led the employer to believe that strict contractual rights would not be insisted upon and he was thereby estopped from making the challenge.

The conflicting decisions of *Jarvis Brent* and *Bell* have subsequently been considered in the case of *J F Finnegan Ltd* v. *Community Housing Association* (1995) by the Court of Appeal.

The appeal concerned only the procedural aspects of the liquidated damages provisions in the contract between Finnegan and Community Housing. The rulings of the judge at first instance (discussed in detail in Chapter 4 above) that the damages clause was not penal were not challenged.

The Court of Appeal, approving *Bell* and disapproving *Jarvis Brent* (in part) held:

1. Written notice from the employer under clause 24.2.1 is a condition precedent to the deduction of liquidated and ascertained damages.
2. Such a notice need not precede the deduction, but can accompany the deduction.
3. The notice need only make clear two things: whether the employer is making any deduction of liquidated and ascertained damages, and what sum is being deducted, the whole or only part of the liquidated and ascertained damages. The notice must be such as would make these two things clear to a reasonably literate and numerate contractor.

Effect of final certificates

A point worth noting here is that under the wording of some standard forms the issue of a final certificate is stated to be conclusive evidence of the fulfilment of the contractor's obligations. The effect of this, subject to any qualifications which are applicable, is that whether or not there has been a previously issued certificate of completion the contractor cannot be held liable for any subsequent default.

Only in exceptional cases will a final certificate be issued at such a time that it has the effect of prematurely cutting off the contractor's liability for liquidated damages but, in the author's experience, it has been known to occur as a result of erroneous administration of a contract.

Of more concern is the effect of final certificates on prematurely cutting off the contractor's liability for latent defects. In two cases, *Colbart Ltd* v. *H Kumar* (1992) and *Crown Estate Commissioners* v. *John Mowlem & Co. Ltd* (1994) the courts held that, on the wording of the contracts, the issue of the final certificates had that effect. As a result of these cases changes were made to the drafting of some standard forms, including JCT 80.

10.4 *Methods of recovery*

For employers, the preferred method of recovery of liquidated damages is by deduction from sums due to the contractor. This avoids the trouble and expense of suing for recovery and it avoids the problem of the insolvent contractor who is not worth suing.

Standard forms of contract usually make express provision for deduction. Note, however, there may also be need for the service of withholding notices – a point covered later in this chapter.

The *Token* v. *Charlton* (1973) case referred to in section 10.3 confirmed that deductions can be made from interim certificates where the contract expressly gives that right and there seems little doubt that phrases such as 'from any moneys due' will be taken to confer it.

Restrictions on deduction

However, it is by no means certain that wording such as that in ICE Minor Works Conditions: '. . . the contractor shall be liable to the employer in the sum stated . . .' gives any right to deduct from interim certificates or, indeed, any contractual right to deduct at all. In such cases it is necessary to look at the payment provisions to see whether they are restrictive on the employer or whether they contemplate deductions of any kind.

The JCT Agreement for Minor Building Works avoids these difficulties with wording similar to the main JCT forms:

> '. . . The employer may deduct such liquidated damages from any moneys due to the contractor under this contract or he may recover them from the contractor as a debt . . .'

Deduction essential in some forms

Hudson at section 10.056 makes the comment that some rare forms of contract make provision for deductions which can be construed as mandatory and exclusive so that failure to make deductions may disentitle the employer from recovering damages. The case of *Baskett* v. *Bendigo Gold Dredging Co.* (1902) is quoted where the phrase '. . . will be deducted from any moneys due to the contractor . . .' was held to prevent the employer from recovering when moneys had been paid without deductions.

Suing for recovery

When the employer finds it necessary to sue for recovery of liquidated damages or finds himself defending a claim for full payment on a certificate where they have been deducted, the employer has available the alternative claim or alternative counter-claim for general damages. In *Temloc* v. *Errill* (1987) the alternative claim failed because the liquidated damages provisions were valid and excluded general damages. In *Rapid Building* v. *Ealing Family Housing Association* (1984) the alternative counterclaim succeeded when the liquidated damages provisions were ruled invalid and there was therefore no exclusion.

Deductions from retentions

There has long been argument on whether employers are entitled to deduct liquidated damages from sums held as retention moneys. The principle that the employer's interest in retention money is fiduciary as trustee for the contractor is expressed in the main building forms and probably implied into other forms. Many of the disputes which reach the courts on retention are on whether or not the employer is obliged to set aside retention money in a separate trust fund, and in a series of cases from *Rayack Construction Ltd* v. *Lampeter Meat Co. Ltd* (1979) to *Wates Construction (London) Ltd* v. *Frantham Property Ltd* (1991) the courts have held that the employer's interest in retention is as a trustee and not as a beneficiary and that a trust fund is implied. That even applied in *Wates* where an express provision for retention to be placed in a separate bank account had been deleted.

Nevertheless, this does not prohibit the employer from deducting sums due under the contract. Mr Justice Vinelott in *Rayack* said:

> 'Lastly, [Counsel for Lampeter Meat Co.] said that there had been delay in completing the contract and that liquidated damages were likely to exceed the retention moneys. However, the contention that delay would give rise to a claim for liquidated damages rests on speculative grounds and in any event, if such a claim were maintainable, the defendants would be entitled under condition 30(4)(a) to withdraw the equivalent sum from the trust account.'

Rayack was followed in *Henry Boot Building Ltd* v. *The Croydon Hotel and Leisure Co. Ltd* (1985) where it was held that although the employer was obliged to set aside retentions in a separate fund, the obligation could not be enforced by injunction at a time when the employer was entitled to deduct a greater sum as liquidated damages.

In the later case of *J F Finnegan Ltd* v. *Ford Sellar Morris Developments Ltd* (1991) an injunction that retention should be placed in a separate account was granted to the contractor, but in this case the contractor had a defence to the claim for liquidated damages, and there was no dispute on the principle of deducting from retention money.

10.5 Time limits on recovery

Legal limits

English law requires that actions for breach of contract are brought within six years of the cause of action for contracts under hand, and within 12 years for contracts under seal or executed as a deed.

Unless there are express provisions in the contract to the contrary, these limits apply to actions for the recovery of liquidated damages either as a main claim or a counterclaim. It is less clear whether these time limits also apply to recovery of liquidated damages by deductions from sums due to the contractor. Were it not for the extraordinary delays in the settlement of some final accounts, the question would appear to be academic, but there are indeed cases where significant sums are still outstanding to contractors many years after the work has been completed.

The probability is that where the contract permits recovery from sums due to the contractor, such recovery can be made at any time after the sums become due unless there is some contractual restriction.

Contractual limits

There are good reasons for imposing such restrictions, not least to bring certainty to final account settlements which may only have been achieved after years of protracted negotiations and compromise. It would be inequitable, and possibly a case for arguing waiver or estoppel, if an employer stayed silent through negotiations on his intention to deduct damages from the final payment to secure a better deal on the final accounts.

JCT 2005 endeavours to eliminate the problem of continuing uncertainty by stating in clause 1.10.3 the effect of the final certificate to be: 'conclusive evidence that all and only such extensions of time, if any, as are under clause 2.28 have been given' and by stating in clause 2.32.2 that the employer's notice of intention to require the payment or allowance of liquidated damages must be given not later than the date of the final certificate.

However, because the architect is *functus officio* after the issue of the final certificate – a point confirmed in *H Fairweather Ltd* v. *Asden Securities Ltd* (1979) – and it is not part of the architect's duty to deal with deductions of damages in certificates, and all his certificates including those for extensions of time can, in any event, be opened up in arbitration, the position of the parties is not as restricted by the issue of the final certificate as might first appear.

ICE 7th Edition requires the employer to pay the balance on the final account, less any deductions for liquidated damages, within 28 days of the issue of the engineer's final certificate. The wording, however, is not restrictive enough to prevent later claims from the employer for such damages, although both contractor and employer can be caught under these conditions with the tight timing restrictions in the arbitration agreement – clause 66.

Payment in full – no barrier to recovery

The general rule that the courts will not bar a claim for liquidated damages simply because payment has already been made in full, was established in *Clydebank Engineering* v. *Yzquierdo y Castaneda* (1905) and contractual wording to exclude this principle would have to be very precise.

10.6 Interest on repayment

Provided liquidated damages are deducted legitimately under the contract, there is no breach of contract if a review of the contractor's entitlement to extensions of time leads to a revised date for completion and such damages have to be repaid in whole or in part to the contractor.

JCT 2005 provides for the repayment of liquidated damages when a later completion date is fixed but there is no mention of payment of interest on the sums so repaid. Were it not for the much-criticised decision of the High Court of Northern Ireland in *Department of the Environment for Northern Ireland* v. *Farrens (Construction) Ltd* (1981) it is unlikely that any argument in favour of interest could be sustained. There is no common law right of interest on sums due, so if there is no contractual right, there is no case other than to claim special damages under the second limb of *Hadley* v. *Baxendale.* But if there is no breach how can there be such a claim?

In *Farrens,* which concerned a JCT 63 contract, Mr Justice Muncy appeared to take the view that the architect could issue only one certificate of non-completion and if the employer deducted damages on the basis of a certificate which was subsequently superseded, he did so at his own risk. This would then amount to breach of contract and interest would be payable as special damages under the second limb of *Hadley* v. *Baxendale* as the foreseeable consequences of the employer's failure to pay on the due date, following the decision in *Wadsworth* v. *Lydall* (1981).

Legal commentators suggested that *Farrens* was wrongly decided on JCT 63 but in any event it would not apply to later editions of JCT because of differences in wording. However, there remains a view that interest is not payable on refunded liquidated damages under forms where there is no express mention of interest.

The position is different under forms such as ICE Conditions which have express provisions for interest to be paid on the reimbursement of liquidated damages.

10.7 *Withholding notices*

The Housing Grants, Construction and Regeneration Act 1996, requires at Section 111 the service of a withholding notice before deductions can be made from sums due under construction contracts covered by the Act. Most UK construction contracts are so covered.

Section 111 of the Act reads:

'111. (1) A party to a construction contract may not withhold payment after the final date for payment of a sum due under the contract unless he has given an effective notice of intention to withhold payment.

The notice mentioned in section 110(2) may suffice as a notice of intention to withhold payment if it complies with the requirements of this section.

(2) To be effective such a notice must specify –
   (a) the amount proposed to be withheld and the ground for withholding payment, or
   (b) if there is more than one ground, each ground and the amount attributable to it,

and must be given not later than the prescribed period before the final date for payment.

(3) The parties are free to agree what that prescribed period is to be.

In the absence of such agreement, the period shall be that provided by the Scheme for Construction Contracts.

(4) Where an effective notice of intention to withhold payment is given, but on the matter being referred to adjudication it is decided that the whole or part of the amount should be paid, the decision shall be construed as requiring payment not later than –
   (a) seven days from the date of the decision, or
   (b) the date which apart from the notice would have been the final date for payment,

whichever is the later.'

Scheme for Construction Contracts

The period referred to in Section 111 of the Act as that provided by the Scheme for Construction Contracts is the 7-day period stated in Paragraph

10 of Part II – Payment – of the Scheme for Construction Contracts Regulations 1998 which reads:

> 'Notice of intention to withhold payment
> 10. Any notice of intention to withhold payment mentioned in section 111 of the Act shall be given not later than the prescribed period, which is to say not later than 7 days before the final date for payment determined either in accordance with the construction contract, or where no such provision is made in the contract, in accordance with paragraph 8 above.'

Application to liquidated damages

The rule which can be derived from the Act is that deductions cannot be made from sums due under the contract unless a withholding notice has been served in compliance with the terms of the contract or, in the absence of any such terms, in compliance with the 7-day period stated in the statutory Scheme.

It is evident from a string of the cases, the most recent being *Avoncroft Construction Limited v. Sharba Homes (CN) Limited* (2008), that the rule applies to deductions for liquidated damages.

What is more, as the *Avoncroft* case makes clear, is that a valid notice is necessary if the deduction is to be made from a payment ordered by an adjudicator. Dealing with arguments on this the judge in *Avoncroft* said:

> 'Validity of withholding notice
> 14. Given my conclusions on these points, it is not necessary to deal with the further point which Mr Thompson raises, but I do so because it is a point of interest and which merits consideration. Mr Thomson submits that the defendant's withholding notice dated 15 February 2008 was not served in time.
> 15. Two clauses in the contract make express provision for service of a withholding notice, namely clause 30.1.1.4 (which deals with payment of Interim Certificates) and clause 30.8.3 (which deals with payment due pursuant to the Final Certificate.) Neither of these is relevant here. The sum which the adjudicator has decided is due to the claimant is not and does not reflect an entitlement to an interim payment, nor does this sum arise out of the provisions of Clause 30.8: the Final Certificate has not been issued so the mechanism provided by Clause 30.8 has not come into operation.
> 16. I accept Mr Thompson's submission that the decision provides for a sum due under the contract. The adjudicator decided that a sum was payable; by reason of the obligation contained in clause 41A.7.2, the parties must comply with the decision reached by the adjudicator; accordingly, the sum is due under the provisions of the contract ie under the contract. The sum which the adjudicator awarded is not due pursuant to the contractual payment mechanisms. The

adjudication provisions of the building contract do not provide for a notice of withholding to be served against a decision. Section 111 of the 1996 Act requires provision to be made for service of a withholding notice against "a sum due under the contract". As the contract has made no provision for service of a withholding notice against the decision of an adjudicator, one must look to section 111(3) of the Act. This provides that the prescribed period for service of a withholding notice shall be that provided by the Scheme for Construction Contracts Regulations 1998. Pursuant to paragraph 10 of Part 2 of the Scheme, that period is "not later than 7 days before the final date for payment".

17. The adjudicator's decision required the defendant to pay £56,380 "peremptorily" and by no later than 4 pm on 21 February 2008. Accordingly, the final date for payment was 21 February 2008. The notice was served only six days before 21 February 2008 and was thus out of time. Accordingly, the defendant would be prohibited, in any event, from withholding any money from the sum awarded by the adjudicator's decision.

18. In response, Mr Maguire submitted that the effective and applicable withholding notice was that dated 19 September 2007. I reject that submission: the notice dated 19 September 2007 is expressly stated to apply to sums to be withheld from the claimant's Application for Payment number 13, and not to sums to be withheld from the payment due pursuant to the adjudicator's decision.

19. Mr Maguire also submitted that, as the decision was made on 14 February, it would be impossible for the defendant to satisfy the provisions of the Scheme. He submits that the court should recognise that impossibility and make an allowance in favour of the defendant for later service of the notice eg on the following day. I reject that submission also: many authorities in this field stress the importance of strict compliance with the time limits provided by the Act, and I see no reason here to depart from that general approach.

20. The defendant has no real prospect of success. No other reason has been advanced why summary judgment should not be given. It follows that judgment should be entered for the claimant for the full sum claimed.

10.8 Decisions of adjudicators

The introduction, by the Housing Grants, Construction and Regeneration Act 1996, of a statutory right to adjudication of disputes in construction contracts led inevitably, although perhaps not intended, to a vast increase in the number of disputes referred for third party determination. Disputes relating to extensions of time and liquidated damages have been high on the list of those so referred.

For the most part decisions of adjudicators remain confidential to the parties and where they are exposed to public scrutiny it is usually in relation to enforcement proceedings. These are principally proceedings seeking court orders on sums found due by adjudicators. Little of substance has been added to the general body of construction law by the hundreds of rulings given by the courts in enforcement actions but there is already a significant body of adjudication-related law of great interest to the construction industry – part of which covers the recovery of amounts claimed as liquidated damages.

One of the issues dealt with by the courts, as can be seen from the previously mentioned *Avoncroft* case, is whether a valid withholding notice has been served. A similar issue was considered in *Edmund Nuttall Ltd* v. *Sevenoaks District Council* (2000) where the conditions of contract stated conditions precedent to the employer's right to deduct liquidated damages but did not cover deductions from amounts found due in adjudication. The employer argued for an implied term that would have allowed set-off but this was rejected by the court, with Mr Justice Dyson saying:

> '[Employer's counsel] contends that a term of the contract should be implied that, where an Adjudicator has made an award in favour of the contractor, the employer should be able to deduct liquidated and ascertained damages from the amount of the award. He submits that such a term is necessary to give business efficacy to the contract. Without it the contract is unworkable where an Adjudicator's award is issued in favour of a contractor, and the employer wishes to make a deduction, for example of liquidated and ascertained damages.
>
> I cannot accept this submission. It seems to me that the contract works perfectly satisfactorily without such a term. Moreover, I think I ought to be extremely wary about implying a term as to the circumstances in which liquidated and ascertained damages may be deducted from a sum due to the contractor when the contract contains detailed express provisions which deal precisely with that issue.
>
> It may be that those who draft these standard forms of contract will decide to enlarge the scope of clause 2.7, so as to admit the deduction of liquidated and ascertained damages by employers from sums awarded by Adjudicators to contractors. That however is not a matter for me. There may well thought to be good policy reasons for rejecting this suggestion in any event.'

Broader issues which have troubled the courts concern the position where the adjudicator has determined the appropriate extension of time but has not dealt with liquidated damages. The early judgments on this appeared to be at odds with *VHE Construction plc* v. *RBSTB Trust Co. Ltd* (2000) indicating that set-off was not permissible and *David McLean Housing Contractors Ltd* v. *Swansea Housing Association Ltd* (2002) indicating that it was. However, these were explained by Mr Justice Jackson as reconcilable in *Balfour Beatty Construction* v. *Serco Limited* (2004) where he said:

'50 There is, in my judgment, no inconsistency between the reasoning in VHE Construction and David McLean. In each case the decision flows from an analysis of what the adjudicator had decided and from the particular circumstances of the case.

51 The manner in which VHE Construction and David McLean can be reconciled has been discussed by His Honour Judge Seymour QC in *Solland International Ltd* v *Daraydan Holdings Ltd* [2002] EWHC 220 (TCC); 83 CONLR 109 at paragraphs 30 to 32. The same matter has been discussed by His Honour Judge Thornton QC in *Bovis Lend Lease Ltd* v *Triangle Development Ltd* [2003] BLR 31 at paragraphs 35 to 36. I note next the decision of the Court of Appeal in *Parsons Plastics (Research and Development) Ltd* v *Purac Ltd* [2002] BLR 334. In that case the contract contained a specific claim as to set off which determined the outcome.'

Having then reviewed further authorities, Mr Justice Jackson went on to say:

'53 I derive two principles of law from the authorities, which are relevant for present purposes.

(1) Where it follows logically from an adjudicator's decision that the employer is entitled to recover a specific sum by way of liquidated and ascertained damages, then the employer may set off that sum against monies payable to the contractor pursuant to the adjudicator's decision, provided that the employer has given proper notice (insofar as required).

(2) Where the entitlement to liquidated and ascertained damages has not been determined either expressly or impliedly by the adjudicator's decision, then the question whether the employer is entitled to set off liquidated and ascertained damages against sums awarded by the adjudicator will depend upon the terms of the contract and the circumstances of the case.'

These rules have been followed in subsequent cases including *William Verry Ltd* v. *London Borough of Camden* (2006) and the *Avoncroft* v. *Sharba Homes* (2008) case mentioned previously.

Chapter 11
Defences / challenges to liquidated damages

11.1 Benefits of precedents

Contractors faced with the deduction of liquidated damages or an action to recover the same do not always accept that deductions are justified or due. This may be because of extension of time disputes; because of perceived legal flaws in the contractual provisions for liquidated damages; or because of alleged maladministration by the contract administrator.

Although it will often be the case that the underlying dispute between the parties is factual in nature, particularly where the defence or challenge to liquidated damages is the amount of extension of time due, there will frequently be some aspect of a dispute where legal precedent can offer some guidance to the parties on the likely outcome of any formal dispute resolution proceedings.

Traditionally arbitration was the principal method of dispute resolution in the construction industry but currently, along with litigation, it is more of a back-stop after mediation or adjudication proceedings. But of these only litigation provides legal precedents.

There has been debate as to whether the huge amount of legal argument and analysis on matters of interest to the construction industry which goes into arbitration proceedings could, in some way, be harnessed to general beneficial effect and accorded secondary status. However, for a variety of reasons, some legal, some practical, it seems unlikely that much will come of this – although it has been known for redacted arbitration awards and adjudication decisions to be put forward in mediations as providing rulings which might sensibly, if not legally, be accorded some weight.

Consequently, as things presently stand and are likely to remain, it is necessary to look to the cases for legal guidance and precedents on the defences and challenges to claims for liquidated damages – save for those defences which found on the facts. In the following analysis summaries only are given where the subject has been covered in depth in other chapters.

11.2 Extension of time due

Probably the commonest plea put forward by contractors when faced with liquidated damages is that extensions of time are still due. This is usually because the contract administrator has not recognised any justifiable delay or has recognised too little. If negotiation fails to produce a satisfactory result the contractor has little choice but to proceed to adjudication, arbitration or litigation.

Review in adjudication

As noted in the previous chapter, there is a statutory right in the United Kingdom for disputes in construction contracts to be referred to adjudication. The effect of this is that all disputes, including those on extensions of time and liquidated damages, can be referred by either party to adjudication under the contractual scheme, or under the statutory scheme if there is no contractual scheme or if it is not compliant with statutory requirements.

An adjudicator's decision is binding on the parties unless and until the dispute is referred to arbitration or litigation. This was explained by Mr Justice Ramsay in the case of *William Verry Ltd* v. *The Mayor and Burgesses of the London Borough of Camden* (2006) as follows:

> 'Whilst adjudication is not arbitration, in my judgment, the phrase "the decision of the adjudicator is binding" is intended to provide a similar degree of compliance by the parties, except that in the case of an adjudicator's decision, the decision is not "final" but is "interim" unless the parties agree to accept it as finally determining the dispute. The intention of Parliament must be that the decision is binding and enforced at interim stage. If the decision were no more than another contractual obligation, which could be breached or could be reduced or diminished by other contractual obligations, then the fundamental purpose of providing cash flow in the construction industry would be undermined. As Lord Justice Mantell said in *Ferson* v. *Levolux* at para. 30, "the contract must be construed so as to the give effect to the intention of Parliament, rather than to defeat it". In my judgment, that can only be done by giving proper effect to the word "binding" by enforcing the decision of adjudicators.'

The judge went on to say later in his judgment:

> 'Equally, Mr. Matthias relies on the decision of His Honour Judge Lloyd QC in *David McLean Housing Limited* v. *Swansea Housing Association Limited* [2003] BLR 125 in which he held that the employer was entitled to deduct liquidated damages from the adjudicator's decision where the adjudicator had determined the appropriate extension of time, but had not dealt with liquidated damages.'

That particular question has been the subject of a subsequent decision by Mr Justice Jackson in *Balfour Beatty Construction Ltd* v. *Serco Limited* [2004] EWHC 3336 in which, having considered the relevant decisions including *David McLean, Bovis Lend Lease, Parsons Plastics* and *Fersons* v. *Levolux*, he derived the following two principles:

'(i) Where it follows logically from an adjudicator's decision that the employer is entitled to recover a specific sum by way of liquidated and ascertained damages, then the employer may set off that sum against monies payable to the contractor, pursuant to the adjudicator's decision provided that the employer has given proper notice (insofar as required).

(ii) Where the entitlement to liquidated and ascertained damages has not been determined either expressly or impliedly by the adjudicator's decision, then the question whether the employer is entitled to set off liquidated and ascertained damages against sums awarded by the adjudicator will depend upon the terms of the contract and the circumstances of the case.

29 The particular issue of whether liquidated damages can be deducted when the adjudicator's decision deals with extensions of time but does not deal with the consequential effect on an undisputed or indisputable claim for liquidated damages raises, I consider, a distinct question of the manner and extent of compliance with the adjudicator's decision. It does not, in my judgment, raise a question as to the ability to set-off sums generally against an adjudicator's decision.'

One of the particular problems of extension of time disputes is that the decision of the adjudicator is not of itself enforceable in court proceedings in the same way that an order for payment of a sum of money can be enforced. Nevertheless, what is clear from the above quotation in the *Verry* case and other adjudication related cases is that the decision is binding on the parties. Some contracts require the contract administrator to recognise the effect of adjudication decisions and to certify accordingly but even if there is no such provision or there is no amending certificate the decision is still effective.

In the event that one or both of the parties is dissatisfied with the decision, the normal way to deprive it of effect on substantive grounds (as distinct from jurisdictional or other procedural grounds) is for one party to refer the dispute to arbitration or litigation. Neither of these is an appeal process. Both entail a new review of the dispute in which the adjudicator's decision plays no part. There is another way by which, in some circumstances, a referring party can seek to improve on an adjudicator's decision. This involves commencing a second adjudication and being able to successfully argue that the second referred dispute is different from the first referred dispute.

This matter was considered by the Court of Appeal in *Quietfield Ltd* v. *Vascroft Construction Ltd* (2006). Lord Justice May had this to say:

'21. The judge concluded in paragraph 42 of his judgment that there were four relevant principles where there are successive adjudications about extension of time and the deduction of damages for delay, as follows:

"(i) Where the contract permits the contractor to make successive applications for extension of time on different grounds, either party, if dissatisfied with the decisions made, can refer those matters to successive adjudications. In each case the difference between the contentions of the aggrieved party and the decision of the architect or contract administrator will constitute the 'dispute' within the meaning of section 108 of the 1996 Act.

(ii)   If the contractor makes successive applications for extension of time on the same grounds, the architect or contract administrator will, no doubt, reiterate his original decision. The aggrieved party cannot refer this matter to successive adjudications. He is debarred from doing so by paragraphs 9 and 23 of the Scheme and section 108(3) of the 1996 Act.

(iii) Subject to paragraph (iv) below, where the contractor is resisting a claim for liquidated and ascertained damages in respect of delay, pursued in adjudication proceedings, the contractor may rely by way of defence upon his entitlement to an extension of time.

(iv)   However, the contractor cannot rely by way of defence in adjudication proceedings upon an alleged entitlement to extension of time which has been considered and rejected in a previous adjudication."

In my judgment, these principles are a correct analysis for the purposes of the present case. Mr Holt, for the appellant, attempted to persuade us, unsuccessfully in my view, that the judge's paragraph (i) was wrong – see later in this judgment.'

and

'31. Section 108(3) of the 1996 Act and paragraph 23 of the Scheme provide for the temporary binding finality of an adjudicator's *decision*. More than one adjudication is permissible, provided a second adjudicator is not asked to decide again that which the first adjudicator has already decided. Indeed paragraph 9(2) of the Scheme obliges an adjudicator to resign where the dispute is the same or substantially the same as one which has previously been referred to adjudication and a *decision* has been taken in that adjudication.

32. So the question in each case is, what did the first adjudicator decide? The first source of the answer to that question will be the actual decision of the first adjudicator. In the present appeal, Mr Holt did not even take us to the first adjudicator's decision, although he was invited more than once by the court to do so. He was conscious, no doubt, that it would show, as it does, that the decision was limited to the grounds for extension of time in the two letters.

33. The scope of an adjudicator's decision will, of course, normally be defined by the scope of the dispute that was referred for adjudication. This is the plain expectation to be derived from section 108 of the 1996 Act and paragraphs 9(2) and 23 of the Scheme. That is also the plain expectation of paragraph 9(4) of the Scheme, which refers to a dispute which varies significantly from the dispute referred to the adjudicator in the referral notice and which for that reason he is not competent to decide. There may of course be some flexibility, in

that the scope of a dispute referred for adjudication might by agreement be varied in the course of the adjudication.'

Lord Justice Dyson said:

'41. The contract contains no express provision limiting the number of such written notices that may be given by the contractor in respect of any particular Relevant Event or the number of times that the contractor may, in respect of any Relevant Event, give particulars of the expected effects or make estimates of the expected delay to the completion of the Works beyond the Completion Date. Nevertheless, I would hold that, upon the true construction of the contract or by necessary implication, the contractor cannot give successive notices of the same material circumstances including the same cause or causes of delay or identify the same Relevant Event as he has given and included in a previous written notice. Nor can he successively give the same particulars of the expected effects of the same Relevant Event or make the same estimates of the expected delay to completion as he has previously given or made. In other words, the contractor cannot merely repeat himself, hoping that the architect may, on a reconsideration of substantially the same material that he has already considered reach a different conclusion. In practice, of course, the contractor is rarely likely to consider that there is any point in doing this.

42. In my judgment, therefore, the contractor must present some new material which could reasonably lead the architect to reach a different conclusion from that on which he based his earlier decision or decisions. The judge did not explain what he meant by "different grounds" in his first principle. I can see no reason to construe clause 25 so as to prohibit the contractor from relying on the same Relevant Event as he relied on in support of a previous application for extension of time, giving materially different particulars of the expected effects and/or a different estimate of the extent of the expected delay to the completion of the Works. If the position were otherwise, the contractor could not make good shortcomings of one application by a later application, and would be obliged to refer the matter to arbitration. That cannot have been intended by the contract. There is nothing in the express language which prevents the contractor from making good the deficiencies of an earlier application in a later application.

43. So much for the position under clause 25. The judge's first principle may appear to suggest that every dispute arising from the rejection of an application for an extension of time may be referred to adjudication. I do not consider that that is necessarily the case. The question whether a contractor may make successive applications for extensions of time depends on the true construction of clause 25 and any term necessarily to be implied. The question whether disputes arising from the rejection of successive applications for an extension of time

may be referred to adjudication depends on the effect of section 108(3) of the 1996 Act and paragraph 9(2) of the Scheme.

44. There are obvious differences between successive applications for extensions of time under the contract and successive referrals of disputes to adjudication. In the real world, there is often a regular dialogue between contractor and architect in relation to issues arising from clause 25. If an architect rejects an application for an extension of time pointing out a deficiency in the application which the contractor subsequently makes good, it would be absurd if the architect could not grant the application if he now thought that it was justified. To do so would be part of the architect's ordinary function of administering the contract. But referrals to adjudication raise different considerations. The cost of a referral can be substantial. No doubt that is one of the reasons why the statutory scheme protects respondents from successive referrals to adjudication of what is substantially the same dispute.

45. Paragraph 9(2) provides that an adjudicator must resign where the dispute is the same or substantially the same as one which has previously been referred to adjudication and a decision has been taken in that adjudication. It must necessarily follow that the parties may not refer a dispute to adjudication in such circumstances.

46. This is the mechanism that has been adopted to protect respondents from having to face the expense and trouble of successive adjudications on the same or substantially the same dispute. There is an analogy here, albeit an imperfect one, with the rules developed by the common law to prevent successive litigation over the same matter: see the discussion about *Henderson* v. *Henderson* (1843) 3 Hare 100 abuse of process and cause of action and issue estoppel by Lord Bingham of Cornhill in *Johnson* v. *Gore Wood & Co. (a firm)* [2002] 2 AC 1, 30H–31G.

47. Whether dispute A is substantially the same as dispute B is a question of fact and degree. If the contractor identifies the same Relevant Event in successive applications for extensions of time, but gives different particulars of its expected effects, the differences may or may not be sufficient to lead to the conclusion that the two disputes are not substantially the same. All the more so if the particulars of expected effects are the same, but the evidence by which the contractor seeks to prove them is different.

48. Where the only difference between disputes arising from the rejection of two successive applications for an extension of time is that the later application makes good shortcomings of the earlier application, an adjudicator will usually have little difficulty in deciding that the two disputes are substantially the same.'

In the case of *H G Construction Ltd* v. *Ashwell Homes (East Anglia) Ltd* (2007) Mr Justice Ramsay, after referring to the rules on *Quitefield*, summarised the position as follows:

'(1) the parties are bound by the decision of an adjudicator on a dispute or difference until it is finally determined by court or adjudication proceedings or by an agreement made subsequently by the parties.

(2) The parties cannot seek a further decision by an adjudicator on a dispute or difference if that dispute or difference has already been the subject of a decision by an Adjudicator.

(3) As a matter of practice, an adjudicator should consider (based either on an objection raised by one of the parties or on his own volition) whether he is being asked to decide a matter on which there is already a binding decision by another Adjudicator. If so he should decline to decide that matter or, if that is the only matter which he is asked to decide, he should resign.

(4) The extent to which a decision or a dispute is binding will depend on an analysis of
  (a) the terms, scope and extent of the dispute or difference referred to adjudication and
  (b) the terms, scope and extent of the decision made by the adjudicator.

(5) In considering the terms, scope and extent of the dispute or difference the approach has to be to ask whether the dispute or difference is the same or substantially the same as the relevant dispute or difference.

(6) In considering the terms, scope and extent of the decision, the approach has to be to ask whether the Adjudicator has decided a dispute or difference which is the same or fundamentally the same as the relevant dispute or difference.'

See also the case of *Emcor Drake & Scull Ltd* v. *Costain Construction Ltd & Skanska Central Europe AB* (2004) dealing with the overlap between successive adjudication decisions.

Review in arbitration

Most standard forms of contract used in construction contain arbitration agreements which empower the arbitrator to open up and review any certificates. So, quite apart from referrals to adjudication, there is no absolute finality in extensions of time awarded by the contract administrator. Even statements in some contracts that the architect's final certificate is conclusive evidence that all such extensions as are due have been given, is usually subject to the proviso of review in arbitration or other proceedings. However, in some contracts, there are time limits on the commencement of arbitration or other proceedings and failure to commence on time leaves the certificates unchallengeable. Then no further review of extensions of time can be made.

Usually where arbitration agreements are written into contracts, those agreements bind the parties to settling their differences by arbitration. If one party starts an action in the courts it is open to the other party to apply for a stay of proceedings on the grounds that there is an effective arbitration agreement. Of course if both parties prefer litigation to arbitration they may choose to ignore their arbitration agreement and go to court. This, however, is not without its difficulties.

Review in litigation

In *Northern Regional Health Authority* v. *Derek Crouch Construction Co. Ltd* (1984), a dispute under a JCT 63 contract which concerned in the first instance the sufficiency of extensions of time granted by the architect, the Court of Appeal ruled that the courts have no power to open up and review certificates where the contract gives the duty of certification to an architect or engineer. If the parties have in their arbitration agreement conferred the power of opening up and reviewing on an arbitrator, that is effective, but the power is not transferred to the courts.

The decision in *Crouch* was followed in *Oram Builders* v. *M J Pemberton* (1985) on a JCT Agreement for Minor Building Works form.

The *Crouch* decision was the cause of some alarm if not dismay to those with actions running at the time, but its effect has now been diminished by section 100 of the Courts and Legal Services Act 1990 which enables the parties, by agreement, to confer the powers of their arbitrator on the courts.

More recently the *Crouch* decision has come under scrutiny in two cases of considerable interest to the subject of extensions of time.

In *Balfour Beatty Civil Engineering Ltd* v. *Docklands Light Railway Ltd* (1996) the Court of Appeal considered a dispute on the amount of time due to the contractor under an amended version of the ICE 5th Edition where the employer acted as the engineer and the arbitration clause was omitted. The judge at first instance had decided, following *Crouch*, that the court had no general power to open up and review the decisions and certificates of the employer and the powers of the court would be limited to special circumstances where such decisions and certificates were proved to have been not in accordance with the provisions of the contract.

The Court of Appeal held that the parties' rights and obligations were governed by the contract they had made. Under the terms of the contract, the contractor's entitlements were to be such as the employer considered due in the employer's judgment. Whilst that judgment was not expressed to be binding and conclusive there was no agreed means of challenging it. Sir Thomas Bingham, Master of the Rolls, said this:

> 'It is not for the court to decide whether the contractor made a good bargain or a bad one; it can only give effect to what the parties agreed.'

However, the decision made clear that the employer was bound to act honestly, fairly and reasonably in arriving at his judgment (even though there was no express obligation in the contract to do so) and if the contractor could prove a breach of this duty it would be entitled to a remedy in damages.

In *John Barker Construction Ltd* v. *London Portman Hotel Ltd* (1996) the court was asked to consider whether it had jurisdiction to entertain a contractor's claim in relation to extensions of time granted by the architect and, if it had such jurisdiction, what was the proper determination of the claim. The applicable conditions of contract were JCT 80 but with the arbitration clause deleted and replaced by the words 'The proper law of the agreement shall be English and the English court shall have jurisdiction'. The judge, reviewing the *Crouch* decision, said:

> 'The essential points of the *Crouch* decision, reading the judgments as a whole, appear to me to be the following:
>
> 1. The contractual machinery established by the parties provided in the first instance for determination of what was a fair and reasonable extension of time by the architect.
> 2. That agreed allocation of responsibility to the architect was subject to two safeguards:
>   (a) implicitly, an obligation on the architect to act lawfully and fairly, and
>   (b) explicitly, the power of review by an arbitrator, who was entitled to substitute his opinion for that of the architect.
> 3. If safeguard (a) failed, the court could declare the architect's decision invalid, but it could not substitute its decision for that of the architect solely because it would have reached a different decision, for that would be to usurp the role of the arbitrator.
> 4. If safeguard (b) failed because the arbitration machinery broke down, the court could substitute its own machinery to ensure enforcement of the parties' substantive rights and obligations that a fair and reasonable extension should be given.'

The judge then went on to say:

> 'In the present case it seems to me that both parties' arguments go too far. Clause 25 provided for the determination of what was a fair and reasonable extension of time by the architect. If the architect made his determination fairly and lawfully, the parties would get what they had bargained for, and I would not accept, as a matter of construction of the contract, that either party would be entitled in those circumstances to ask the court to substitute its opinion for that of the architect. By lawfully I mean acting within his power and properly directing himself as to the terms of the contract.
>
> Nor would I agree, on the other hand, with the Defendants' argument that the grounds on which a decision of the architect under clause 25 may be challenged are limited to bad faith or manifest excess of jurisdiction. I find quite unacceptable the suggestion that the parties can have intended that a decision on a matter of such potential importance should be entrusted to a third person, who was himself an agent of one party, without that person being under any obligation to act fairly. It seems to me to go without saying that the parties must have intended the

decision-maker to be under such an obligation, the imposition of which is necessary to give efficacy to the contract.'

Later on in his judgment, the judge went on to rule that the extension of time granted by the architect was fundamentally flawed in that it was based on an impressionistic assessment rather than a logical analysis of delay. Consequently the contractual machinery had broken down to the extent that the court was required to determine on the evidence what was a fair and reasonable extension of time.

For further comment on this aspect of the judgment see the comment in Chapter 12.

11.3 Completion achieved earlier than certified

Such a defence, as with extensions of time, amounts to a challenge of a certificate. There are two main lines of argument. One that the employer occupied or used the works prior to the date on the completion certificate; the other, that the works were completed and ready for hand-over but the architect or engineer was over-zealous in demanding perfection or absolute completion.

Both are matters which can be decided on findings of fact but regard will be given to the wording of the particular contract and, as mentioned in Chapter 10, whilst occupation and use does not constitute completion under JCT forms, there is a strong presumption that it does so under the wording of ICE forms.

As to the over-zealous architect or engineer the problem for contractors is that most standard forms incorporate in clauses on completion the phrase 'in the opinion of' the architect or engineer. Providing the architect or engineer is applying his professional judgment, is doing so fairly and lawfully, and is not subject to interference or pressure from the employer, then it will take good evidence and clear facts to overturn his decision.

Action against certifiers by contractors

It is not unknown for contractors who feel aggrieved by the content and consequences of certificates to question whether they have any course of action against the certifier. In the absence of a contractual link, this would, of course, be in tort. The possibility of such action was given some life by the comments of Lord Salmon in *Arenson v. Arenson* (1977) when he said:

> '. . . The architect owed a duty to his client, the building owner, arising out of the contract between them to use reasonable care in issuing his certificates. He also, however, owed a similar duty of care to the contractor arising out of their proximity: see *Hedley Byrne & Co. Ltd* v. *Heller & Partners Ltd* (1964). In *Sutcliffe* v. *Thackrah* (1974) the architect negligently certified more money was due than was in fact due, and he was

successfully sued for the damage which this had caused his client. He might, however, have negligently certified that less money was payable than was in fact due and thereby starved the contractor of money.

In a trade in which cash flow is especially important this might have caused the contractor serious damage for which the architect could have been successfully sued . . .'

In *Michael Salliss & Co. Ltd* v. *Calil* (1987) Judge Fox-Andrews also held that a supervisory officer held a duty of care to the contractor when certifying although he did appear in these comments to be indicating bias as a reason rather than contrary opinion. He said:

'. . . it is self-evident that a contractor who is party to a JCT contract looks to the architect or supervising officer to act fairly as between him and the building employer in matters such as certificates and extensions of time. Without a confident belief that that reliance will be justified, in an industry where cash flow is so important to the contractor, contracting would be a hazardous operation. If the architect unfairly promotes the building employer's interest by low certification or merely fails properly to exercise reasonable care and skill in his certification it is reasonable that the contractor should not only have the right as against the owner to have the certificate reviewed in arbitration but also should have the right to recover damages against the unfair architect . . .'

However, the Court of Appeal in *Pacific Associates* v. *Baxter* (1988) declined to follow either of the above and ruled that in considering a duty of care, it was necessary to look at all the circumstances, including the provisions of the relevant contract. On the wording in that case, and the fact that the contract had an arbitration clause for dealing with disputes on certificates, there was no duty of care. See, however, the comments in Chapter 12 on the liability of certifiers.

11.4 Certificates not valid

If it can be shown that certificates for extension of time or completion are not valid, the courts will hold, applying the *contra proferentem* rule, that there is no date from which liquidated damages can run and the employer will be left to sue for whatever general damages he can prove.

Thus in *Miller* v. *London County Council* (1934) the court held that the phrase 'to assign such other time or times for completion' invalidated an extension of time granted after completion by the engineer and the exercise of the power too late prevented the employer from recovering liquidated damages.

A certificate will not be invalid if the fault amounts to no more than incompetence by the certifier – a point borne out by *Lubenham* v. *South Pembrokeshire* (1986) discussed in Chapter 10 – but if there is disregard for the express rules of the contract or there is evidence of collusion with the employer, fraud or similar wrongdoing, the employer will not be able to rely on the certificate.

Breach of express rules

There are various challenges to validity under this heading:

(i)   certificates not given on time;
(ii)   certificates not in correct form;
(iii)   certificates given by wrong person;
(iv)   no named certifier;
(v)   decisions improperly delegated;
(vi)   certificates not given fairly;
(vii)   contractual machinery not applied.

Certificates not given on time – *certifier functus officio*

Standard forms of contract usually have some point after which the certifier becomes *functus officio*. In JCT 2005 it is the issue of the final certificate; in ICE 7th Edition it is, on any particular issue, the giving of a clause 67 decision. Clearly any certificate which is issued after this point is invalid.

See *H. Fairweather Ltd* v. *Asden Securities Ltd* (1979) for a building case and *Monmouthshire County Council* v. *Costelloe & Kemple Ltd* (1965) and *ECC Quarries Ltd* v. *Merriman Ltd* (1988) for civil engineering cases.

Certificates not given on time – specified time requirements

Failure to meet time requirements specifically stated in extension of time clauses is not a straightforward issue. In *Temloc* v. *Errill* (1987) it was ruled that the 12-week period specified in JCT 80 for the architect's review of extensions after completion was directory only as to time, and failure did not invalidate the liquidated damages provisions. Where the time requirements apply before completion – for example, 12 weeks from the contractor's application in JCT 2005 and 'forthwith' in ICE 7th Edition – it is thought even less likely that failure by the certifier would invalidate liquidated damages having regard to the obligation on certifiers to conduct final reviews. The contractor's remedy in such circumstances would appear to lie in a claim for constructive acceleration.

On balance it seems that the *functus officio* rule holds good but other contractual time requirements are less significant. This perhaps follows the line developed by Lord Denning in *Amalgamated Building Contractors Ltd* v. *Waltham Holy Cross UDC* (1952) when he declared that *Miller* v. *London County Council* (1934) turned on the very special wording of the contract and upheld an extension granted after completion.

Certificates not given in correct form

Disputes on this usually centre on whether a certificate has been given in writing and served in the specified manner. In the *Token* v. *Charlton* (1973) it was held that an architect's letter to the employer did not constitute a certificate and general advice is that certificates and notices should have clear identity as to their purpose. It is doubtful that minutes of meetings could be said to constitute 'notices in writing' in themselves but it is possible, but not recommended, that if delivered with an accompanying letter of appropriate wording they could serve as notices.

In connection with the problems posed by modern technology, it is worth noting the definition in ICE 7th Edition of communications in writing:

'Communications which under the contract are required to be "in writing" may be hand-written, type-written or printed and sent by hand, post, telex cable or facsimile or other means resulting in a permanent record.'

Certificates given by wrong person

It is fundamental that where a contract names an individual or a post-holder as the architect, engineer or contract administrator who has the duty of certifying, then that person or post-holder only can exercise that duty.

Hudson gives two old cases in support of this, *ESS* v. *Truscott* (1837) and *Lamprell* v. *Billericay Union* (1849), so the problem is by no means a new one and the local government reorganisation of 1974, which coincided with a high level of construction activity, showed how easy it is for wrong names to end up on certificates.

In *Hounslow London Borough Council* v. *Twickenham Garden Developments Ltd* (1971) one of the attacks on the architect's certificate was that it was not given by the architect. The named architect in the contract was a firm, Matthew Ryan Simpson. Certificates were issued and signed by Mr Matthews, the senior partner, and phrased throughout in terms 'I' and not 'we'. It was said that one partner is not the firm. Mr Justice Megarry, after observing that there was nothing in the contract which required the certificate to be signed, held that:

'a partner who has power to bind the firm has power to give a notice on behalf of the firm.'

Named individual

Where there is a named individual in the contract and he retires or is properly replaced then a new individual will have to be appointed by the employer. There is no need to re-appoint in the case of a designated post-holder where there is a change by succession unless, of course, the first post-holder was also named as an individual. Where there is a take-over or amalgamation of firms or a management structure re-organisation such that designated posts disappear – the chief architect, perhaps becoming deputy director of technical services – then the employer should always clarify the position with a new appointment.

Power to re-appoint

It is essential that the contract should give the employer power to re-appoint with words such as 'or other such person so appointed from time to time by the employer and notified in writing as such to the contractor' or similar. Without such a power the employer may suffer the embarrassment of being unable to change a named individual even after he had been dismissed or retired. In the event of death there would probably be an implied term permitting a new appointment to give the contract business efficacy.

For a legal view on the renomination of certifiers *Croudace Ltd* v. *London Borough of Lambeth* (1986) is instructive. Here, the chief architect of Lambeth retired and no one was appointed in his place. This, in part, led to delay in dealing with the contractor's claim. When the case went to the Court of Appeal this is how Lord Justice Balcombe summarised the findings of Judge Newey which were approved:

'(1) Croudace's application for payment for loss and expense under Conditions 11(6) and 24 was made within a reasonable time.
(2) If "the Chief Architect" could be regarded as if he were a sort of corporation sole, then in the absence of any individual bearing the title any qualified member of Lambeth's Architect's Department might act as such. If that were the correct view, then the Chief Architect by the Department failed to comply with the obligation to ascertain Croudace's loss and expense and, since he or they were acting as Lambeth's agent, Lambeth were liable for the breach of contract.
(3) If, on the other hand, the Chief Architect must be an individual, Lambeth's failure since 1 June 1983 to nominate one had prevented performance of Lambeth's contract with Croudace, since there had been no one legally entitled to ascertain or direct ascertainment. The agreement gave Lambeth the power to nominate a successor to the "Chief Architect" if he should cease to act and there must be implied a reciprocal obligation requiring them to do so. On that alternative view Lambeth were again in breach of contract.'

No named certifier

This is a common problem with small works forms of contract when the employer for reasons of economy has attempted to avoid the services of a contract administrator. It is a matter of argument how the provisions of the contract work in such circumstances when there is apparently no one to certify extensions, variation, payments, completion and the like.

On one view it is suggested that the contractor by entering into the contract with no certifier has implicitly accepted that the employer will carry out the certifier's functions. Another view is that those specific provisions of the contract relying on a certifier no longer apply. The consequences of this, if correct, would be that no extensions of time could be granted; variations would be outside the contract; and the contractor's entitlement to interim payments would be in jeopardy.

Contractors and employers who enter into contracts which require a certifier, but do not have one, do so at their peril.

Decisions improperly delegated

Some forms of contract such as ICE Conditions expressly permit the engineer to delegate some of his functions under the contract whilst making it clear that he cannot delegate his functions as the contract certifier. Other forms, such as JCT 2005, describe the powers of the architect without any reference to delegation.

The general rule is that the contract administrator cannot delegate any of his functions under the contract unless he is expressly empowered to do so and the issue of any certificate which is improperly delegated is invalid.

Nature of delegation

There is often thought to be a problem in that unofficial 'delegation' appears to be commonplace in so far as architects and engineers rely on their subordinate staff to analyse disputes, prepare reports and draft letters. This is, however, to misunderstand the nature of delegation. A power which has been delegated has been passed to another; it no longer resides with the original power holder. The process of arriving at a decision with the assistance of others is not delegation.

The point was considered in *Anglian Water Authority* v. *RDL Contracting Ltd* (1988) where an arbitrator had found that an Engineer's decision under clause 66 of ICE 5th Edition had been invalidly given. In his award, the arbitrator gave his reasons as follows:

'12. The Engineer's decision letter of the 5th September 1986 states that the matter is being handled by Bob Baxter, whose initials are given in the reference to the letter.
13. Mr. Baxter was the Project Engineer not the Engineer under the contract, and notwithstanding the fact that the letter was signed by Mr. Rouse, having been addressed on this subject, I do not consider this letter is his decision under the contract, but has been formulated by Mr. Baxter. I therefore find the decision given in his letter invalid.'

In allowing an application to appeal against the award, Judge Fox-Andrews said this:

'... In the commercial world many decisions are made by people such as Mr. Rouse, who append their signatures to letters drafted by others. It would require compelling evidence to establish in such circumstances that the decision was not that of the signatory. The facts that Mr. Baxter was the Project Engineer and had taken an active part previously in the contract had no probative value. Some, albeit limited, guidance is to be found on this aspect in *Clemence* v. *Clarke* (1880).

I find the arbitrator erred in law in holding that such decisions (if any) made in the letter of 5 September were not those of Mr. Rouse ...'

In *Clemence* v. *Clarke* (1880) a contract provided for extras to be certified by the architect. The architect employed a measuring surveyor and when certifying the extras stated 'as certified by the measuring surveyors'. The employer declined to pay on the grounds that the certificate was invalid. It was held that the architect had not abdicated his duties.

It should perhaps be added that in both the above cases the correct signature was used on the relevant document. The position would be significantly different if a subordinate purported to use the name of the proper certifier by signing 'for and behalf of' or similar since there could, in such circumstances, be genuine doubt whether the views of the proper certifier coincided with those expressed on the certificate.

Certificates invalid because of pressure or fraud

It goes without saying that fraud on the part of the certifier invalidates his certificates. As Lord Justice Denning said in *Lazarus Estates Ltd* v. *Beasley* (1956):

'Fraud unravels everything ... It vitiates judgments, contracts and all transactions whatsoever.'

But fraud in a deliberate sense is, or appears to be, fortunately rare. A far more common problem is alleged bias, interest or lack of independence because of the certifier's close relationship with the employer. Sometimes the dividing line between what is proper and improper will be fine, in other cases less so.

In *Hickman & Co.* v. *Roberts* (1913) an employer in financial difficulties instructed his architect not to issue any further interim certificates. It was held that the architect had lost his independence by allowing himself to be influenced by the employer.

In *Ranger* v. *Great Western Railway* (1854) it was held that the contractor could not object to the engineer's decisions on grounds of bias where the engineer was the employers' servant since he had been aware of the situation when he entered into the contract. Nor could he object on the fact, unbeknown to the contractor at the time of tendering, that the engineer was also a shareholder of the employing company.

Commercial interest

The problem of shareholding is a difficult one and probably comes down to a matter of scale. Modest holdings of shares in a public company would not, it is suggested, normally cast doubt on the independence of a professional acting for that company but substantial ownership in a private company would be a different proposition, whether the share ownership be in the contractor's business or the employer's business.

Various property booms in the 1980s and 1990s did give rise to some unorthodox situations where professional practices and individuals were closely involved as entrepreneurs in development projects and at the same time were acting as architects or engineers under the development contracts.

Paid officials

A more common situation is that of the paid official. The difficulties of his independence have long been recognised. He has to act as agent of the employer in designing and controlling the project, including administration of the contract but he has to act independently as certifier under the contract. He is a man with two hats with the difficulty that they must both be worn at once.

The relationship of paid officials to their employers was considered at length in the Australian case of *Perini Corporation* v. *Commonwealth of Australia* (1969), particularly in regard to whose interests the engineer should consider in granting extensions of time. A succinct summary of the long and complex judgment of Mr Justice Macfarlan is given by the editors of *Building Law Reports* at page 82 of 12 BLR:

'(1) The Director of Works was a certifier under the contract and as such had certain duties imposed on him by the contract.
(2) The Director of Works had a discretion as to whether or not he would grant an extension of time.
(3) The Director of Works could rely on other persons to supply him with the information on which he would exercise his discretion.
(4) The Director was bound to give his decision on any application for extension of time within a reasonable time which in the circumstances plainly meant that he should give a decision as soon as his investigation into the facts was completed and was not entitled to defer his decision.
(5) In making his decision, the Director was entitled to consider departmental policy but would be acting wrongfully if he were to consider himself as controlled by departmental policy.
(6) There was an implied term in the contract that the Commonwealth would not interfere with the Director of Works' duties as certifier.

(7) There was also an implied term in the contract that the Common-wealth would ensure that the Director of Works did his duty as certifier.'

Certificates not given fairly

It is well settled that the validity of a certificate which is not given fairly is open to challenge. See, for example, the various legal references and extracts from judgments given under the heading 'Review in litigation' in Section 11.2 above.

The effect of a successful challenge could be not only to deprive the employer of any rights to liquidated damages which might other-wise be due but also to leave the employer liable to the contractor for damages.

Contractual machinery not applied

Complaints from contractors that certifiers have either been dilatory in dealing with extensions of time or casual in assessing the amount of time properly due are commonplace. Many certifiers might well admit to taking a broad brush approach to their task – a practice which some years ago was commonly considered acceptable.

However, attitudes have changed and detailed delay analysis is now a skill which anyone undertaking the assessment of extensions of time will have to apply. The legal consequences of failure to do so are evident from the case of *John Barker Construction Ltd* v. *London Portman Hotel Ltd* (1996) discussed under the heading 'Duty of certifiers' in Chapter 12.

It is clear that if it can be shown that the certifier has failed to properly operate the machinery of the contract, by, for example, making an impres-sionistic assessment of delay instead of a logical analysis, then the validity of the certificate can be challenged.

11.5 Conditions precedent not observed

The effectiveness of challenging liquidated damages by claiming non-observance of conditions precedent to their deduction depends exclusively on the wording of the particular contract.

In *Aoki Corporation* v. *Lippoland (Singapore) Pte Ltd* (1994) there was a con-tractual requirement that the architect should reply within one month of an application for extension giving a decision in principle on the contractor's right to an extension. The court decided that the architect's failure to do so did not affect the employer's right to deduct liquidated damages but sug-gested that the contractor could claim damages arising from the failure which could include costs of accelerated working.

If there are express requirements making the consideration of extensions of time, the issuing of certificates of completion or non-completion, and the giving of notices of intention to deduct, conditions precedent, they may be effective, as shown in a previous chapter. But this is not an area where the courts are likely to imply terms. It might seem inequitable that liquidated damages should be deducted before extensions of time have been considered but this is not prohibited by some standard forms and the courts will not improve the contract for the parties.

Challenges on conditions precedent will normally arise after deduction of liquidated damages have been made and there is a strong probability that an action to recover damages would fail if the default in procedure was rectified before the action came to trial. Moreover, if there is an arguable alternative defence for withholding damages that might succeed.

11.6 No date for commencement

Most arguments on time relate to the date for completion with the dispute between the parties being whether sufficient extensions of time been granted. It is easy enough therefore to overlook the point that to have a date for completion from a specified time it is first necessary to have a date for commencement.

In *Kemp* v. *Rose* (1858) where the date for commencement was omitted from a written contract, the court declined in the face of conflicting oral evidence to set a date.

Similar circumstances still occur with surprising regularity with modern standard forms. With ICE Conditions, where the date for commencement is left to be set by the engineer 'a reasonable time after the date of acceptance of tender', it is usual for the contractor and engineer to agree at a pre-start meeting what the date for commencement should be and it is not uncommon for written confirmation of the date to be overlooked.

JCT contracts provide in the appendix for both the date for possession and the date for completion to be stated but at tender stage these are often left blank since the date for possession has not been decided and the contractor's own proposals on time for completion are to be considered. It only needs a slip then in administrative procedure in documentation to set the contract running with no formal date for commencement and possibly none for completion.

Letters of intent

One particular cause of confusion are letters of intent which in various ways lead to work starting on site before a formal contract date has been established or could legally apply. It is not unknown for the contract works to be completed on a succession of such letters. The problem for the employer is how to retain the time requirements originally envisaged and the liquidated

damages provisions when part of the work has been completed before the
date for commencement.

11.7 Prevention

The principle of prevention has been discussed at length in other chapters.
In short it acts as a defence to liquidated damages where there is no contractual provision to extend time for any act or breach by the employer
which has impeded the contractor in his obligation to finish within time.
    Prevention defences fall into two categories:

(i) where the contract has no provision to extend time;
(ii) where there are contractual provisions but they do not cover the par-
     ticular alleged prevention.

The following is a list of some of the better known cases:

(i)   *Holme* v. *Guppy* (1838) – delay in possession and by employer's
        workforce;
(ii)   *Dodd* v. *Churton* (1897) – additional works;
(iii)   *Wells* v. *Army & Navy Co-operative Society* (1902) – delay in possession
        and late information;
(iv)   *Miller* v. *London County Council* (1934) – extras and interference by
        employer's contractors;
(v)   *Neodox Ltd* v. *Borough of Swinton & Pendlebury* (1958) – late supply of
        information;
(vi)   *Perini Pacific Ltd* v. *Greater Vancouver Sewerage* (1966) – employer's
        supply of defective machinery;
(vii)   *Peak Construction (Liverpool) Ltd* v. *McKinney Foundations Ltd* (1970) –
        employer's delay in re-nomination;
(viii) *Percy Bilton Ltd* v. *Greater London Council* (1982) – employer's delay in
        re-nomination;
(ix)   *Rapid Building Group Ltd* v. *Ealing Family Housing Association* (1984) –
        late possession of site;
(x)   *SMK Cabinets* v. *Hili Modern Electrics Pty* (1984) – additional works and
        employer's contractors;
(xi)   *McAlpine Humberoak Ltd* v. *McDermott International Inc.* (1992) – instruc-
        tions given after the completion date;
(xii)   *Balfour Beatty Building Ltd* v. *Chestermount Properties Ltd* (1993) – instruc-
        tions given after the completion date;
(xiii) *Multiplex Constructions (UK) Ltd* v. *Honeywell Control Systems Ltd* (2007)
        – useful summary of the law on prevention.

11.8 Penalties

Liquidated damages held by the courts to be penalties will not be
enforced.

Penalties, as shown in Chapter 4, are not confined to sums which are patently stipulated *in terrorem*. Lord Dunedin in *Dunlop Pneumatic Tyre* v. *New Garage* (1915) stated two propositions:

(i) the essence of a penalty is a payment of money stipulated as *in terrorem* of the offending party;
(ii) the essence of liquidated damages is a genuine covenanted pre-estimate of damage.

Lord Dunedin went on to suggest three tests for penalties:

(i) extravagant in comparison with loss;
(ii) a sum greater than an amount due;
(iii) a single sum covering several events.

The courts have tended to describe as penalties any stipulated sum which is either a penalty in the light of the above tests or can otherwise be shown not to meet the rule for liquidated damages as a genuine pre-estimate of damage. In both cases the effect is the same and the employer can recover only the loss he can prove and probably only up to the amount of the stipulated sum.

Most of the leading cases on challenges to liquidated damages as penalties have been covered in detail in other chapters, but for summary purposes some of the best known are listed here:

(i) *Kemble* v. *Farren* (1829) – non-payment of a single sum;
(ii) *Ranger* v. *The Great Western Railway Co.* (1854) – retention of all moneys due to contractor;
(iii) *Clydebank Engineering Co. Ltd* v. *Yzquierdo y Castaneda* (1905) – pre-estimate of damage;
(iv) *Public Works Commissioner* v. *Hills* (1906) – retention money as liquidated damages;
(v) *Dunlop Pneumatic Tyre Co. Ltd* v. *New Garage & Motor Co. Ltd* (1915) – rules for penalties;
(vi) *Ford Motor Company (England) Ltd* v. *Armstrong* (1915) – single sum for different breaches;
(vii) *Widnes Foundry (1925) Ltd* v. *Cellulose Acetate Silk Co. Ltd* (1933) – under-liquidation;
(viii) *Stanor Electric Ltd* v. *R. Mansell Ltd* (1987) – single sum for two houses;
(ix) *BFI Group Ltd* v. *DCB Integration Systems Ltd* (1987) – loss need not be suffered;
(x) *Arnhold & Co. Ltd* v. *Attorney General of Hong Kong* (1989) – stop figure on proportioning down;
(xi) *Multiplex Constructions Pty Ltd* v. *Abgarus Pty Ltd* (1992) – review of the law on liquidated damages and penalties and rejection of various challenges;
(xii) *Philips Hong Kong Ltd* v. *Attorney General of Hong Kong* (1993) – rejection of challenge based on hypothetical calculations and restatement of general case for upholding liquidated damages;

(xiii) *Finnegan (JF) Ltd* v. *Community Housing Association Ltd* (1993) – rejection of challenges relating to the use of a formula for pre-estimation of loss and the effects of funding on the employer's actual loss;
(xiv) *Lordsvale Finance Plc* v. *Bank of Zambia* (1996) – interest rate for default in syndicated loan contract not a penalty;
(xv)   *Alfred McAlpine Capital Projects Ltd* v. *Tilebox Ltd* (2005) – stipulated sums for late completion not a penalty;
(xvi) *Murray* v. *Leisureplay Plc* (2005) – provisions for payment on termination of an employment contract not a penalty.

Damages per week or part thereof

It is arguable that damages expressed as 'per week or part thereof' could fall foul of the penalty rules on the basis that loss for one day must be less than loss for seven days. This is not necessarily true but the trap should be avoided by stating damages as per week or per day.

Where damages are expressed per week it is probable that they apply only to each full week late and cannot be apportioned on a daily basis.

11.9 Provisions void for uncertainty

By application of the *contra proferentem* rule, provisions for liquidated damages which can be shown to be uncertain or inconsistent will be held to be unenforceable. Consider *Bramall & Ogden* v. *Sheffield City Council* (1983) and *Arnhold* v. *Attorney-General of Hong Kong* (1989), as discussed in Chapters 4 and 6.

Where the figure for liquidated damages is not clearly stated in the appendix, the employer will not be able to correct this after the contract is made. Nor will the employer be able to correct the omission of any sum in the appendix except by agreement with the contractor before the breach occurs.

11.10 Waiver / estoppel

Contractors frequently find that liquidated damages are enforced notwithstanding some assurance or understanding given to the contrary. The question then is, how good are the defences of waiver or estoppel?

Waiver

The law on waiver is exceedingly complex and the terminology confused enough to strain the best lawyers. In *Tool Metal Manufacturing Co. Ltd* v. *Tungsten Electric Co. Ltd* (1955) Lord Justice Denning said:

'... If the defendant, as he did, led the plaintiffs to believe that he would not insist on the stipulation as to time and that if they carried out the work he would accept it, and they did it, he could not afterwards set up the stipulation as to time against them. Whether it be called waiver or forbearance on his part, or an agreed variation or substituted performance, does not matter. It is a kind of estoppel. By his conduct he evinced an intention to affect their legal relations. He made, in effect, a promise not to insist on his strict legal rights. That promise was intended to be acted on, and was in fact acted on. He cannot afterwards go back on it ...'

Many years earlier in *Birmingham and District Land Co.* v. *London and North Western Rail Co.* (1888) Lord Justice Bowen had given this view on the doctrine of waiver:

'... If persons who have contractual rights against others induce by their conduct those against whom they have such rights to believe that such rights will either not be enforced or will be kept in suspense or abeyance for some particular time, those persons will not be allowed by a court of equity to enforce the rights until such time has elapsed, without at all events placing the parties in the same position as they were in before ...'

Waiver, therefore, occurs when one party expressly or implicitly, indicates to the other his intention to forgo certain rights under a contract and it is effective in law when the other party changes their position in reliance on the waiver.

Estoppel

Estoppel is a rule of evidence which acts as a defence in preventing one party alleging facts necessary to a claim where he has previously by his conduct represented the contrary.

Applied to liquidated damages it would seem to amount to this. If an employer assured a contractor he did not intend to deduct damages for late completion and the contractor finished at his own pace instead of accelerating to avoid damages, the employer could be estopped from suing for liquidated damages and the contractor could rely on the doctrine of waiver to recover damages deducted against the assurance given.

Variation

Clearly, if there is any consideration given for the waiver, such as a promise by the contractor not to press claims for loss and expense or extra cost, then that amounts to a variation of the terms of the contract and is legally binding. Where there is a waiver without consideration it may be withdrawn by reasonable notice as in *Rickards (Charles)* v. *Oppenheim* (1950) mentioned in Chapter 2.

Contract administrator no power to vary terms

In practice in the construction industry true waiver is unusual if only because direct communication between the contractor and employer is itself fairly unusual. For most matters the contractor deals with the contract architect or engineer whose capacity to vary the terms of the contract, either officially or by waiver, is restricted to his powers under the contract. If the architect or engineer did purport to act as agent of the employer with wider powers, including those to waive the terms of the contract, that would be a different matter.

   Consequently comments by architects or engineers at site meetings or the like of the kind 'don't worry about damages', would be of little value to the contractor in a defence against damages.

   To avoid any possibility of interfering with contractual requirements and procedures and to avoid the charge of waiver being raised, it is the policy of some employers that they do not attend meetings with or deal directly with the contractor. This is sound policy because it is very easy at a site meeting if the employer is present for him to be drawn into concessions on contractual performance which he may later regret.

11.11 Unfair Contract Terms Act 1977

The question is sometimes asked, does the Unfair Contract Terms Act 1977 apply to liquidated damages? Usually the question comes from someone who thinks that paying liquidated damages is unfair. Ironically, however, if the Act does apply, unfairness under the Act relates to the limitation effect of liquidated damages. And if the Act offers protection to anyone in this matter, which is doubted, such protection might well be in favour of the recipient and not the payer of liquidated damages.

   The Act was introduced to limit the use of exclusion or exemption clauses. It commences:

> 'An Act to impose further limits on the extent to which under the law of England and Wales and Northern Ireland civil liability for breach of contract, or for negligence or for other breach of duty, can be avoided by means of contract terms and otherwise and under the law of Scotland civil liability can be avoided by means of contract terms.'

In broad terms, the Act governs contract terms which:

(i)   exclude or restrict liability for negligence;
(ii) exclude or restrict liability for breach of contract;
(iii) permit different contractual performance from that expected or permit no performance at all;

    (iv) require indemnities against the other party's negligence or breach of
          contract;
    (v) exclude liability for breach of terms implied into contracts by the Sale
          of Goods Act and the Supply of Goods and Services Act;
    (vi) exclude liability in respect of misrepresentation.

Negligence liability under the Act applies only to liability arising from business activities.
    The Act governs liability arising in contract when one party either:

    (i) deals as a consumer; or
    (ii) deals on the other's written standard terms of business.

Test of reasonableness

In relation to a contract term there is a test of reasonableness to the effect that a term:

> 'shall have been a fair and reasonable one to be included having regard to the circumstances which were, or might reasonably have been, known to or in the contemplation of the parties when the contract was made.'

Application to liquidated damages

To bring liquidated damages in a construction contract within the scope of the Act it would be necessary to show:

    (i)   an intention to exclude or restrict liability for breach of contract;
    (ii) that the party whose rights were excluded or restricted dealt on the
           other's written standard terms of business;
    (iii) that the exclusion or restriction terms failed the test of
           reasonableness.

The first requirement above is, perhaps, met. Liquidated damages are intended to restrict liability for breach of contract in the sense that they exclude general damages. The third requirement would be difficult to meet since if the liquidated damages are a genuine pre-estimate of loss, or a lesser sum, they must evidently meet the reasonableness test. But even if it was possible to overcome this, the second requirement on dealing on the other party's standard terms would present a difficulty.

Standard terms

Firstly, there is the general question of whether or not standard conditions of contract could be said to represent one party's terms. As *Emden's Construction Law* issue 22 says at III 176:

'There is, however, considerable debate as to whether the standard forms of building contract (such as JCT 80) fall within the Unfair Contract Terms Act 1977. The principal point of controversy is whether such contracts can accurately be said to represent one party's written standard terms of business (i.e. in normal cases, those of the contractor) when the standard forms go to substantial lengths to define and balance the interests of both parties, and have been evolved by long processes of consultation. The question is a difficult one, but two points can perhaps be usefully made. The first is that there is nothing explicit in the Act to prevent a set of terms from being treated as the standard terms of one party merely because they might also be described as the standard terms of the other party. The second, and more serious, point is that the mere generosity or evenhand-edness of a set of terms should not, by itself, disqualify them from clas-sification as the standard terms of a particular party; such matters are better adjudged according to the statutory reasonableness test than by pre-empting that test outright. If the relevant standard form is one which the party in question habitually employs in his commercial relations, there would seem to be no decisive difficulty in treating that form as embodying his written standard terms of business, however sympathetic and responsive those terms may be to the legitimate commercial interests of the other contracting party.'

Other party's terms

Then there remains the final hurdle for any party seeking to apply the Act to liquidated damages in a construction contract – the need to show that they were dealing on the other party's terms. That is to say, an employer claiming that his rights were restricted by liquidated damages would have to show that he was dealing on the contractor's terms – an unusual reversal of normal contractual practice but not impossible in specialist or small works contracts.

Application to construction contracts

The Unfair Contract Terms Act has not had a great deal of application to construction contracts but in *Rees Hough Ltd* v. *Redland Reinforced Plastics Ltd* (1984) the court did consider the 'reasonableness test' in connection with the suppliers' terms of sale for special concrete pipes for pipe-jacking.

Redland's terms, as is not unusual, excluded all liability in respect of loss or damage suffered by the customer as a result of any defect in the goods or lack of fitness for their purpose. During construction various defects occurred in the pipes and the pipe-jacking was abandoned in favour of segmental tunnelling. In an action to recover damages, Judge Newey had to consider, amongst other things, whether Redland's terms were reasonable under the Unfair Contract Terms Act. These are extracts from his judgment:

'Considerations which in my view support Redland's contention that their standard conditions are reasonable are: that the contract was between two companies and, while Redland were the larger, RH were capable of looking after themselves; all or most other concrete pipe manufacturers contract on the basis of standard terms; the PJA [Pipe-Jacking Association] did not protest at such terms, although they had with regard to statements of limitation on loading; RH had long been aware that Redland had standard terms; the terms would have been understandable by any intelligent businessman; and RH never attempted to negotiate alterations in the terms.'

'Considerations which in my opinion are against Redland's contention that the terms are reasonable are that: RH were regular customers of Redland and major purchasers of their jacking pipes; when in the past pipes had been defective, Redland had not relied on the terms but had paid compensation to RH; Redland did not refer to the terms during negotiation of the contract; the sums to be paid by RH for the pipes were substantial; Redland was likely to gain from the development of a new pipe, which after successful use could be marketed generally; the remedies of repair and replacement provided by the terms were if defects were liable to result in pipe jacking having to be abandoned when a pipeline was incomplete; the terms had not been negotiated between the PJA and the CPA [Concrete Pipe Association] or any other trade association; and Redland could, as it had before 1968, have maintained product liability insurance.'

'Doing the best I can, I reach the conclusion that Redland has failed to prove that their standard terms were reasonable; indeed I think that the balance of the considerations is strongly against the terms being reasonable.'

'I therefore, hold that Redland cannot rely upon the standard terms of sale to invalidate the express and implied terms between themselves and RH.'

Chapter 12
Extensions of time

12.1 *Purposes of extension provisions*

A contractor is under a strict duty to complete on time except to the extent that he is prevented from doing so by the employer or is given relief by the express provisions of the contract. The effect of extending time is to maintain the contractor's obligation to complete within a defined time and failure by the contractor to do so leaves him liable to damages, either liquidated or general, according to the terms of the contract. In the absence of extension provisions, time is put at large by prevention and the contractor's obligation is to complete within a reasonable time. The contractor's liability can then only be for general damages but first it must be proved that he has failed to complete within a reasonable time.

Extension of time clauses, therefore, have various purposes:

(i)   to retain a defined time for completion;
(ii) to preserve the employer's right to liquidated damages against acts of prevention;
(iii) to give the contractor relief from his strict duty to complete on time in respect of delays caused by designated neutral events.

It is a common belief in the construction industry that extensions of time are solely for the benefit of the contractor. At face value by giving the contractor more time to complete the works and by reducing his liability for liquidated damages they do appear to be one-sided. This view is reinforced by the drafting of extension clauses which require the contractor to apply for and to substantiate his case and by contractor's traditional linkage of extensions of time with claims for loss and expense or extra cost.

There seems to be little in all this for the employer.

Overcoming prevention

As shown in earlier chapters, it is not the contractor who has most need of extension of time provisions, it is the employer. A string of well-documented cases from *Holme* v. *Guppy* (1838) to *Rapid Building* v. *Ealing* (1984) confirm that the courts will not uphold liquidated damages where the employer has prevented completion on time unless there is express provision in the contract to extend time for the employer's default. Lord Fraser's comment in *Percy Bilton Ltd* v. *Greater London Council* (1982) sums it up:

'. . . The general rule is that the main contractor is bound to complete the work by the date for completion stated in the contract. If he fails to do so, he will be liable for liquidated damages to the employer. That is subject to the exception that the employer is not entitled to liquidated damages if by his acts or omissions he has prevented the main contractor from completing his work by the completion date: see, for example, *Holme* v. *Guppy* (1838) and *Wells* v. *Army and Navy Co-operative Society* (1902). These general rules may be amended by the express terms of the contract . . .'

Relief to contractor

If overcoming prevention was the sole purpose of extension clauses they would deal with such matters as: late possession of the site; late supply of drawings and information; interference by the employer's workmen; variations and extras. There would be no need to include for: neutral events which are the fault of neither party; force majeure; war or riots; nor for events which inherently are contractors' risk, such as weather and strikes. By these additional events the contractor is given relief from his otherwise strict duty to complete on time and in this respect extension of time clauses do operate to the contractor's benefit.

The extent of the benefit varies considerably from form to form. With standard forms the need for consensus between the parties in the drafting process may allow the introduction of events which go well beyond what most would regard as neutral in ordinary circumstances. Thus, in some contracts the contractor can seek an extension for inability to obtain labour or materials for reasons beyond his control; although many employers regard such clauses as going too far and delete them. Where the form is employer drafted it is a matter of commercial judgment on the allocation of risk. The employer has no need to include provisions for exceptional adverse weather nor any other neutral events but if he omits such grounds he has to consider how much the contractor might add to his tender price to cover the greater risk of incurring liquidated damages.

12.2 Notices, applications and assessments

Extension of time clauses of standard forms may at first reading give the impression that an application by the contractor is a condition precedent to any extension being granted. This comes from the use of phrases of the type 'the contractor shall forthwith give written notice' and 'the engineer shall upon written request'. Closer examination of the clauses will often reveal that the contractor's obligation is to give notice only of delay and that there is a defined duty on the contract administrator to consider the contractor's entitlement to an extension whether or not an application has been made.

The facility for contract administrators to grant extensions if they are due is essential to maintain the extension provisions. If an application by the contractor was a condition precedent to an extension, it is arguable that it would be at the option of the contractor whether or not the provisions were effective, and by choosing not to apply for an extension to cover acts of prevention the contractor could render the liquidated damages provisions inoperative. See, for example, the comments of Mr Justice Jackson in *Multiplex* v. *Honeywell* (2007) quoted in Chapter 5 above.

Duty of certifiers

It follows that the contract administrator cannot merely take a passive role in the extension process; he must consider his duty to the employer. In *Holland Hannan & Cubitts (Northern) Ltd* v. *Welsh Health Technical Services Organisation* (1981) design defects in windows supplied by a nominated sub-contractor led to delays but the architect, in the mistaken belief that the problem lay between the main contractor and the sub-contractor, declined to issue a variation order or an extension of time. The architect was joined in the action which followed and sued by the employer for negligence and breach of contract. This is what Judge Newey had to say:

> '... PTP's [Percy Thomas Partnership] failure to issue a Variation Instruction when defects in design had become apparent, when they had come to believe that Crittall's remedies were overcoming the defects and when they had no alternative proposals of their own, may possibly be excused on the grounds that they were labouring under a mistake of law as to Cubitt's responsibilities. However, I find it impossible to believe that architects in charge of a great building project, which has been brought to a stop by an unexpected difficulty, are entitled to adopt a passive attitude, as PTP did in this case. PTP's failures were ones of omission rather than of commission, but I think that they nonetheless amounted to breach of contract.
>
> The same conclusion as I have reached in regard to the issue of a variation instruction applies, I think, to the grant to Cubitts of an extension of time ...'

Failure by a certifier to act fairly can lead to invalidation of his certificates. See the comments in Chapter 11. But even where the certifier acts fairly that will not be enough to sustain his decisions if it can be shown that he has failed to apply the machinery of the contract to his decision making process.

This point comes out very strongly in the case of *John Barker Construction Ltd* v. *London Portman Hotel Ltd* (1996) where the judge held that the effect of the architect making an impressionistic assessment instead of a logical analysis of delay rendered his extension of time fundamentally flawed. The judge said:

'I accept that [the architect] believed, and believes, that he made a fair assessment of the extension of time due to the Plaintiffs. It is fairly apparent that the Defendants were concerned by the overrun of the contract in time and costs, and I have no doubt that [the architect] was conscious of this, but I believe also that he endeavoured to exercise his judgment independently. However, in my judgment his assessment of the extension of time due to the Plaintiffs was fundamentally flawed in a number of respects, namely:

(1) [The architect] did not carry out a logical analysis in a methodical way of the impact which the relevant matters had or were likely to have on the Plaintiffs' planned programme.
(2) He made an impressionistic, rather than a calculated, assessment of the time which he thought was reasonable for the various items individually and overall. (The Defendants themselves were aware of the nature of [the architect's] assessment, but decided against seeking to have any more detailed analysis of the Plaintiffs' claim carried out unless and until there was litigation.)
(3) [The architect] misapplied the contractual provisions, as more particularly set out above. Because of his unfamiliarity with SMM7 he did not pay sufficient attention to the content of the bills, which was vital in the case of a JCT contract with quantities.
(4) Where [the architect] allowed time for relevant events, the allowance which he made in important instances (such as the items relating to the walls or the cutting of pockets in the bathroom screeds) bore no logical or reasonable relation to the delay caused.

I recognise that the assessment of a fair and reasonable extension involves an exercise of judgment, but that judgment must be fairly and rationally based.

All in all, I am satisfied that the Plaintiffs have established that, although there was no bad faith or excess of jurisdiction on the part of the architect, his determination of the extension of time due to the Plaintiffs was not a fair determination, nor was it based on a proper application of the provisions of the contract, and it was accordingly invalid.'

This case provides a clear warning to everyone whose duty it is to assess and certify extensions of time that anything less than a thorough delay analysis may invalidate the certificate. The consequences of that could well be that the certifier could find himself in breach of his professional contract and liable to the employer for breach of duty.

Also of interest on the duty of certifiers to act impartially is the unusual case of *Costain Ltd & Others* v. *Bechtel Ltd* (2005). Costain was part of a consortium of contractors, known as Corber, engaged to carry out part of the Channel Tunnel Rail Link project. Bechtel was part of a consortium engaged to act as project manager. Costain was concerned that Bechtel was deliberately adopting a policy of administering the contract in an unfair and adverse manner. Costain sought interim injunctions restraining the project manager's conduct. A key issue in the case was whether in assessing sums payable to the contractor, the project manager was under a duty to act impartially between the employer and the contractor or merely to act in the interests of the employer. Costain relied on the principles established by the House of Lords in *Sutcliffe* v. *Thackrah* (1974) that a certifier has a duty to act fairly and impartially. Bechtel argued:

- the terms of the contract were specific and detailed and that they conferred no discretion on the project manager – there was, therefore, no need to imply any term on impartiality
- the decisions of the project manager could be challenged in adjudication – thereby excluding the need for an implied term on impartiality
- the position of the project manager under the contract was analogous to that of the project manager in *Royal Brompton Hospital NHS Trust* v. *Hammond & Others* (2002) where the project manager had been specifically employed to look after the interests of the employer
- there were terms in the contract (these were additional conditions) which excluded terms implied by custom.

The judge, whilst declining to grant the interim injunctions sought by Costain, expressed the views that:

- the principles of *Sutcliffe* v. *Thackrah* did apply to the contract
- the provisions for adjudication did not affect any duty to act fairly and impartially
- the project manager's position under the contract was not analogous to that in the *Royal Brompton* case
- the additional conditions excluding terms implied by custom had no impact since the implied duty of a certifier to act fairly and impartially was a matter of law not custom.

Is the contractor's application a condition precedent?

The direct question of whether a contractor's application was a condition precedent for granting an extension was one of the many issues considered in *London Borough of Merton* v. *Stanley Hugh Leach Ltd* (1985). The issue was put as follows:

'Upon the true construction of clause 23 is the contractor entitled to an extension of time in respect of any cause of delay falling within sub-clauses (a) to (k) if he fails to give written notice thereof forthwith upon it becoming reasonably apparent that the progress of the works is delayed?'

Mr Justice Vinelott held that giving notice was not a condition precedent. In the course of his judgment, he said:

'. . . The case for *Merton* is that the architect is under no duty to consider or form an opinion on the question whether completion of the works is likely to have been or has been delayed for any of the reasons set out in clause 23 unless and until the contractor has given notice of the cause of a delay that has become "reasonably apparent" or, as it has been put in argument, that the giving of notice by the contractor is a condition precedent which must be satisfied before there is any duty on the part of the architect to consider and form an opinion on these matters. The arbitrator's answer to this question was that "a written notice from the contractor is not a condition precedent to the granting of an extension of time under clause 23 . . ."

I think the answer to *Merton's* contention is to be found in a comparison of the circumstances in which a contractor is required to give notice on the one hand and the circumstances in which the architect is required to form an opinion on the other hand. The first part of clause 23 looks to a situation in which it is apparent to the contractor that the progress of the works is delayed, that is, to an event known to the contractor which has resulted or will inevitably result in delay. The second part looks to a situation in which the architect has formed an opinion that completion is likely to be, or has been delayed beyond the date for completion. It is possible that the architect might know of events (in particular "delay on the part of artists, tradesmen or others engaged by the employer in executing work not forming part of this contract") which is likely to cause delay in completion but which has not caused an actual or prospective delay in the progress of the work which is apparent to the contractor. If the architect is of the opinion that because of an event falling within sub-paragraphs (a) to (k) progress of the work is likely to be delayed beyond the original or any substituted completion date he must estimate the delay and make an appropriate extension to the date for completion. He owes that duty not only to the contractor but also to the building owner . . .'

Mr Justice Vinelott went on to say that failure by the contractor to give notice of delay was itself breach of contract and this could have some effect on his right to an extension of time. Drawing from views expressed in *Keating* he made the point that if the architect, because of failure by the contractor to give notice of delay, was unable to avoid or reduce a delay to completion, the contractor should have no greater extension than if he had given notice.

The legal principle here is that no one should benefit from his own breach of contract but it is closely allied to the duty to mitigate. Both JCT and ICE forms encompass these issues by reference to 'fair and reasonable' tests and JCT goes further by expressly requiring the contractor to use his best endeavours to prevent delay.

Meaning of delay

As to what is meant by 'delay' in an extension of time clause, there are differing views. Much depends upon the particular words used. The natural meaning would be delay to completion if the clause is concerned only with extensions of time. It is difficult to see what purpose would be served by requiring the contractor to give notice of delays to non-critical activities or to give notice when adequate float time remained to permit completion within the time allowed. If there is a wider meaning than delay to completion and the contractor fails to give notice of subordinate delays it is unlikely that the employer would suffer any damage as a result of the breach. The contractor alone might suffer by some diminution of loss and expense or extra cost payments in appropriate circumstances.

However, in some forms it is not clear whether delay means 'delay in completion' or simply 'delay in progress' and whether the contractor is entitled to an extension of time for any delay whether or not completion is likely to be achieved before or after the due date. Recent versions of ICE Conditions resolved the problem by requiring the engineer to deal separately with delays and extensions.

Application for neutral events

It is possible that the contractor could disadvantage himself by failing to give notice of delay or making application for extension of time in respect of neutral events. Indeed there is a school of thought that says that whilst notices and applications may not be conditions precedent for acts of prevention, express requirements for notices and applications are conditions precedent for neutral events.

On legal principles there is some logic in this, since it is clearly up to the contractor to make his case and to obtain whatever benefit he can from the contractual provisions. In practice, however, there is a complication. Contractors much prefer extensions for reimbursable rather than neutral events and will look for the possibility of gaining the first at the loss of the second. So it is not uncommon for a contractor to play down the delaying effects of his own problems and of neutral events, particularly adverse weather, and to attribute all delay to acts of prevention.

In such circumstances contract administrators may well feel they have a duty to the employer to consider all the facts known to them and to grant extensions for neutral events even when no application has been made. This is not strictly necessary because if the true cause of delay is a neutral event and the contractor has forfeited his right by non-application, that should rightly be the end of the matter. The contractor's rights of claim for loss and expense or extra cost if he has any are not affected by this. It would be odd if the contract placed a duty on the contract administrator to rectify the omissions of the contractor except where by doing so he was serving his duty to the employer.

12.3 *Time for granting extensions*

There are two issues to consider here. Firstly, that extensions of time might be granted too late to be effective in keeping liquidated damages provisions alive and secondly, that extensions might be granted too late to allow the contractor to re-programme his work. In the first case it is the employer who is the loser; in the second case the contractor, except to the extent that he might be successful in recovering his acceleration costs from the employer.

The case of *Miller* v. *London County Council* (1934) has already been mentioned in Chapter 11. In that case it was held that the phrase 'to assign such other time for completion' contemplated exercise of the power within a reasonable time of the delay and a retrospective extension came too late to be effective.

In *Amalgamated Building Contractors* v. *Waltham* (1952), Lord Justice Denning declined to follow *Miller* which he said turned on its particular wording and he had this to say on the contractor's argument that the architect must give a completion date at which they could aim in the future:

> '. . . I do not agree with this contention. It is only necessary to take a few practical illustrations to see that the architect as a matter of business, must be able to give an extension even though it is retrospective – in such a case, seeing that the cause of the delay operates until the last moment, when the works are completed, it must follow that the architect can give a certificate after they are completed . . .'

Commenting on this case which concerned extensions of the neutral type relating to difficulties in the supply of labour and materials, *Keating* suggests that the power to extend time retrospectively would not apply to delay caused by the employer unless very clear words are used.

Standard forms such as JCT 2007 and ICE 7th Edition do expressly provide for a review of extensions after completion although both have time limits – 12 weeks after the date of practical completion in JCT 2005 and 28 days after the issue of the certificate of substantial completion in ICE 7th Edition. But in any event, following the decision in *Temloc* v. *Errill* (1987), mentioned in Chapter 11, that failure by the architect to observe the 12 week requirement did not invalidate liquidated damages, these time limits are perhaps to be taken as directory only.

Contractor's opportunity to re-programme

The question of whether failure to grant an extension in time for the contractor to re-programme is a breach, and the related question of what is the consequence if it is a breach, again must hinge in part on whether the cause of delay is employer's fault or a neutral event. Mr Justice Roper in *Fernbrook Trading Co. Ltd* v. *Taggart* (1979) put the matter this way:

> '. . . I think it must be implicit in the normal extension clause that the contractor is to be informed of his new completion date as soon as reasonably practicable. If the sole cause is the ordering of extra work then in the normal course extensions should be given at the time of ordering, so that the contractor has a target for which to aim. Where the cause of delay lies beyond the employer, and particularly where its duration is uncertain,

then the extension order may be delayed, although even then it would be a reasonable inference to draw from the ordinary extension clause that the extension should be given a reasonable time after the factors which will govern the exercise of the [architect's] discretion have been established. Where there are multiple causes of delay, there may be no alternative but to leave the final decision until just before the issue of the final certificate . . .'

In *Perini Corporation v. Commonwealth of Australia* (1969) Mr Justice Macfarlan devoted a major part of his lengthy judgment to this issue, but in finding that the certifier was bound to give his decision in a reasonable time he did not appear to distinguish between employer's fault and neutral events. The three following extracts from the judgment show his line of thought.

'It was also submitted that when an application was made the Director of Works was obliged to give his decision promptly. It was argued that he did not have any discretion as to whether he should grant or refuse an application; he should it was said, once he found the facts of cause and delay in the erection of the buildings, automatically grant an extension.

The argument continued that it was quite irrelevant, and indeed wrong, for the Director to distinguish between rain which was normal and therefore, as he said, to be expected, and rain which was abnormal; to refuse an extension because he was of the opinion that the contractor could make up the lost time during the period that remained for completion of the contract. It was also, so it was argued, quite wrong for him to defer his decision on an application and to inform the contractor that he had done so and that the facts upon which the deferred application had been based had been noted and would be considered with other facts which might thereafter occur to determine then whether or not an extension was justified at a later time. At other times the Director gave his reason for refusing an application that the policy of the Department was that applications, made in circumstances such as were being considered, should be refused.

It was argued on behalf of the plaintiff that this attitude to his duty was unauthorised. All these arguments submitted on behalf of the plaintiff stemmed in a fundamental sense from the consideration that with an agreement as large and complex as this one was, it was of the utmost importance that the progress of the work should be planned and that if by reason of a refusal of an application there had to be a change in the plan it was important for the purposes of the contract that the change should be made as early as possible. It was also important so it was argued, that the change should be made early because of the loss to the plaintiff if it were not then done.'

'Two other points remain for consideration. The first is one to which I have already referred and is whether or not the Director must, as was alleged on behalf of the plaintiff, make his decision on an application promptly. Clause 35 is silent on this point. It provides that the application of the contractor shall be within 14 days of the happening etc. but does not provide any time within which the Director must give his decision. I have already described the general character of this agreement and I will not repeat the description, but in my opinion it is clear that the exigencies of this particular agreement, as exemplified by all its provisions, require that a decision shall not be deferred or delayed. I do not quite appreciate how condition 35 can be construed as an obligation to decide promptly, but I am clear that both the exigencies of this agreement as well as the words of clause 35 require that the decision must be given within a reasonable time. The measurement of a reasonable time in any particular case is always a matter of fact. Plainly the Director must not delay, nor may he procrastinate, and in my opinion he is not entitled simply to defer a decision. On the other hand he is, in my opinion, and this follows from the nature of his obligation to give his own personal decision on the point, necessarily obliged to have available for that consideration such time as is necessary to enable him to investigate the facts which are relevant to making it. When that investigation is complete I am of the opinion that his decision should then be made.

I cannot accept all the arguments submitted by learned counsel for the plaintiff that the Director is bound to investigate every dependent fact himself; this conclusion would, I think, be to ignore the realities of the situation. I am of the opinion, though, that by this agreement and by his mandate he may act upon the findings and opinions of other persons, be they subordinates or independent persons such as architects or meteorological observers; he may also consider and pay attention to the recommendations of subordinates with respect to the very application he is considering. I do agree though that the actual decision must be one which flows from the volition of his own mind and I am of the opinion that it is quite irrelevant that that decision is expressed by the placing of his initials upon the recommendation of a subordinate officer.'

'Paragraph 7 of the Statement of Claim alleges that the term to be implied obliges the defendant to ensure that the Director of Works decides an application for extension promptly. I have already expressed the opinion that the proper implied term in the circumstances is to make a decision within a reasonable time. I am also of the opinion that the obligation created by an implied term of the defendant is to insure that the Director of Works gives his decision within a reasonable time in the manner in which I have already explained . . .'

Constructive acceleration

There is no suggestion in *Perini* that time would be put at large and liquidated damages lost by failure to give extensions within a reasonable time and indeed the case is sometimes quoted as authority for the proposition that the contractor has a remedy for the costs of enforced or constructive acceleration. Bearing in mind that *Perini* was decided on implied terms, the answers to the questions on breach and consequences arising from late granting of extensions prior to completion may be as follows:

(i)   if there are express time requirements on the certifier, the liquidated damages provisions may be invalidated – *Temloc* v. *Errill* (1987) it should be noted was decided on the time requirement after completion;

(ii) if there are express or implied time requirements on the certifier and the contractor accelerates to avoid liquidated damages he may be entitled to his costs;

(iii) it remains open whether employers' fault and neutral events carry the same status in (i) and (ii) above – it would be wise for the certifier to assume that they do.

Demand after tender

A small point from *Perini*, wholly unrelated to extensions of time but one which will surprise many, is that the contractor was unable to recover the cost of a bank guarantee required by the employer after formal acceptance of the tender. The judge said:

> '... The reality of the situation in my opinion is that the plaintiff and the defendant agreed to the provision by the plaintiff of an additional guarantee but that they did not make any agreement at all with respect to the liability of one side or the other side for the cost of doing so. It is in my opinion simply a matter upon which the parties have not expressed any agreement and for that reason the claim of the plaintiff on this point must fail ...'

12.4 *Application to claims*

The link between extension of time clauses and the recovery of loss and expense or extra cost by the contractor has long been a controversial issue.

Many contractors work to the simple maxim 'get the time first and the money will follow' and many contract administrators are nervous about granting extensions because they anticipate exactly what will follow. Although extension of time clauses are rarely drafted with a view to providing grounds for loss and expense claims, there is probably more time spent in the construction industry on relating extensions to claims than to relief from liquidated damages.

Reimbursable and non-reimbursable

Usually there will be no link at all between extension of time clauses of a contract and provisions for additional payments, but just as the contractor looks for something tangible on which to base his claims, so the contract supervisor looks for something tangible to justify his certification of extra payments. Tacitly, if not expressly then, there is a link between extensions and claims and there is general recognition that there are two types of extension: (i) 'reimbursable' extensions which are based on employer's fault; and (ii) 'non-reimbursable' extensions which are based on neutral events. In the Society of Construction Law's Delay and Disruption Protocol these are referred to as 'compensable' and 'non-compensable' events.

Neutral events – loss lies where it falls

In *Henry Boot Construction Ltd* v. *Central Lancashire Development Corporation* (1980) on a JCT 63 contract regarding a dispute on whether 'statutory undertakers' were within the definition of tradesmen or others engaged by the employer in laying mains for a development site, Judge Fay expressed the matter this way:

'... Now if there is delay in carrying out a construction contract there is, of course, loss. There is in the first instance, loss to both parties. To the employer, the owner, there is loss of the return upon his investment. The day when he starts getting return by way of rent from his property is postponed and he may well have an extended period of expenditure upon supervision and the like. Equally, there is loss upon the contractor owing to the prolongation of the period for which he has to supply matters falling within overheads as well as other expenditure, and indeed possibly idle time as well...'

'... The broad scheme of these provisions is plain. There are cases where the loss should be shared, and there are cases where it should be wholly borne by the employer. There are also cases which do not fall within either of these conditions and which are the fault of the contractor. But in the cases where the fault is not that of the contractor, the scheme clearly is that in certain cases, the loss is to be shared; the loss lies where it falls. But in other cases the employer has to compensate the contractor in respect of the delay, and that category, where the employer has to compensate the contractor, should one would think, clearly be composed of cases where there is fault upon the employer or fault for which the employer can be said to bear some responsibility...'

The reference here to loss being shared is made in the sense that the employer loses his right to damages for late completion and the contractor stands his own costs. This indeed is the essence of the non-reimbursable extension – as Judge Fay said – the loss lies where it falls.

Extensions not conditions precedent to recovery of cost

The difficulty for the contractor is that unless the extension provisions require the certifier to indicate the grounds and apportionment of any extension granted, when the contractor has made application under more than one heading, he may not know whether he has been granted reimbursable or non-reimbursable extensions. On the basis that loss and expense or extra cost provisions stand alone, this should not matter. As Judge Fox-Andrews said in *Fairweather* v. *Wandsworth* (1987), obtaining an extension of the time is not a condition precedent to recovering loss and expense, but the judge went on to recognise the realities of the situation when he said:

> '. . . Neither this part of the judgment nor the terms of the contract itself point to an extension of time under condition 23(f) being a condition precedent to recovery of direct loss and expense under condition 24(1)(a). However, the practical effect ordinarily will be that if the architect has refused an extension under the former, the contractor is unlikely to be successful with the architect on an application under condition 24(1)(a) . . .'

Apportionment of extensions

JCT 2005, although not making an extension of time a condition precedent to recovery of loss and expense, does have in its loss and expense clause a sensible requirement that the architect shall state in writing what extensions have been made in respect of reimbursable events if that is necessary for the ascertainment of loss and expense. ICE Minor Works form has a similar linkage in its additional payments clause. ICE 7th Edition tackles the issue in a different way by separately identifying events which can give rise to both claims for extra costs and extensions of time.

For further comment on apportionment see Chapter 14 and the case of *John Doyle* v. *Laing* (2004) later in this chapter.

12.5 Proof of entitlement

Standard forms usually require the contractor to give notice of delay or make his application within reasonable time of the happening of the delaying event and to provide such details and particulars as are necessary to assist the contract administrator in making his decision.

Need for records

The burden of proof of delay rests on the contractor whether the application be made before or after completion and whether the delaying event is the employer's fault or neutral. If the contractor intends to challenge liquidated damages on the grounds of delay or aims to avoid liquidated damages by obtaining an extension from the contract administrator or an arbitrator, the contractor will have to produce evidence that delay occurred and that the cause of the delay gave an entitlement to an extension.

Consequently it is up to the contractor to keep records as evidence of both delay and cause. The strength of the contractor's case will depend on the quality of his records.

Some delays can be shown to flow naturally from the cause; thus an order to suspend work for a defined period would give both a readily identifiable event and a readily measurable delay. With variation orders and extras, however, the position is not as straightforward. There may be the argument that the contractor has a contractual obligation to accommodate variations and extras and has made no attempt to do so. And there will often be the suspicion that the delays the contractor is trying to pass off as due to relevant events, have other causes of the contractor's own making, or within his control for which there is no entitlement to extension.

Monthly progress meetings

It is not only the contractor who must keep records. A similar burden rests with the contract supervisor to protect the employer's position. Ideally, the contractor and the contract supervisor should be of the same mind as the contract progresses on the causes and extent of any delays. They may not be able to agree, but at least they should make some attempt to do so. There is probably no effective substitute for the monthly progress meeting at which the contractor makes his report, the contract administrator makes his, and they jointly agree by discussion what should be placed on record as the true state of affairs.

Usefulness of programmes

How effective is the contractor's programme as a scale for the measurement of delay? Much depends on the quality of the programme and whether the contractor was achieving the planned rate of progress before the alleged delay occurred. There is no doubt that a comprehensive programme marked up on a regular basis to show actual start dates, durations, and completion dates of significant activities, is at least a credible record of progress even if it says nothing on the actual causes of any delays. But by adding to the programme indicators on when variations, revisions and the like were ordered; instructions were given; and other events were encountered; the programme can provide a detailed picture which should satisfy the tests of good evidence.

The contractor who fails to produce an effective and realistic programme puts himself at a disadvantage on proof of entitlement to extensions. First of all he must show that he has suffered delay and that this delay has affected completion of the works; then he must show that the delay was caused by a relevant event and not his own deficiencies and difficulties.

To do this without a programme showing what was intended by way of orderly progress and output, and records showing how progress and output were comparing with the programme before the alleged delay, is a near impossible task. As for the contract administrator having to form his own view in such circumstances, he may have so few definite facts on which to base his decision that the contractor can hardly complain of unfair treatment. This later point came up in *Hounslow v. Twickenham Garden Developments* (1971) as a side issue on whether or not the contractor's refusal to provide a programme had improperly influenced the architect's decision not to give an extension of time for a strike and his decision to issue a notice under the determination provisions. Mr Justice Megarry in considering the architect's duty on these matters in the absence of a programme said this:

'It will be seen from this that provided the contractor has given written notice of the cause of delay, the obligation to make an extension appears to rest on the architect without the necessity of any formal request for it by the contractor. Yet he is required to do this only if in his opinion the completion of the works "is likely to be or has been delayed beyond the Date of Completion", or any extended time for completion previously fixed. If a contractor is well ahead of his works and is then delayed by a strike, the architect may nevertheless reach the conclusion that completion of the works is not likely to be delayed beyond the date of completion. Under condition 21(1), the contractor is under a double obligation: on being given possession of the site, he must "thereupon begin the Works and regularly and diligently proceed with the same", and he must also complete the works "on or before the Date for Completion", subject to any extension of time. If a strike occurs when two-thirds of the work has been completed in half the contract time, I do not think that on resuming work a few weeks later the contractor is then entitled to slow down the work so as to last out the time until the date for completion (or beyond, if an extension of time is granted) if thereby he is failing to proceed with the work "regularly and diligently".'

12.6 Global claims

All claims whether for loss and expense, extra cost, or for extension of time, should meet the legal requirement of linking damage with cause. A claim for loss and expense or extra cost is not effective without a cause and a claim for extension of time is not effective without a relevant event. Standard forms of contract display this principle by requiring each and every claim to stand on its own merits.

Contractors may well say there are circumstances where it is not possible to isolate individual delaying events. If there has been severe disruption of planned activities by changes, variations and late instructions, it may be impracticable to attribute loss and expense or extra cost, to individual heads of claim and it may be difficult, if not impossible, to isolate the effects of delay for each relevant event. Is it possible in such circumstances for the contractor to make a 'global' approach to the presentation of his claim?

For loss and expense or extra cost there is authority for the proposition that the global approach can be used when the contractual machinery has been exhausted. In *J. Crosby & Sons Ltd* v. *Portland Urban District Council* (1967) consideration had to be given on a pipeline contract which had been delayed by 46 weeks by a combination of matters to whether the arbitrator was entitled to award a lump sum by way of compensation to the contractor. The arbitrator had found:

> '... The result, in terms of delay and disorganisation, of each of the matters referred to above was a continuing one. As each matter occurred its consequences were added to the cumulative consequences of the matters which had preceded it. The delay and disorganisation which ultimately resulted was cumulative and attributable to the combined effect of all these matters. It is therefore impracticable, if not impossible, to assess the additional expense caused by delay and disorganisation due to any one of these matters in isolation from the other matters ...'

On appeal to the High Court against the arbitrator's award the argument for the Council was put as follows:

> '... The respondents say that the contract provides a most elaborate code whereby prices and rates can be varied or prescribed in almost every eventuality. They say that this code is intended to operate in relation to each piece of work separately and no provision is made for variation of the contract price generally. Whilst they conceded that an arbitrator at the end of the day may make an award of a lump sum, they insist that this lump sum must be ascertained simply by adding together the individual amounts which he finds to be due under each head of claim. This results, they say, from the fact that the code in the contract provides different bases of assessment for different claims ...'

And for the contractor the argument was put in this way:

> '... Since, however, the extent of the extra cost incurred depends upon an extremely complex interaction between the consequences of the various denials, suspensions and variations, it may well be difficult or even impossible to make an accurate apportionment of the total extra cost between the several causative events. An artificial apportionment could of course have been made; but why, they ask, should the arbitrator make such an apportionment which has no basis in reality ...?'

Mr Justice Donaldson giving judgment in favour of the contractor accepted the global approach with reservations. He said:

> '... so long as the arbitrator does not make any award which contains a profit element, this being permissible under clauses 51 and 52 but not under clauses 41 and 42, and provided he ensures that there is no duplication, I can see no reason why he should not recognise the realities of the situation and make individual awards in respect of those parts of

individual items of the claim which can be dealt with in isolation and a supplementary award in respect of the remainder of these claims as a composite whole. This is what the arbitrator has done . . .'

The ruling in *Crosby* was followed by Mr Justice Vinelott in *London Borough of Merton* v. *Leach* (1985). He said:

'. . . In *Crosby* the arbitrator rolled up several heads of claim arising under different heads and indeed claims for which the contract provided different bases of assessment. The question accordingly is whether I should follow that decision. I need hardly say that I would be reluctant to differ from a judge of Donaldson J's experience in matters of this kind unless I was convinced that the question had not been fully argued before him or that he had overlooked some material provisions of the contract or some relevant authority.

Far from being so convinced, I find his reasoning compelling. The position in the instant case is, I think, as follows. If application is made (under clause 11(6) or 24(1) or under both sub-clauses) for reimbursement of direct loss or expense attributable to more than one head of claim and at the time when the loss or expense comes to be ascertained it is impracticable to disentangle or disintegrate the part directly attributable to each head of claim, then, provided of course that the contractor has not unreasonably delayed in making the claim and so has himself created the difficulty, the architect must ascertain the global loss directly attributable to the two causes, disregarding, as in *Crosby,* any loss or expense which would have been recoverable if the claim had been made under one head in isolation and which would not have been recoverable under the other head taken in isolation.

To this extent the law supplements the contractual machinery which no longer works in the way in which it was intended to work so as to ensure that the contractor is not unfairly deprived of the benefit which the parties clearly intend he should have . . .'

Neither *Crosby* nor *Merton* gave anything but qualified approval to the global approach. Both required the use of contractual procedure to its limit and then evidence of the impossibility of accurate apportionment for the remainder. *Crosby* and *Merton* were certainly not, as some contractors believe, authority for the proposition that a post-completion global claim submission is a legitimate alternative to timely and individual applications.

Later cases

In the case of *Wharf Properties Ltd* & *the Wharf (Holdings) Ltd* v. *Eric Cumine Associates* (1991) an action by a developer against his architect was struck out because it did not disclose a reasonable cause of action. The employer who claimed that the architect had failed properly to manage and co-ordinate the project had pleaded that the complexity of the project made it impossible to isolate individual delays and their effects.

Crosby, Merton and *Wharf* were all considered by Recorder Tackaberry in *Mid Glamorgan County Council* v. *Williams* (1991) where an employer claimed on a global basis against his architect following claims settled with the contractor for late supply of information. An application for the claim to be struck out was rejected and it was held that claims formulated in a global manner could be pursued if it was impossible or impracticable to break down the interaction of events.

In the case of *British Airways Pension Trustees Ltd* v. *Sir Robert McAlpine & Sons Ltd* (1994) the contractor had been successful in having an action against him struck out on the grounds that the statement of claim did not set out the remedial cost of each alleged defect and did not particularise for each defect the alleged diminution in value. The Court of Appeal reversed the striking out order, holding that although the pleading as to damages was embarrassing in that it was open to further particularisation it was not seriously prejudicial to the defendants. Lord Justice Saville said:

'The basic purpose of pleadings is to enable the opposing party to know what case is being made in sufficient detail to enable that party properly to prepare to answer it. To my mind it seems that in recent years there has been a tendency to forget this basic purpose and to seek particularisation even when it is not really required. This is not only costly in itself, but is calculated to lead to delay and to interlocutory battles in which the parties and the court pore over endless pages of pleadings to see whether or not some particular point has or has not been raised or answered, when in truth each party knows perfectly well what case is made by the other and is able properly to be prepared to deal with it. Pleadings are not a game to be played at the expense of the litigants, nor an end in themselves, but a means to the end, and that end is to give each party a fair hearing. Each case must of course be looked at in the light of its own subject matter and circumstances.'

In the case of *GMTC Tools and Equipment Ltd* v. *Yuasa Warwick Machinery Ltd* (1994) the Court of Appeal held that it was not the function of the courts to require a party to establish causation and loss by any particular method. Lord Justice Leggatt said:

'I have come to the clear conclusion that the plaintiffs should be permitted to formulate their claims for damages as they wish, and not be forced into a straightjacket of the judge's or their opponents' choosing.'

Comment

The common rule which can be taken from the above-mentioned cases from *Crosby* to *GMTC Tools* is that generally the law requires cause and effect to be particularised for each item of claim but in circumstances where such particularisation is impossible then globalisation of some or all of the effects may be permissible. However, underlying this rule is another rule to the effect that for a claimant to succeed with a global claim it cannot include within its claim matters for which it carries responsibility. Lord Macfadyen (the judge of first instance) in *John Doyle* v. *Laing* put it this way:

'36. The logic of a global claim demands, however, that all the events which contribute to causing the global loss be events for which the defender is liable. If the causal events include events for which the defender bears no liability, the effect of upholding the global claim is to impose on the defender a liability which, in part, is not legally his. That is unjustified. A global claim, as such, must therefore fail if any material contribution to the causation of the global loss is made by a factor or factors for which the defender bears no legal liability . . . The point has on occasion been expressed in terms of a requirement that the pursuer should not himself have been responsible for any factor contributing materially to the global loss, but it is in my view clearly more accurate to say that there must be no material causative factor for which the defender is not liable.

37. Advancing a claim for loss and expense in global form is therefore a risky enterprise. Failure to prove that a particular event for which the defender was liable played a part in causing the global loss will not have any adverse effect on the claim, provided the remaining events for which the defender was liable are proved to have caused the global loss. On the other hand, proof that an event played a material part in causing the global loss, combined with failure to prove that that event was one for which the defender was responsible, will undermine the logic of the global claim. Moreover, the defender may set out to prove that, in addition to the factors for which he is liable founded on by the pursuer, a material contribution to the causation of the global loss has been made by another factor or other factors for which he has no liability. If he succeeds in proving that, again the global claim will be undermined.'

Added to the above-described difficulties of sustaining global claims, there is the point that for claims made under construction contracts, whether for time or for money, there are usually prescriptive rules limiting, and sometimes going so far as to prohibit, global claims. For this reason perhaps, as much as anything else, there is very little by way of legal authority on the globalisation of extension of time claims.

To the extent that extensions of time are seen as entitlements it is for the contractor to comply with the applicable procedural rules and few modern contracts, if any, allow for global applications or assessments. However, in contracts where there are frequent delaying events the contractual machinery for dealing with extensions of time sometimes breaks down raising the interesting question of whether the consequence is that globalisation should be allowed or whether the employer's right to liquidated damages is lost. There does not appear to be any definitive legal answer to this but in practice it is the globalisation point which is usually argued rather than the loss of right to damages point.

The *Doyle* v. *Laing* case (2004)

Since its publication in 2004 the opinion of the Scottish Inner House in the case of *John Doyle Construction Ltd* v. *Laing Management (Scotland) Ltd* (2004) has received enormous attention from lawyers, contractors and employers anxious to establish how far it has changed the law on global claims not only in Scotland but in the rest of the United Kingdom. On this latter point the case of *London Underground Ltd* v. *Citylink Telecommunications Ltd* (2007) is of interest. In considering a challenge to an arbitrator's award, Mr Justice Ramsey said in that case:

> 'In deciding whether there was serious irregularity, I consider that the proper approach to global claims is relevant. The approach set out in the decision in [*Doyle* v. *Laing*] is not challenged on this application and I accept that approach.'

Note also the endorsement of the approach in *Skanska Construction UK Ltd* v. *Egger (Barony) Ltd* (2004) and the new comment on global claims in *Keating*:

> 'On the other hand, if a global claim fails, that does not mean that no claim whatsoever could succeed, since there may be sufficient evidence to apportion individual losses to individual events . . .'

The extent to which the law has changed, simply stated, is that prior to *Doyle* v. *Laing* global claims tended to stand or fall in their entirety whereas post *Doyle* v. *Laing* it is permissible to consider apportionment.

The reasoning in *Doyle* v. *Laing* was stated by Lord Drummond Young as follows:

> '[10] For a loss and expense claim under a construction contract to succeed, the contractor must aver and prove three matters: first, the existence of one or more events for which the employer is responsible; secondly, the existence of loss and expense suffered by the contractor; and, thirdly, a causal link between the event or events and the loss and expense. (The present case involves a works contract concluded between a management contractor and a works contractor; in such a case the management contractor is obviously in the position of the employer and the works contractor in the position of the contractor). Normally individual causal links must be demonstrated between each of the events for which the employer is responsible and particular items of loss and expense. Frequently, however, the loss and expense results from delay and disruption caused by a number of different events, in such a way that it is impossible to separate out the consequences of each of those events. In that event, the events for which the employer is responsible may interact with one another in such a way as to produce a cumulative effect. If, however, the contractor is able to demonstrate that all of the events on which he relies are in law the responsibility of the

employer, it is not necessary for him to demonstrate causal links between individual events and particular heads of loss. In such a case, because all of the causative events are matters for which the employer is responsible, any loss and expense that is caused by those events and no others must necessarily be the responsibility of the employer. That is in essence the nature of a global claim. A common example occurs when a contractor contends that delay and disruption have resulted from a combination of the late provision of drawings and information and design changes instructed on the employer's behalf; in such a case all of the matters relied on are the legal responsibility of the employer. Where, however, it appears that a significant cause of the delay and disruption has been a matter for which the employer is not responsible, a claim presented in this manner must necessarily fail. If, for example, the loss and expense has been caused in part by bad weather, for which neither party is responsible, or by inefficient working on the part of the contractor, which is his responsibility, such a claim must fail. In each case, of course, if the claim is to fail, the matter for which the employer is not responsible in law must play a significant part in the causation of the loss and expense. In some cases it may be possible to separate out the effects of matters for which the employer is not responsible.

[11] The expression "global claim" has normally been used in Scotland, England and other Commonwealth countries to denote a claim calculated in the foregoing manner. In the United States the corresponding expression is "total cost claim". In the American cases before the Court of Claims, however, a further category is recognised, that of a "modified total cost claim". In the American terminology, a total cost claim involves the contractor's claiming that the whole of his additional costs in performing the contract have been the result of events for which the employer is responsible. A modified total cost claim is more restrictive, and involves the contractor's dividing up his additional costs and only claiming that certain parts of those costs are the result of events that are the employer's responsibility. This terminology has the advantage of emphasising that the technique involved in calculating a global claim need not be applied to the whole of the contractor's claim. Instead, the contractor can divide his loss and expense into discrete parts and use the global claim technique for only one, or a limited number, of such parts. In relation to the remaining parts of the loss and expense, the contractor may seek to prove causation in a conventional manner. This may be particularly useful in relation to the consequences of delay, as against disruption. The delay, by itself, will invariably have the consequence that the contractor's site establishment must be maintained for a longer period than would otherwise be the case, and frequently it has the consequence that foremen and other supervisory staff have to be engaged on the contract for longer periods.

Costs of that nature can be attributed to delay alone, without regard to disruption. Moreover, because delay is calculated in terms of time alone, it is relatively straightforward to separate the effects of delay caused by matters for which the employer is responsible and the effects of delay caused by other matters. For example, delay caused by late instructions and delay caused by bad weather can be measured in a straightforward fashion, subject only to the possibility that the two causes operate concurrently; we discuss concurrent causes below at paragraphs [15]–[19].

[12] Perhaps the most detailed description of total cost claims is found in *John Holland Construction & Engineering Pty Ltd* v. *Kvaerner RJ Brown Pty Ltd*, a decision of the Supreme Court of Victoria. In that case, Byrne J. stated (at 82 BLR 85–87):

"The claim as pleaded . . . is a global claim, that is, the claimant does not seek to attribute any specific loss to a specific breach of contract, but is content to allege a composite loss as a result of all of the breaches alleged, or presumably as a result of such breaches as are ultimately proved. Such claim has been held to be permissible in the case where it is impractical to disentangle that part of the loss which is attributable to each head of claim, and this situation has not been brought about by delay or other conduct of the claimant . . .

Further, this global claim is in fact a total cost claim. In its simplest manifestation a contractor, as the maker of such claim, alleges against a proprietor a number of breaches of contract and quantifies its global loss as the actual cost of the work less the expected cost. The logic of such a claim is this:

(a) the contractor might reasonably have expected to perform the work for a particular sum, usually the contract price;
(b) the proprietor committed breaches of contract;
(c) the actual reasonable cost of the work was a sum greater than the expected cost.

The logical consequence implicit in this is that the proprietor's breaches caused that extra cost or cost overrun. This implication is valid only so long as, and to the extent that, the three propositions are proved and a further unstated one is accepted: the proprietor's breaches represent the only causally significant factor responsible for the difference between the expected cost and the actual cost. In such a case the causal nexus is inferred rather than demonstrated . . . The understated assumption underlying the inference may be further analysed. What is involved here is two things: first, the breaches of contract caused some extra cost; secondly, the contractor's cost overrun is this extra cost. The first aspect will often cause little difficulty but it should not, for this reason, be ignored . . . It is the second aspect of the understated assumption, however, which is likely to cause the more obvious problem because it involves an allegation that the breaches of contract were the material cause of all of the contractor's cost overrun. This involves an assertion that, given that the breaches of contract caused some extra cost, they must have caused the whole of the extra cost because no other relevant cause was responsible for any part of it."

Byrne J went on to consider the claim made by the plaintiffs in the case before him, and pointed out that, because it was a total cost claim, it was necessary to eliminate any causes of inadequacy in the tender price other than matters for which the employer was responsible. It was also necessary to eliminate any causes of overrun in the construction cost other than matters for which the employer was responsible.

[13] In *Boyajian v United States*, 423 F 2d 1231 (1970), the Court of Claims approved of the following passage commenting on the total cost method of calculation (at 1243):

"This theory has never been favoured by the court and has been tolerated only when no other mode was available and when the reliability of the supporting evidence was fully substantiated . . . The acceptability of the method hinges on proof that (1) the nature of the particular losses make it impossible or highly impracticable to determine them with a reasonable degree of accuracy; (2) the plaintiff's bid or estimate was realistic; (3) its actual costs were reasonable; and (4) it was not responsible for the added expenses."

In that case it was held that any suggestion that there was a presumption that the contractor's expenditure was reasonable must be rejected.

[14] We agree with the foregoing statements of the law by Byrne J. and the Court of Claims. It is accordingly clear that if a global claim is to succeed, whether it is a total cost claim or not, the contractor must eliminate from the causes of his loss and expense all matters that are not the responsibility of the employer. This requirement is, however, mitigated by the considerations discussed by the Lord Ordinary at paragraphs [38] and [39] of his opinion. In the first place, it may be possible to identify a causal link between particular events for which the employer is responsible and individual items of loss. On occasion that may be possible where it can be established that a group of events for which the employer is responsible are causally linked with a group of heads of loss, provided that the loss has no other significant cause. In determining what is a significant cause, the "dominant cause" approach described in the following paragraph is of relevance. Determining a causal link between particular events and particular heads of loss may be of particular importance where the loss results from mere delay, as against disruption; in cases of mere delay such losses as the need to maintain the site establishment for an extended time can often readily be attributed to particular events, such as the late provision of information or design changes. We note that in the United States the Court of Claims has approved of an approach along the foregoing lines. In *Boyajian v United States*, the

Court of Claims approved of the following passage (at 423 F 2d 1244):

"In situations where the court has rejected the 'total cost' method of proving damages, but where the record nevertheless contained reasonably satisfactory evidence of what the damages are, computed on an acceptable basis, the court has adopted such other evidence . . .; or where such other evidence, although not satisfactory in and of itself upon which to base a judgment, has nevertheless been considered at least sufficient upon which to predicate a 'jury verdict' award, it has rendered a judgment based on such a verdict . . . However, where the record is blank with respect to any such other alternative evidence, the court has been obliged to dismiss the claim for failure of damage proof, regardless of the merits."

[15] In the second place, the question of causation must be treated by "the application of common sense to the logical principles of causation": *John Holland Construction & Engineering Pty Ltd* v. *Kvaerner RJ Brown Pty Ltd, supra*, at 82 BLR 84I per Byrne J.; *Alexander* v. *Cambridge Credit Corporation Ltd*, (1987) 9 NSWLR 310; *Leyland Shipping Company Ltd* v. *Norwich Union Fire Insurance Society Ltd*, [1918] AC 350, at 362 per Lord Dunedin. In this connection, it is frequently possible to say that an item of loss has been caused by a particular event notwithstanding that other events played a part in its occurrence. In such cases, if an event or events for which the employer is responsible can be described as the dominant cause of an item of loss, that will be sufficient to establish liability, notwithstanding the existence of other causes that are to some degree at least concurrent. That test is similar to that adopted by the House of Lords in *Leyland Shipping Company Ltd* v. *Norwich Union Fire Insurance Society Ltd*, where a ship was torpedoed by a German submarine and taken into the harbour of Le Havre. When a gale sprang up she was moved to a berth inside the outer breakwater, where she took the ground at each ebb tide. Ultimately her bulkheads gave way and she sank. She was insured against perils of the sea, but excluding the consequences of hostilities. It was held that the proximate cause of the loss was the damage inflicted by the torpedo, which fell within the exclusion. In our opinion the same approach should be taken to cases such as the present. If an item of loss results from concurrent causes, and one of those causes can be identified as the proximate or dominant cause of the loss, it will be treated as the operative cause, and the person responsible for it will be responsible for the loss.

[16] In the third place, even if it cannot be said that events for which the employer is responsible are the dominant cause of the loss, it may be possible to apportion the loss between the causes for which the employer is responsible and other causes. In such a case it is obviously necessary that the event or events for which the employer

is responsible should be a material cause of the loss. Provided that condition is met, however, we are of opinion that apportionment of loss between the different causes is possible in an appropriate case. Such a procedure may be appropriate in a case where the causes of the loss are truly concurrent, in the sense that both operate together at the same time to produce a single consequence. For example, work on a construction project might be held up for a period owing to the late provision of information by the architect, but during that period bad weather might have prevented work for part of the time. In such a case responsibility for the loss can be apportioned between the two causes, according to their relative significance. Where the consequence is delay as against disruption, that can be done fairly readily on the basis of the time during which each of the causes was operative. During the period when both operated, we are of opinion that each should normally be treated as contributing to the loss, with the result that the employer is responsible for only part of the delay during that period. Unless there are special reasons to the contrary, responsibility during that period should probably be divided on an equal basis, at least where the concurrent cause is not the contractor's responsibility. Where it is his responsibility, however, it may be appropriate to deny him any recovery for the period of delay during which he is in default.

[17] Apportionment in this way, on a time basis, is relatively straight-forward in cases that involve only delay. Where disruption to the contractor's work is involved, matters become more complex. Nevertheless, we are of opinion that apportionment will frequently be possible in such cases, according to the relative importance of the various causative events in producing the loss. Whether it is possible will clearly depend on the assessment made by the judge or arbiter, who must of course approach it on a wholly objective basis. It may be said that such an approach produces a somewhat rough and ready result. This procedure does not, however, seem to us to be funda-mentally different in nature from that used in relation to contribu-tory negligence or contribution among joint wrongdoers. Moreover, the alternative to such an approach is the strict view that, if a contrac-tor sustains a loss caused partly by events for which the employer is responsible and partly by other events, he cannot recover anything because he cannot demonstrate that the whole of the loss is the responsibility of the employer. That would deny him a remedy even if the conduct of the employer or the architect is plainly culpable, as where an architect fails to produce instructions despite repeated requests and indications that work is being delayed. It seems to us that in such cases the contractor should be able to recover for part of his loss and expense, and we are not persuaded that the practical difficulties of carrying out the exercise should prevent him from doing so.

[18] An apportionment procedure of this nature has been used with apparent success in the United States in cases before the Court of Claims. Thus in *Lichter v. Mellon-Stuart Company*, 305 F 2d 216 (1962), the plaintiffs' total cost claim on one contract was rejected on the ground that a substantial amount of their loss was the consequence of factors other than breaches of contract by the defendants. The court could find no basis for allocation of the plaintiff's claim, which was for a lump sum, between those causes which were actionable and those which were not, with the result that the entire claim was rejected. Nevertheless, the Court of Claims allowed a claim based on another contract between the same parties to succeed in part, and its decision was upheld by the United States Court of Appeals for the Third Circuit. The Court of Claims had held that part of the plaintiff's extra cost on this contract was attributable to the fault of the defendant and part was attributable to other non-compensable factors. The Court of Appeals stated the result of that finding as follows (at 305 F 2d 221):

"Once it had thus been established that only part of the . . . claim represented extra cost chargeable to Mellon, the one question remaining was whether a reasonable allocation of part of the total sum was possible. The court undertook such an allocation, guided by evidence concerning the extra time required for the performance of the stone contract as the result of the improper shelf angles. We cannot say that this was an arbitrary method of allocation. Indeed, [the plaintiff] is not in position to complain that the allocation was imprecise since it bore the burden of proving how much of the extra cost resulted from Mellon's improper conduct. [The plaintiff] risked the loss of its entire claim, as occurred with reference to the masonry contract, if the court should not have been able to make a reasonable allocation."

The important points that emerge from this decision are, first, that the Federal courts in the United States are willing to undertake an apportionment exercise and, secondly, that any such apportionment must be based on the evidence and carried out on a basis that is reasonable in all the circumstances. In our opinion a similar procedure should be available in Scots law. We stress, however, that the allocation must be based on the evidence, and that under Scottish procedure the evidence must be based on a foundation in the pleadings.

[19] In *Phillips Construction Co. Inc v. United States*, 394 F 2d 834 (1968), the plaintiff undertook the construction of a large housing project connected with an air force base. During construction, heavy rainfall and extensive flooding were encountered. Under the parties' contract, the plaintiff assumed the risks incident to abnormal rainfall as such. Nevertheless, it claimed that its difficulties were greatly compounded by the inadequacy of the government-designed drainage system for the project, and it sued for the loss that it said resulted from the defective drainage system. The Armed Services

Board of Contract Appeals, the body charged with determining the dispute at first instance, rejected a total cost claim by the plaintiff, because the plaintiff's total loss was caused partly by matters for which the government were responsible and partly by the exceptional rainfall, for which neither party was responsible. Nevertheless, the Board agreed with the plaintiff's contention about the inadequacy of the drainage system, and apportioned the plaintiff's additional costs between flooding caused by defective drainage and other factors. That exercise was upheld by the Court of Claims, which observed that "It represented the best judgment of the fact trier on the record before it", and this is all "that the parties have any right to expect". In our opinion a broadly similar apportionment exercise is possible in a Scottish case, for the reasons discussed above.

[20] The present case is concerned with the relevancy of the pursuers' pleadings, and the argument for the defenders in large measure consisted of a detailed and sustained attack on the overall structure of those pleadings. Nevertheless, it must be borne in mind that the present case involves a commercial action, and in the Commercial Court elaborate pleading is unnecessary. All that is required is that a party's averments should satisfy the fundamental requirements of any pleadings, namely that they should give fair notice to the other party of the facts that are relied on, together with the general structure of the legal consequences that are said to follow from those facts. In doing that, the pleadings of one party should disclose sufficient to enable the other party to prepare its own case and to enable the parties and the court to determine the issues that are actually in dispute. The relevancy of pleadings must always be tested against these fundamental requirements. In a case involving the causal links that may exist between events having contractual significance and losses suffered by the pursuer, it is obviously necessary that the events relied on should be set out comprehensively. It is also essential that the heads of loss should be set out comprehensively, although that can often best be achieved by a schedule that is separate from the pleadings themselves. So far as the causal links are concerned, however, there will usually be no need to do more than set out the general proposition that such links exist. Causation is largely a matter of inference, and each side in practice will put forward its own contentions as to what the appropriate inferences are. In commercial cases, at least, it is normal for those contentions to be based on expert reports, which should be lodged in process at a relatively early stage in the action. In these circumstances there is relatively little scope for one side to be taken by surprise at proof, and it will not normally be difficult for a defender to take a sufficiently definite view of causation to lodge a tender, if that is thought appropriate. What is not necessary is that averments of causation should be over-elaborate, covering

every possible combination of contractual events that might exist and the loss or losses that might be said to follow from such events.'

12.7 The liability of certifiers

In Chapter 11 the possibility of an aggrieved contractor bringing an action against a certifier for alleged under-certification of extension of time is considered. And in Section 12.2 above, comment is made on the duties of certifiers in undertaking their role in the extension of time process.

Also worth noting as an indication of the potential liabilities which certifiers face if they fail to exercise due skill and care in the performance of their duties are the following:

- *Edgeworth Construction Ltd* v. *N D Lea & Associates* (1993) where the Supreme Court of Canada, taking a different view than the English Court of Appeal in *Pacific Associates* v. *Baxter,* held that a consulting engineer's duty of care to a contractor was not negated by either the existence or the terms of the contract between the contractor and the employer. The case did not concern extensions of time but nevertheless it serves as an indicator of how the law may develop in actions by contractors against certifiers.
- *West Faulkener Associates* v. *London Borough of Newham* (1994) where the Court of Appeal held that an architect who had failed to understand the obligation imposed on the contractor by a contractual requirement that he should proceed regularly and diligently and consequently had failed to serve a notice of default entitling the employer to terminate the contract was in breach of his duty of skill and care to the employer.
- *Wessex Regional Health Authority* v. *HLM Design Ltd* (1995) where it was held that an employer had independent and concurrent causes of action arising out of over-certification of extensions of time against both the contractor and the architect.
- *Royal Brompton Hospital NHS Trust* v. *Hammond & Others* (2000) where His Honour Judge Seymour stated the tests for professional negligence established in *Bolam* v. *Friern Hospital Management Committee* (1957):

'(i) what, at the material time, were the standards of ordinarily competent members of the relevant profession in relation to whatever it is which it is alleged that the defendant should have done, but failed to do, or did, but should not have done;

(ii) what it is that the defendant actually failed to do, or did, as the case may be;

(iii) by a comparison of (i) and (ii) above, that the defendant fell below the standards of the ordinarily competent member of matter or his profession in respect of the matter or matters of complaint.'

The judge then said:

'However, for the reasons which I have set out above, in my view, in practical terms the burden shouldered by a claimant who contends that an architect or a project manager has been negligent in granting, or being involved in the grant of, an extension of time for completion of works governed by a contract in the Standard Form is a heavy one: unless the case is very obvious it is most unlikely to succeed.'

before going on to find that the architect had been negligent in respect of one of the extensions of time granted.

Chapter 13
Relevant events

13.1 Force majeure

The expression 'force majeure' is of French origin. Under the French Civil Code force majeure is a defence to a claim for damages for breach of contract. It needs to be shown that the event:

(i)   made performance impossible;
(ii) was unforeseeable;
(iii) was unavoidable in occurrence and effects.

In English law there is no doctrine of force majeure. Before 1863 and the case of *Taylor* v. *Caldwell* it was a rule of the law of contract that the parties were absolutely bound to perform any obligations they had undertaken and the fact that performance had become impossible did not provide relief from damages. In *Taylor* v. *Caldwell* a music hall which was to be hired for a concert was destroyed by fire the day before the performance; the court of Queen's Bench held the hirer not liable for damages by implying a term on impossibility of performance. Mr Justice Blackburn said:

> '... in contracts in which the performance depends on the continued existence of a given person or thing, a condition is implied that the impossibility of performance arising from the perishing of the person or thing shall excuse the performance ...'

From this case developed the doctrine of frustration extending the sphere of impossibility to other instances of frustration. On basic legal principles, therefore, it is frustration and not force majeure which must be pleaded as a defence in English contract law.

Force majeure does, however, have a place in English law where it is expressly introduced as a contract term – as for example, in MF/1 where it provides grounds for extension of time. There can also be an oblique application through the medium of EU law.

EU law

In *Dairyvale Foods Ltd* v. *Intervention Board of Agricultural Produce* (1982) Mr Justice Parker, in considering whether industrial action came within the definition of 'force majeure' in EEC regulations, held that to be effective the occurrence had to be:

(i) an external event beyond the control of the party relying on it;

and

(ii) had to have consequences which could not be avoided.

Force majeure excludes fault

Contractually based force majeure to be effective has to meet the same tests and has to conform with the doctrine of frustration in that there must be no fault attaching to the party using force majeure as a defence or a ground for claim. In *Sonat Offshore SA* v. *Amerada Hess Development Ltd* (1987) a force majeure clause entitled Sonat, an oil rig operation, to payment in certain circumstances. The clause applied '. . . when performance is hindered or prevented by strikes (except contractor induced strikes by contractor's personnel) or lockout, riot, war (declared or undeclared), act of God, insurrection, civil disturbances, fire, interference by any Government Authority or other cause beyond the reasonable control of such party . . .'. Arising from the fault of Sonat there was an explosion and severe fire. The Court of Appeal held that '. . . other cause beyond the reasonable control . . .' did not include for negligence.

Contractual application

On the question of scope of the phrase 'force majeure' when used in a contract there are few English examples for judicial guidance. It would seem from *Lebeaupin* v. *Crispin* (1920) that the phrase could cover wars, epidemics and strikes amongst other things.
The point was made in that case that:

> '. . . a force majeure clause should be construed in each case with close attention to the words which precede or follow it and with due regard to the nature and general terms of the contract . . .'

In JCT 2005 most of the events which might be considered as force majeure are covered elsewhere in the contract as grounds for extension – e.g. war, strikes, riot, fire, storms and exceptional weather and the application of the term is obviously restricted. In ICE forms the phrase is not used at all.
Force majeure by its nature is a neutral event between the parties and is therefore a non-reimbursable event as far as extensions of time are concerned.
Amongst the few definitions of force majeure in construction contracts is the following from FIDIC Conditions of Contract:

> 'In this Clause, "Force Majeure" means an exceptional event or circumstance:

> (a) which is beyond a Party's control,

(b) which such Party could not reasonably have provided against before entering into the Contract,

(c) which, having arisen, such Party could not reasonably have avoided or overcome, and

(d) which is not substantially attributable to the other Party.'

13.2 Adverse weather

Adverse weather of itself does not give any grounds for non-performance of contractual obligations. Unless there are provisions in the contract offering relief, the contractor is deemed to have taken all risks from weather. *Hudson* quotes the early case of *Maryon* v. *Carter* (1830) where a pavement was to be laid by a certain date. Due to bad weather the contractor did not complete on time and forfeited his right to payment.

The question of whether or not adverse weather should be a relevant event is purely a matter of risk distribution in the contract. The case for the risk being allocated to the contractor is that it follows the general principle that risks should be allocated to the party best able to control them. There is also the general point that the contractor can spread the risk over his various contracts whereas if the employer takes the risk there may be no corresponding scope for spreading the risk. Consider, for example, an employer engaging in a single project.

It is noticeable that in recent years a trend appears to be developing, led perhaps by central government–drafted contracts such as GC/Works/1 Edition 3 and the Highways Agency Design and Build Contract, of firmly placing all weather risks on the contractor and in such contracts there is, of course, no relevant event for adverse weather.

Standard forms

Nonetheless many standard forms of construction contract do make some provision for extension of time in respect of exceptional adverse weather and some, such as the New Engineering Contract, go further by allowing some of the risk of the contractor's costs to fall on the employer. Experimental contracts have even been run on the basis that the employer carries the risk for all time lost due to weather, but there is no evidence that either contractors or employers would welcome the general application of such a radical approach.

The extent to which adverse weather applies as a relevant event depends, therefore, almost exclusively on the wording of the contract. In JCT 2005, it is given as delay due to 'exceptionally adverse weather conditions'. The old JCT 63 said 'exceptionally inclement weather' but this was considered to be too restrictive to include for heat. ICE forms use the phrase 'exceptional adverse weather' conditions.

Weather records

It is not just the phrase which has to be considered but also its context. The contractor's entitlement, under JCT forms, to an extension is in respect of delay to the progress of the works by exceptionally adverse weather conditions. The starting point then, is that there has to be delay, not just exceptionally adverse weather. This may seem rather theoretical but it has become so common for contractors to apply for extensions on the grounds that the weather has been worse than average that sight can become lost of the need for proof of delay. The practice of obtaining local weather records and comparing them on a year to year basis, or on a particular year against average, may show that the weather has been exceptional but it says nothing about delay.

The point came up in the case of *Walter Lawrence & Son Ltd* v. *Commercial Union Properties (UK) Ltd* (1984) where a contractor was suing for return of amounts deducted as liquidated damages. Judge Hawser held that:

> '. . . When considering an extension of time under clause 23(b) of JCT 63, on the ground of "exceptionally inclement weather" the correct test for the architect to apply is whether the weather itself was "exceptionally inclement" so as to give rise to delay and not whether the amount of time lost by the inclement weather was exceptional . . .'

Assessment applies to time work carried out

Another matter of significant interest arose in the *Walter Lawrence* case in respect of the time at which the weather should be assessed. The architect in correspondence had said: '. . . It is our view that we can only take into account weather conditions prevailing when the works were programmed to be put in hand, not when the works were actually carried out . . .'

The contractor refuted this and claimed that his progress relative to programme was not relevant to his entitlement to an extension. It was held that the effect of the exceptionally inclement weather is to be assessed at the time when the works are actually carried out and not when they were programmed to be carried out even if the contractor is in delay.

Weather after due date for completion

It should be added that the judge in *Walter Lawrence* drew a distinction between delays which occur during the original or extended time for completion and delays after the due date where the contractor is in culpable delay. He said:

> '. . . These letters do raise the issue as to whether the plaintiffs can legitimately claim in respect of delays which occurred after the date when the

contract plus any proper extension of time ought to have been completed. I think that there is clearly an issue on that aspect of the matter, but it would appear to me that the plaintiffs have a claim of substantial character in respect of the period to the end of the contract – as properly extended . . .'

This question of whether the contractor can get an extension of time for a neutral event which occurs after the due date for completion but when the contractor is still proceeding in culpable delay cannot be generally settled. It depends on the wording of the particular contract. In the case of *Amalgamated Building Contractors v. Waltham Holy Cross UDC* (1952) Lord Justice Denning, on an early building form, said:

'. . . Take a simple case where contractors, near the end of the work, have overrun the contract time for six months without legitimate excuse. They cannot get an extension for that period. Now suppose that the works are still uncompleted and a strike occurs and lasts a month. The contractors can get an extension of time for that month . . .'

This apparent acceptance of entitlement to an extension for neutral events after the due date for completion does not give a general rule and the probability is that where extension provisions refer to the contractor being 'fairly' entitled, as most modern standard forms do, then the contractor would have some difficulty in establishing his case.

Is adverse weather 'beyond the contractor's control'?

A question frequently asked in respect of standard forms which have no specific relevant event for adverse weather but which contain a relevant event worded 'beyond the contractor's control' or similar is whether such a broadly worded event can cover delays caused by adverse weather.

The answer to that, applying the factual test which would seem to follow the decision in the *Scott Lithgow* case (see the comment in Chapter 6) is that it has to be decided as a matter of fact in each case whether the delay caused by the adverse weather was beyond the contractor's control. That may depend on the steps taken, or which could have been taken, by the contractor to alleviate the effects of the weather.

The question sometimes arises in respect of process and plant contracts under model forms such as MF/1 and the IChemE Red Book. In such contracts the probability is that susceptibility to weather conditions has not been considered sufficiently important to justify a specific relevant event for weather. And, having regard to the balance of risk in those contracts, which is generally favourable to the contractor, the case for allowing extensions of time for delays caused by adverse weather as being matters beyond the contractor's control is arguably stronger than would be the case in an ordinary construction contract which was silent on weather.

13.3 Civil commotion, strikes, etc.

These events may be included by express reference in detail or may not be included at all. Where not included they may in appropriate circumstances be covered by generalised events such as 'force majeure' or 'other special circumstances'.

Civil commotion has been defined as 'an insurrection of the people' for general purposes and for insurance purposes as 'a stage between a riot and a civil war'. 'Local combination of workmen' is an old-fashioned expression, the exact meaning of which is uncertain. It might cover events falling short of a strike such as a mass picket or perhaps even an organised 'go slow' or 'work-to-rule'.

Strikes

Strikes can be difficult matters in considering extensions. Do they apply solely to the site of the works and the contractor's own workforce or is there a more general application to sub-contractors and suppliers and if so how far removed? And should the contractor get an extension of time if his own bad employee relationship has caused the strike or, taking the matter a stage further, if it is suspected that the contractor has engineered the strike?

JCT 2005 answers the first question by its wide application of the clause: '. . . affecting any of the trades employed upon the works or any of the trades engaged in the preparation, manufacture, or transportation of any of the goods or materials required for the works . . .'. The contract is silent on the question of culpability but once again, the fair and reasonable test seems to have some relevance although strong evidence against the contractor would be needed to avoid giving him the benefit of any doubt.

ICE forms say nothing on strikes, etc. and it is up to the contractor to argue his cause under 'other special circumstances'.

An example of the difficulty in applying provisions in standard forms to strikes and the like can be found in *Boskalis Westminster Construction Ltd* v. *Liverpool City Council* (1983). Delays arose on the contract from strikes on works by statutory undertakers which came within the category of 'others engaged by the respondents in executing work not forming part of the contract'. The architect gave the contractor an extension under the prevention provisions; the employer claimed it should have been given under strikes. At the back of this, of course, was the question of recovery of loss and expense. The court upheld the arbitrator's decision that the delay did not come within the scope of the strike clause.

Duty to mitigate

One area of difficulty is whether the contractor can avoid the effects of strikes etc. by changing suppliers and sub-contractors, albeit at extra cost.

The general duty to mitigate may not apply since the contractor would rarely be in a position to recover his extra costs from the employer but in so far as there is an express contractual requirement for the contractor to use his best endeavours to prevent delay, howsoever caused, then he will have an obligation to consider alternatives if available.

13.4 *Damage to the works*

Few subjects in construction are as complicated as damage to the works and insurance and when it comes to whether delays caused by insurable risks are grounds for extensions of time, there are few definitive answers.

Starting from the position that the contractor's obligation is to complete the works within the specified time, it would seem not unreasonable to suggest that the contractor should bear the consequences of, or insure against, all losses and delays arising from damage to the works. But that is an over-simplification. Damage to the works can arise from any of three causes:

(i)   contractor's negligence;
(ii) circumstances outside the contractor's control;
(iii) employer's fault through use, occupation or design.

On general principles, the contractor would have no case for an extension in the first instance; a possible case in the second according to the risk sharing aspects of the contract; and a definite entitlement to an extension in the third on the grounds of prevention.

Where the contract is silent on extensions of time for damage to the works, the contractor has no entitlement to an extension except to the extent that general provisions such as 'force majeure' or 'other special circumstances' might be held to apply. It is difficult to see how either could apply to contractor's negligence and, on rules outlined in Chapters 5 and 6 developed from prevention and *contra proferentem*, they should not apply to employer's fault unless such fault is specifically included in the provisions. This leaves circumstances outside the contractor's control as the only application.

It is not always seen or treated this way. ICE forms are silent on the subject and as users of the forms will know there is very little consistency in the granting of extensions by engineers or by arbitrators. Most, but not all, will exclude contractor's negligence; some take a broad view of 'other special circumstances' and will allow for everything but obvious negligence including such causes as vandalism, site accidents and circumstances outside the contractor's control, whilst others take a hard line and exclude the lot. Many do not accept that an extension cannot be granted under 'any special circumstances' for employer's fault.

JCT forms expressly deal with some aspects of damage to the works by making 'the specified perils' relevant events. These are defined as:

'fire, lightning, explosion, storm, tempest, flood, bursting or overflowing of water tanks, apparatus or pipes, earthquake, aircraft and other aerial devices or articles dropped therefrom, riot and civil commotion, but excluding Excepted Risks.'

Nothing is said about contractor's negligence and it may not be correct to exclude it on the wording. Moreover, there is an optional clause for the employer to insure against loss of liquidated damages in respect of the specified perils. Nothing is said about vandalism, accidental impact damage and the like and with no other events covering these matters they are not obvious grounds for extensions.

13.5 Sub-contractors

Domestic sub-contractors

On basic principles, delays caused by domestic sub-contractors do not give grounds for extensions of time. Unless there are express provisions in the contract to cover delays so caused or there are other provisions for extensions which can be interpreted to cover sub-contractors, the problems of sub-contractor default will rest between the main contractor and sub-contractor.

If the contractor is required by the terms of the contract to obtain approval to his sub-contractors from the contract administrator and that approval is unreasonably delayed, that could be a breach of contract with the potential to defeat the liquidated damages provisions unless there are extension clauses covering employer's acts of prevention.

Nominated sub-contractors

Nominated sub-contractors are the cause of many complex disputes, as shown in Chapter 9, and however much forms of contract attempt to place responsibility for such sub-contractors on main contractors, it is very difficult for the employers to avoid sharing some of the responsibility for their delays and defaults. The burden of re-nomination after default is a heavy one as shown in *North West Metropolitan Regional Hospital Board* v. *Bickerton* (1970), *Peak* v. *McKinney* (1970), *Percy Bilton* v. *GLC* (1982) and *Fairclough* v. *Rhuddlan Borough Council* (1985), with the employer bound to avoid delay in re-nominating and to allow time for rectification of faulty work.

Details of the various standard forms are considered in the concluding chapters of this book but few modern forms now include provisions for nominated sub-contractors. ICE 7th Edition is one that does but it allows extensions only for delays arising from determination of the nominated sub-contractors' employment.

Beyond contractor's control

The possibility of general phrases in extension provisions being interpreted to cover sub-contractor delay arises from the interpretation of 'a cause beyond the contractor's control' in *Scott Lithgow* v. *Secretary of State for Defence* (1989). Although that case related to a right of payment, the same decision would probably have been given in relation to extension of time. See the comment in Chapter 6. The case does indicate the need for great caution in drafting general phrases which give rights and obligations.

Thus in early JCT Minor Building Works forms the contractor was entitled to an extension if it became apparent that the works would not be completed by the due date for 'reasons beyond the control of the contractor'. The form was amended in 1991 to take note of the *Scott Lithgow* decision so as to include 'within' the control of the contractor any default of sub-contractors or suppliers.

However, there are still other well-used standard forms of contract, including MF/1, which use such phrases as 'beyond the reasonable control of the contractor' in defining relevant events. In such contracts it may well be open to the contractor to argue that he is entitled to an extension of time if he is delayed by sub-contractors who, as a matter of fact, are beyond his reasonable control.

An example of the difficulties which can result from imprecise drafting is found in the case of *John Mowlem & Company plc* v. *Eagle Star Insurance Company Ltd* (1995). The contract contained an extension of time clause which read (in part):

> '. . . if, in the opinion of the architect, the completion of any section is likely to be or has been delayed . . . by any cause beyond the control of the Management Contractor, his subcontractors or materials suppliers . . . the architect shall allow and certify a fair and reasonable extension of time . . .'

The preliminary question considered by the Court of Appeal was as follows:

> 'On the true construction of the management contract would the management contractor ever be entitled to extensions of time for causes beyond its control but within the control of subcontractors or materials suppliers, and if so in what circumstances?'

One submission put to the court was that so long as the cause of delay was beyond the control of any of the entities, the management contractor, his sub-contractors, or his materials suppliers then the management contractor was automatically entitled to an extension of time. That submission required the extension of time clause to be read as 'any cause beyond the control of the management contractor or his sub-contractors or material suppliers'. The effect would be that any cause beyond the control of the management contractor but within the control of a sub-contractor, or equally within the control of the management contractor but beyond the control of the sub-contractor would be a qualifying event. Lord Justice Hirst, after briefly reviewing the *Scott Lithgow* decision and deciding that the facts of the case were quite different, rejected the submission and held the answer to the preliminary question to be in the negative.

13.6 Non-availability of resources

As with sub-contractors, the full burden of problems in obtaining the resources necessary for constructing the works should, on basic principles, fall on the contractor. If, however, the employer decides to share the risk by allowing extensions of time for problems so arising, that is a concession or a matter of commercial judgment between the parties.

Some standard forms do give the contractor an entitlement to extension for delays caused by his inability to secure labour or materials but generally the inability must be for reasons beyond the control of the contractor and which he could not have reasonably foreseen at the time of tender. This was the position in JCT 80. Similar provisions in JCT 63 were optional and were frequently deleted by the employers.

Apart from the point of principle as to which party should carry the risks of obtaining resources, there is frequently much scope for interpretative argument on the wording of such clauses and on the facts of any case. It may not be the intention of the draftsmen that every upsurge in the industry's workload with its attendant shortages of labour and materials should relieve contractors of their obligations to complete on time, but it is this type of situation which has often led to problems.

Qualified tenders

Perhaps because the risk of obtaining resources is so obviously that of the contractor there are few cases of note on the subject. One of interest, but not directly concerned with extensions of time, was *Davis Contractors Ltd* v. *Fareham UDC* (1956). With its form of tender for the construction of 78 houses, Davis attached a letter stating the price was subject to adequate supplies of labour being available. In the event, they were not and the works took 22 months to complete instead of eight months. Davis claimed that their tender was qualified and that the contract had been frustrated. It was held that Davis's letter did not form part of the contract and that the contract had not been frustrated.

A cautionary point needs to be made here for contractors who think that by submitting letters of qualification with their tenders they have automatically included the qualifications in any contract which follows. The qualification becomes effective only if it is included in a document which becomes bound-in with the contract documents.

13.7 *Statutory undertakers' works*

Gas, water, electricity and telephone installations are likely to be a part of any building project; and in any urban roadworks project there will be services to be diverted or new services to be laid. The question is, who pays the bill when they cause delay?

This was one of the matters which faced Judge Fay in *Henry Boot* v. *Central Lancashire Development Corporation* (1980). He had to decide whether mains laid to a housing site were within the description of 'work being done by others engaged by the employer', in which case any delay caused gave rise to an extension of time and recovery of loss and expense; or whether they were works executed under statutory powers, in which case the contractor got only an extension of time but no costs. Judge Fay decided in this instance that the works were in the first category and he had this to say about where the loss should fall.

> '. . . The broad scheme of these provisions is plain. There are cases where the loss should be shared, and there are cases where it should be wholly borne by the employer. There are also those cases which do not fall within either of these conditions and which are the fault of the contractor, where the loss of both parties is wholly borne by the contract. But in the cases where the fault is not that of the contractor the scheme clearly is that in certain cases the loss is to be shared: the loss lies where it falls. But in other cases the employer has to compensate the contractor in respect of the delay, and that category, where the employer has to compensate the contractor, should, one would think, clearly be composed of cases where there is fault upon the employer or fault for which the employer can be said to bear some responsibility.'

It follows that where the delay caused by statutory undertakers' work is an act of prevention the usual rules apply, but where the delay is neutral, the contractor's entitlement to an extension will depend on the provisions of the contract.

Are undertakers' delays prevention or neutral?

Standard building and civil engineering forms cover both prevention and neutral situations through various clauses so arguments tend to be more on money that on extensions of time. Contractors will wish to prove that prevention has occurred and this leads to disputes on programmed rates of work, times allowed for service diversions, and availability or possession of the site. It may be possible to show that delays have been caused but these will not necessarily amount to prevention or give entitlement to extension of time.

Only if the delays have impeded the contractor in fulfilment of his obligation to complete on time will there be prevention. Disruption of the contractor's planned activities and impediment of the contractor's plans to finish ahead of time do not in themselves give the contractor a remedy against the employer for recovery of loss and expense or extra cost. The employer does not warrant that statutory undertakers whether working on a contractual basis or in exercise of statutory powers will fit in with the contractor's plans.

If the employer has provided information in the tender documents on expected duration times of statutory undertakers' works, and these times are exceeded, the contractor will obviously have better grounds for claim for both costs and extensions than if the employer has left it to the tenderers to make their own enquiries on such times. It is possible for the employer to provide information without warranting its accuracy, but unless there are good commercial reasons for including information on any matters which are outside the employer's control, such information is best omitted.

Delays to the contractor's progress caused by proximity to, or physical contact with, statutory undertakers' apparatus come into a different category from delays caused by the presence on site of statutory undertakers or their own contractors engaged on diverting mains or laying new ones. The question of whether delays caused by proximity or contact qualify for costs or extensions of time depends firstly, on how that risk is covered in the contract and secondly, on the information provided at the time of tender. At one end of the scale there may be no information given and no provision in the contract for the contractor to recover for unforeseen physical conditions or artificial obstructions. In this case the risk is on the contractor. At the other end of the scale there may be detailed, but possibly inaccurate, information given on locations of apparatus and also provisions in the contract for the contractor to get both extensions of time for delays and recovery of extra costs for unforeseen conditions. In this case the risk is on the employer.

13.8 Other special circumstances

The limitations of this phrase as a catch-all extension provision have been discussed in Chapter 6. There is a strong legal view that it cannot be used to cover the employer's acts of prevention unless express words are added to that effect.

The phrase is to be found in ICE forms and is widely used to cover any delay deemed unforeseen and beyond the control of the contractor.

13.9 Statutory powers

Under some standard forms, extensions of time can be granted for delays caused by restrictions imposed by central government using its statutory powers on the use of labour, the supply of materials, or power and energy. Such provisions may be useful to the employer as well as to the contractor for without it under some forms the contractor can, in the event of a long stoppage, determine his own employment under the contract.

Changes in the law imposed during the currency of the contract which affect the contractor's rate of production or otherwise restrict his activities – such matters as reduced noise levels or stricter safety regulations – would not normally give any rights to extensions unless deemed to be 'force majeure' or 'other special circumstances'.

13.10 Possession and access

The question of whether or not failure by the employer to give possession of the site at a time convenient to the contractor is an act of prevention will depend on the wording of the contract. If the contract simply states a date for possession of the site, the contractor is entitled to the whole of the site from the outset; but if the contract specifies phased release of the site, the contractor is obliged to accommodate the restrictions this will impose.

However, the employer does need to make clear his intentions. In *Whittal Builders Co. Ltd* v. *Chester-le-Street D.C.* (1987), the second of two cases between the same parties, it was held that in giving piecemeal possession of the 90 houses to be refurbished, the council was in breach of contract. Judge Fox-Andrews said:

> '... I am satisfied that in a contract of this kind it would be unusual for the contractor to be given possession of all houses at the same time.
>
> Usually, however, in such a contract where in the nature of things the local authority does not know precisely when occupation of the various houses will be given up to them some procedure is laid down whereby the local authority gives the best information about likely possession dates to the contractor to enable him to plan his work.
>
> However, I do not consider, in interpreting a written contract it is proper to take into consideration what normally will occur, particularly where the contract in such circumstances ordinarily contains express provisions dealing with that state of affairs ...'

Contractual provisions

Until amendment 4 of JCT 80, failure to give possession of the site was not expressed as a relevant event and, as happened in *Rapid Building* v. *Ealing Family Housing Association* (1984), any extension granted by the architect for a delay so caused was invalid. JCT 2005 provides for deferred possession and for extension of time for late possession.

ICE forms have always expressly included late possession as grounds for extension but this recognises perhaps that there are special problems in civil engineering works on possession, not least the amount of land sometimes involved and the continuing rights of other road users.

If the employer intends that the contractor should not have sole occupation of the site this also needs to be well expressed, and there certainly need

to be included in the contract, provisions for extending time in the event of
other authorised site users engaged by the employer causing delay.

Access and egress

The employer's obligations with regard to access and egress apply only to
his own land or routes otherwise promised in the contract. The contractor
has no claim for an extension in the event of delays caused by his own
inability to secure access routes or to keep them in satisfactory operating
order.

Traffic restrictions can pose problems for contractors, particularly if
imposed unexpectedly during the construction period. The position is complicated if the employer is the authority which has imposed the regulations.
But in any event probably the most the contractor can hope for is an extension under special circumstances if such a provision is included.

13.11 Late issue of drawings and instructions

Contractors complain much about the late issue of drawings and instructions and certainly many claims would be avoided if all schemes were
prepared to the last detail before the contractor started work. However,
unless there is an express term to the contrary it is not a breach of contract
for the employer through his agent to supply drawings and information as
the works proceed providing the contractor is not impeded in the performance of his obligations.

Implied terms

But clearly, in the absence of express terms relating to necessary information
there must be implied terms. And it is misunderstandings or disputes on
these which frequently cause conflict. The contractor will probably argue for
implied terms which suit his convenience and profitability. The employer
will probably argue for implied terms which did no more than require him
to avoid prevention. As a compromise they might both settle for what is
reasonable in all the circumstances.

In *Roberts* v. *Bury Improvement Commissioners* (1870) it was said:

> '... The contractor, also, from the nature of the works, could not begin
> the work until the commissioners and their architect had supplied plans
> and set out the land, and given the necessary particulars; and, therefore
> in the absence of any express stipulation on the subject, there would be
> an implied contract on the part of the commissioners to do their part
> within a reasonable time; and, if they broke that implied contract, the
> contractor would have a cause of action against them for any damages
> he might sustain . . .'

In *Wells* v. *Army & Navy Co-operative Society* (1902) it was said that:

'The plaintiffs (the contractors) must within reasonable limits be allowed to decide for themselves at what time they are to be supplied with detail.'

More recently in *J and J Fee Ltd* v. *The Express Lift Company Ltd* (1993), Judge Bowsher QC held that it was an implied term of a sub-contract under DOM/2 terms that the main contractor would provide the sub-contractor with correct information concerning the works in such a manner and at such times as was reasonably necessary for the sub-contractor to have in order to fulfil its obligations under the sub-contract.

Reasonable time for supply

Mr Justice Diplock in *Neodox Ltd* v. *Borough of Swinton and Pendlebury* (1958) had to decide what was a reasonable time in a contract with express provisions for the supply of information. He said:

'. . . It is clear from these clauses which I have read that to give business efficacy to the contract, details and instructions necessary for the execution of the works must be given by the engineer from time to time in the course of the contract and must be given in a reasonable time. In giving such instructions, the engineer is acting as agent for his principals, the Corporation, and if he fails to give such instructions within a reasonable time, the Corporation are liable in damages for breach of contract.

What is a reasonable time does not depend solely upon the convenience and financial interest of the claimants. No doubt it is to their interest to have every detail cut and dried on the day the contract is signed, but the contract does not contemplate that. It contemplates further details and instructions being provided, and the engineer is to have a time to provide them which is reasonable having regard to the point of view of him and his staff and the point of view of the Corporation, as well as the point of view of the contractors.

In determining what is a reasonable time as respects any particular details and instructions factors which must obviously be borne in mind are such matters as the order in which the engineer has determined the works shall be carried out (as he is entitled to do under clause 2 of the specification), whether requests for particular details or instructions have been made by the contractors, whether the instructions relate to a variation of the contract which the engineer is entitled to make from time to time during the execution of the contract, or whether they relate to part of the original works, and also the time, including any extension of time, within which the contractors are contractually bound to complete the works.

In mentioning these matters, I want to make it perfectly clear that they are not intended to be exhaustive, or anything like it. What is a reasonable time is a question of fact having regard to all the circumstances of the case, and the case stated does not disclose sufficient details of the circumstances relating to any particular details or instructions to make it possible for me to indicate what would be all the relevant factors in determining what was a reasonable time within which such details and instructions should have been given. What I have mentioned are merely some examples of factors which may or may not be relevant to any particular details or instructions given which the arbitrator has considered . . .'

Nature of implied terms

In *London Borough of Merton* v. *Leach* (1985) when asked to imply a term that Merton would take all steps necessary to enable the contractor to execute the works in a regular and orderly manner, Mr Justice Vinelott would go no further than saying:

'. . . However, the courts have not gone beyond the implication of a duty to co-operate whenever it is reasonably necessary to enable the other party to perform his obligations under a contract. The requirement of "good faith" in systems derived from Roman law has not been imported into English law . . .'

Contractual requirements

Clearly as with so many other matters much depends on the wording of the contract and its express terms, if any, on the supply of drawings and information. ICE forms use the phrase 'any failure or inability of the engineer to issue at a time reasonable in all the circumstances'; JCT forms refer to 'the contractor not having received in due time'. The contractor might well find it easier to argue a case for delay under ICE forms given the construction above as to what is reasonable.

Applications for information

Standard forms usually place some obligation on the contractor to make application for any additional drawings or information he needs to construct the works and the extent to which the contractor satisfies this obligation will have some bearing on his entitlement to an extension in the event of delay. If by not applying, or not applying in good time, he is in part responsible for the delay, that will be reflected in any assessment of the 'reasonableness' test in respect of time of supply and the 'fairness' test in relation to entitlement to an extension.

One of the many points at issue in *Merton* was whether a programme submitted in diagrammatic form at the start of the contract and marked with signs showing the date on which information was required throughout the contract satisfied the contractual requirements of JCT 63 as a specific application in writing, not unreasonably distant from the date required. Mr Justice Vinelott held that it did. This is part of his judgment:

'... As to the first of these two questions, I can see no reason why a document which like Programme 515 sets out in diagrammatic form the planned programme of work and indicates by conventional signs the dates by which instructions, drawings, details and levels are required, should not be a sufficiently specific application to meet the requirements of these clauses. What is called for is a document which indicates whether by words or by the use of conventional signs or in any other form, what the contractor requires and when he requires it and which does so in sufficient detail to enable the architect to understand clearly what is required of him. The arbitrator held that Programme 515 meets that requirement and I see no reason to differ from him ...'

13.12 *Variations and extra works*

When variations or extra works cause delay to completion, the contractor will have a clear case for extension of time, or if there is no express provision to extend time, the employer will have lost his right to liquidated damages.

It is the following types of questions which give rise to arguments, e.g. have the variations actually caused delay to completion? Are the extra works really extra? Could the contractor have accommodated them in his programme? Some of these matters have been considered previously and others are considered under particular forms but it is worth just noting here that extra quantities are not necessarily the same as extra works and in some forms of contract, it is the contractor and not the employer who takes the risks on quantities.

13.13 *Compliance with instructions*

In the absence of express terms, instructions, whether from the employer or the contract supervisor, can readily upset contractual intentions by causing prevention or defeating the contract altogether. Express terms giving contract supervisors power to make orders and instructions, and requiring the contractor to comply are useful therefore in maintaining contractual stability, but it is essential that they are linked to the extension provisions of the contract.

There are difficulties sometimes in distinguishing those instructions which give the contractor rights under the contract and those which are intended to correct or advise him in relation to performance and obligations. All that can be said is that it is up to the contract administrator to make his intentions clear. The case of *Simplex Concrete Piles Ltd* v. *Metropolitan Borough of St*

Pancras (1958) shows how the employer can easily end up with a surprise bill when the contract states that the final account is to include for all instructions and variations. In that case the specified piling system failed on testing and Simplex proposed an alternative system. The architect accepted the proposal in a letter which was subsequently held by the court to be an instruction for a variation entitling Simplex to recover the extra cost of the alternative system. As Mr Justice Edmund-Davies said of Simplex's position: 'On any view of the case, they were accordingly very fortunate in finding so amenable and co-operative an architect.'

13.14 Unforeseen physical conditions

Unforeseen conditions are generally taken to be, in the context of construction contracts, unexpected ground conditions on the site. It is something of an oddity, if not a matter of concern, that some contracts, including ICE conditions refer in their wording neither to the ground nor the site. Thus, clause 12 of ICE conditions refers to physical conditions (and artificial obstructions) encountered during the carrying out of the works.

The consequences of this wording were revealed in the case of *Humber Oil Trustees Ltd* v. *Harbour & General Works (Stevin) Ltd* (1991). In that case the question arose whether the contractor had encountered physical conditions within the scope of clause 12. As a 300-tonne crane on a jack-up barge was placing precast soffit units on piles, the barge became unstable and collapsed, causing extensive damage to the works, plant and equipment. The barge was a total loss and had to be replaced. There was much delay and extra cost.

The contract was under the ICE Fifth Edition conditions and the contractor claimed under clause 12 that collapse of the barge, and its consequences, was due to encountering physical conditions which could not have been foreseen by an experienced contractor. The dispute went to arbitration.

The arbitrator gave an award in favour of the contractor, finding that although the soil conditions were foreseeable, clause 12 contains no limitation on the meaning of 'physical condition'; that a combination of strength and stress, although transient, can fall within the terms; and that in this case an unforeseeable condition had occurred.

The employer appealed, maintaining that the question should be not whether the collapse could have been foreseen, which it clearly could not, but whether physical conditions could reasonably have been foreseen. The judge upheld the arbitrator's award but gave leave to appeal. The arguments advanced for the employer before the Court of Appeal would certainly have found favour with many engineers – namely that a physical condition is something material, such as rock or running sand, and that applied stress is not a physical condition nor is it something which can be encountered. The Court of Appeal, however, rejected the arguments and dismissed the appeal.

Lord Justice Parker dealt with the arguments as follows:

'Mr Dyson [Counsel for the employer] submits that the physical condition of the soil, which was found by the arbitrator to be foreseeable, really concludes the matter and that applied stress is not and cannot be any part of the physical condition.

Attractive as his argument appears to be at first sight, I cannot accept it. The arbitrator was in my view saying that the general soil conditions were foreseeable and well able to stand the applied loads and stresses. There was, however, here a peculiar characteristic which could not have been reasonably foreseen, namely a liability to shear at a much lower loading than had already been withstood.

The matter may perhaps be put in this way. General soil conditions were known and were foreseeable and foreseen. Such soil conditions would not have resulted in a shear failure. There was thus an unforeseeable condition.

Suppose that the Contractor, just before the event, had been informed by the engineer that some further information had just arrived showing that Stevin 73 would collapse because of a special feature of the soil under the leg, which nobody had hitherto known about. In my view the Contractor would then have encountered a physical condition which was not reasonably foreseeable. He would then have made proposals under clause 12, which the engineer might have approved, or indeed even directed the Contractor to take.

As to his submission that applied stress cannot be a part of a physical condition, this in my view is not so. The soil conditions which prevail at any moment when one is considering operations such as foundations or any other operation which puts weight on the soil is in effect the load-bearing capacity of the soil conditions. A particular condition of soil may, for example, be well known safely to sustain without shear 1,000 tonnes. If in fact there is settlement at a load of 300 tonnes what does it show? In my view, surely, that there was an unknown, foreseeable fault which was plainly a physical condition.'

Lord Justice Nourse agreeing said:

'The arbitrator found that there must have been a very unusual combination of soil strength and applied stresses around the base of leg 2 of the barge just before the failure occurred. He found as a fact that that state of affairs could not reasonably have been foreseen by an experienced contractor. That finding cannot be re-opened in this court. Accordingly the first question which we have to decide is whether this very unusual combination of soil strength and applied stresses was, as both the arbitrator and Judge Fox-Andrews have held, a physical condition encountered by the Contractors within clause 12(1) of the ICE Conditions.

The principal submissions of Mr Dyson, for the Employers, are to this effect. He says that a physical condition is something with a material, intransient existence, such as rock or running sand. An applied stress is not a physical condition nor, moreover, is it something which can be encountered. Accordingly, the only physical condition which here fell within clause 12(1) was the soil itself, whose nature could, as the arbitrator found, reasonably have been foreseen by an experienced contractor.

I reject these submissions for the following reasons. First, I agree with Mr Blackburn, for the contractors, that there is nothing to restrict the application of clause 12(1) to intransient, as distinct from transient, physical conditions. Indeed the express reference to weather conditions, albeit by way of exclusion, suggests the contrary. Secondly, while I would agree that an applied stress is not of itself a physical condition, we are not concerned with such a thing in isolation, but with a combination of soil and an applied stress. Thirdly and most significantly, as Butler-Sloss LJ pointed out during the course of the argument, it is impossible to speak of a contractor encountering any form of ground, be it rock, running sand, soil or whatever, without recognising that stress of one degree or another will have to be applied, at any rate notionally, to the ground, which will in turn behave, at any rate notionally, in one way or another; no doubt passively in the case of rock, actively in the case of running sand and perhaps unpredictably in the case of soil. In other words, for the purpose of clause 12(1), you cannot dissociate the nature of the ground from an actual or notional application of some degree of stress. Without such an application you cannot predict how the ground will behave. In the present case I would say that the condition encountered by the contractors was soil which behaved in an unforeseeable manner under the stress which was applied to it, and that that was a physical condition within clause 12(1).'

The decision in *Humber Oil* attracted a good deal of attention – from lawyers concerned as to its logic; from employers wishing to limit its consequences; and from contractors hoping to exploit its opportunities. But the cause of the problem lies not in the legal decision but in the wording of clause 12. Some employers now amend clause 12, substituting 'ground conditions on site' for 'physical conditions'.

Chapter 14
Causation and concurrency

14.1 Introduction

Disputes in construction contracts about extensions of time and liability for liquidated damages are, more often than not, disputes about the amount of extra time due to the contractor who has finished late or is running late rather than disputes about the relevant contractual provisions in themselves. Such disputes are frequently complex, particularly so on larger projects, with the parties likely to be at odds on the causes of delays, the effects of concurrency, criticality and the appropriate method of delay analysis.

In this chapter the general principles of causation and concurrency are examined. Criticality and delay analysis are considered in Chapter 16.

Although it might be thought, given the importance of these matters and the regularity with which they come to be decided in formal dispute resolution procedures, that there would by now be firm sets of rules emanating from the courts or from the drafting of standard forms, there remains, however, a great deal of uncertainty. That is not because of lack of effort; it is because the circumstances of particular contracts and the scenarios which can be contemplated are so diverse, and often so complex, that it is difficult to devise rigid rules of general application. But there is, perhaps, one such rule which comes out of the cases and that is the retention of common sense in the assessment of extensions of time.

Causation

Simply defined causation is the relationship between cause and effect. Lord Wright in *Monarch Steamship Co.* v. *Karlshamns Oljefabriker* (1949) put it as follows:

> 'Causation is a mental concept, generally based on inference or induction from uniformity of sequence as between two events that there is a causal link between them.'

The significance of causation generally is that in order to recover damages for breach of contract or in negligence it is necessary to establish both cause and cost effect. And, in claims for extension of time it is not enough to show that a relevant event occurred – it is also necessary to establish the delaying effect.

Concurrency

Concurrency, simply defined, has the same meaning as simultaneous –
namely that things are taking place, or have taken place, at the same time.

In one of the many *Royal Brompton Hospital* cases (2001) it was suggested
that concurrency only occurs in construction contracts when competing
delaying events have identical start and finish dates but that approach has
not been followed in other cases.

Generally, in construction cases, concurrency is taken to refer to competing events which overlap in their consequences. In that sense it is the timing
of the effects of the causes which are said to be concurrent – not the timing
of the causes in themselves.

Culpable delay

The expression 'culpable delay' is usually applied to the situation when the
contractor has failed to complete the works by the due date and has no
entitlement to an extension of time.

'Culpable' is an unfortunate description as it implies fault, guilt, or, as in
the *Shorter Oxford Dictionary*, 'deserving of punishment'. The fact that the
contractor has no entitlement to an extension does not necessarily make him
culpable in this sense. The delay may have been caused by circumstances
outside the contractor's control which do not qualify as relevant events –
vandalism to the works is an example. What is truly meant by culpable
delay, therefore, is delay with liability for liquidated damages.

The phrase 'culpable delay' patently does not apply to situations where
there is delay arising from prevention but there are no express grounds for
extension. Time is then at large and the contractor has no liability for liquidated damages.

'Culpable delay' before completion date

Culpable delay is sometimes used to describe the situation prior to the date
for completion where the contractor has fallen behind programme or schedule without cause for extension. This might be a reasonable application of
the phrase in respect of allegations of failing to proceed with due diligence
and the like but it has no relevance to liquidated damages and extensions
of time. See the case of *Walter Lawrence & Son Ltd* v. *Commercial Union* (1984)
in Chapter 13, which confirmed that the effects of delays are to be assessed
with regard to the time the works were carried out, not when they were
programmed to be carried out.

14.2 *Causation generally*

The analysis of causation is as old as history but it remains relentlessly
attractive to academics and inescapable in the pursuit of numerous profes-

sions. In the construction industry, the study of causation and the application of computers to the process is a thriving business with ever-developing skills in retrospective delay analysis (RDA). But, despite the studies and despite the skills, for the ordinary practitioner in construction there are still no firm rules of certain application to the everyday problems on building sites. Perhaps that is how it will always be. As Mr Justice Steyn said in the case of *Banque Financière de la Cité* v. *Westgate Insurance Co. Ltd* (1990):

> 'There is no more difficult area in our law than causation. Scientific precepts and philosophical notions are frequently invoked. Ultimately, it seems to me, a judge is on safe ground if he puts his trust in precedent, or in its absence, in common sense.'

Some years earlier, in the case of *Heskell* v. *Continental Express Ltd* (1950), Mr Justice Devlin had made a similar point on the application of common sense when saying:

> 'The cause of the loss has to be ascertained by the standard of common sense of the ordinary man. Common sense is a blunt instrument not suited for probing into minute points, and I cannot believe that if the ordinary man thinks that two causes are of equal efficacy, he cannot say so without being interrogated on fine distinctions.'

However, despite the difficulties that is not to say that delay analysis related to extensions of time and claims for loss and expense or extra cost should not be as thorough as circumstances permit. In the *McAlpine Humberoak* case mentioned in Chapter 2 Lord Justice Lloyd said:

> 'The judge dismissed the defendant's approach to the case as being a retrospective and dissectional reconstruction by expert evidence of events almost day by day, drawing by drawing . . . In our view the defendant's approach is just what the case required.'

And in the *John Barker* case, mentioned in Chapters 11, 12 and 16 the judge held that the architect's impressionistic rather than calculated assessment of the consequences of delaying events invalidated his award of extension of time.

In the *John Doyle* v. *Laing* (2004) case it was said:

> 'the question of causation must be treated by "the application of common sense to the logical principles of causation": *John Holland Construction & Engineering Pty Ltd* v. *Kvaerner RJ Brown Pty Ltd, supra*, at 82 BLR 84I per Byrne J.; *Alexander* v. *Cambridge Credit Corporation Ltd* (1987) 9 NSWLR 310; *Leyland Shipping Company Ltd* v. *Norwich Union Fire Insurance Society Ltd* [1918] AC 350, at 362 per Lord Dunedin. In this connection, it is frequently possible to say that an item of loss has been caused by a particular event notwithstanding that other events played a part in its occurrence. In such cases, if an event or events for which the employer is responsible can be described as the dominant cause of an item of loss, that will be sufficient to establish liability, notwithstanding the existence of other causes that are to some degree at least concurrent. That test is

similar to that adopted by the House of Lords in *Leyland Shipping Company Ltd* v. *Norwich Union Fire Insurance Society Ltd . . .'*

A further neat summary of the law on causation was given in the judgment of Judge Wilcox in the case of *Great Eastern Hotel Company Ltd* v. *John Laing Construction Ltd* (2005) where Laing, as the construction manager, was successfully sued by the employer for breaches of performance. The judge said:

'313. On the question of causation, the law is conveniently set out in *Chitty on Contracts* (29th Edition) at volume 1 at paragraph:
"The courts avoided laying down any formal test for causation; they have relied on commonsense to guide decisions as to whether a breach of contract is a sufficiently substantial cause of the Claimant's loss. The answer to whether the breach was the cause of the loss, or merely the occasion for loss must in the end depend on the court's commonsense in interpreting the facts."

314. If a breach of contract is one of the causes both co-operating and of equal efficiency in causing loss to the Claimant the party responsible for breach is liable to the Claimant for that loss. The contract breaker is liable for as long as his breach was an "effective cause" of his loss. See *Heskell* v. *Continental Express Ltd* [1995] 1 All Eng 1033 at page 1047A. The Court need not choose which cause was the more effective. The approach of Devlin J in *Heskell* was adopted by Steyn J (as he then was) in *Banque Keyser SA* v. *Skandia* [1991] QB page 668 at page 717 and accepted by the Court of Appeal see page 813A to 814C.'

The judge went on to say:

'315. Each claim or group of claims must be examined on their own facts and in the context of the specific contractual provisions such as variations which may give rise to a consideration of the comparative potency of causal events and to apportionment. In the absence of such provision the appropriate test is that if GEH prove that Laing were in breach and the proven breach materially contributed to the loss then it can recover the whole loss, even if there is another effective contributory cause provided that there is no double recovery.'

Approaches to analysis

The complexity of causation in legal matters can be seen from the old case of *Leyland Shipping Co.* v. *Norwich Union Fire Insurance Society* (1918) regarding a ship which, having been torpedoed by a German submarine, was towed to a deepwater harbour at Le Havre and then, when a gale blew up, towed out to an anchorage near a breakwater where she sank when the tide went out. The court had to decide the cause of the sinking because the ship was insured against the perils of the sea but not against war risks. It was held that the cause was the torpedoing and that the unsuccessful attempts to save the ship did not break the chain of causation. Lord Shaw said:

> '... When various factors or causes are concurrent and one has to be selected ... the choice falls upon the one to which may variously be ascribed the qualities of reality, predominance, efficiency ...'

The search for a dominant cause, however, is by no means the only method of analysing causation. Two well-reported medical cases illustrate alternative possibilities.

In *Baker* v. *Willoughby* (1970) the House of Lords took what is sometimes known as the first cause approach. The plaintiff was injured in the leg as a result of a motor accident caused by the defendant's negligent driving. The injury caused pain and suffering and affected the plaintiff's earning capacity. Later, the plaintiff was involved in an attempted robbery during which he was shot in the injured leg. The leg then had to be amputated. The question for the courts was whether the defendant was liable to the plaintiff for a lifetime's worth of pain, suffering and loss of earnings or just for that which occurred prior to the amputation. The House of Lords said the defendant was liable for the full lifetime.

By contrast, in *Jobling* v. *Associated Dairies* (1982), the House of Lords took what may be called the ultimately critical approach. In that case the plaintiff suffered an accident at work which left him with a back injury and a reduced earning capacity. A few years later, but before settlement of damages for the back injury, he was found to be suffering from spinal disease which made him totally unfit for work. The disease had no connection with the injury. The House of Lords held there could be no recovery for loss of earnings from the time of total incapacity. The disease could not be disregarded because it was a supervening cause of the plaintiff's condition which would have overtaken him in any event.

14.3 Concurrency generally

The leading authority on the simpler aspects of causation and concurrency is the judgment of Mr Justice Devlin in the shipping case of *Heskell* v. *Continental Express Ltd* (1950). Having found that there were two separate operative causes he said:

> 'Where the wrong is a tort, it is clearly settled that the wrongdoer cannot excuse himself by pointing to another cause. It is enough that the tort should be a cause and it is unnecessary to evaluate competing causes and ascertain which of them is dominant ... In the case of breach of contract the position is not so clear ...
>
> Whatever the true rule of causation may be I am satisfied that if a breach of contract is one of two causes, both co-operating and both of equal efficacy, as I find in this case, it is sufficient to carry judgment for

damages. *Reischer* v. *Borwick* [1894] 2 QB 548 establishes that for the purposes of a contract of insurance it is sufficient if an insured event is, in this sense, a co-operating cause of the loss. I do not think that *Yorkshire Dale SS Co. Ltd* v. *Minister of War Transport* [1942] AC 691, with its insistence on the ascertainment of "the cause", disapproved this principle. The case decided that the cause of a loss has to be ascertained by the standard of common sense of the ordinary man. Common sense is a blunt instrument not suited for probing into minute points, and I cannot believe that if the ordinary man thinks that two causes are of approximately equal efficacy, he cannot say so without being interrogated on fine distinction.'

The Devlin approach

From the above case comes what is known as 'the Devlin approach' – namely, that if a breach of contract is one of two causes, both co-operating and both of equal efficacy, either is sufficient to establish liability for loss.

The Devlin approach was considered in the case of *Plant Construction Plc* v. *Clive Adams Associates* (2000) where, having noted previous approvals of the approach, the judge said:

'13. In *Heskell* v. *Continental Express Ltd* Devlin J had referred to the two causes as being equally operative "in that if either had ceased the damage would have ceased" (page 1047B), and it would seem that the same was true in the *Banque Keyser Ullmann* case. These were therefore cases like the "head-on-collision" example in paragraph 7 above. Nevertheless the statements of principle, even if not explicitly directed to the question posed in paragraph 7 above, are wide enough to cover it, nor do I see any reason why they should not do so. Moreover, as I understand it, Devlin J's references to "equality" are not directed to any contribution issue but to the question of whether both causes under consideration are "efficacious".

14. On the basis of principle, therefore, with such limited assistance as can be derived from the authorities, I conclude that Plant should not have to show that JMH's breach was the sole cause of the collapse, or that in the absence of that breach there would have been no collapse even if Plant's negligence (including that of Mr Adams as its agent) had remained. The true question, in my view, should be whether JMH's breach was causative, whether alone or as being one of concurrent causes with Plant's negligence, such that but for the concurrence of those causes the collapse would not have occurred.'

Notwithstanding the apparent attempt by the judge in the above case to widen the application of the Devlin approach the point is made by many legal commentators that the Devlin approach does not operate satisfactorily in construction cases where there are counterclaims nor where the competing causes are not of equal efficiency.

Other approaches

Keating, 8th Edition, at pages 272 to 275 gives a detailed analysis of causation as it applies to contracts and shows how unclear the law is when there are claims and counterclaims and competing causes of fault – some the claimant's, some the defendant's, and some neutral. *Keating* comes down broadly in favour of the dominant cause approach for contractors' claims for payment for delay arising from employer's breach where there are competing causes of delay such as neutral events or contractors' own fault.

Keating suggests that there is legal authority for a number of propositions which it lists as follows:

'(a) the Devlin approach. If a breach of contract is one of two causes of a loss, both causes co-operating and both of approximately equal efficacy, the breach is sufficient to carry judgment for the loss.

(b) the dominant cause approach. If there are two causes, one the contractual responsibility of the defendant and the other the contractual responsibility of the plaintiff, the plaintiff succeeds if he establishes that the cause for which the defendant is responsible is the effective, dominant cause. Which cause is dominant is a question of fact, which is not solved by the mere point of order in time, but is to be decided by applying common sense standards.

(c) the burden of proof approach. If part of the damage is shown to be due to a breach of contract by the plaintiff, the claimant must show how much of the damage is caused otherwise than by his breach of contract, failing which he can recover nominal damages only.

(d) the tortious solution. The claimant recovers if the cause on which he relies caused or materially contributed to the loss.'

Other lines of thought are variously described as:

- apportionment
- the 'but-for' test
- the first-in-line approach
- the first-past-the-post approach.

14.4 *Dominant cause approach*

The dominant cause approach as outlined by Lord Shaw in *Leyland Shipping* as stated above is widely taken to be the correct approach to dealing with concurrent causes. Nevertheless it appeared to receive something of a rebuff in the case of *H Fairweather & Co. Ltd* v. *London Borough of Wandsworth* (1987). Judge Fox-Andrews remitted for reconsideration an arbitrator's award which held where there was more than one cause of a delay the extension had to be granted for the dominant reason. The arbitrator had declared the architect correct in granting 81 weeks for strikes. The contractor wanted 18 weeks re-allocated to reimbursable causes.

Judge Fox-Andrews said this:

'... "Dominant" has a number of meanings: "Ruling, prevailing, most influential." On the assumption that condition 23 is not solely concerned with liquidated or ascertained damages but also triggers and conditions a right for a contractor to recover direct loss and expense where applicable under condition 24 then an architect and in his turn an arbitrator has the task of allocating, when the facts require it, the extension of time to the various heads. I do not consider that the dominant test is correct ...'

Although this judgment has been widely reported as signalling the end of the dominant cause test, the judge was not addressing the question of causation as it relates to financial claims but was simply taking a practical approach to the relationship between extensions of time and such claims. He went on to say:

'Neither this part of the judgment nor the terms of the contract itself points to an extension of time under condition 23(f) being a condition precedent to recovery of direct loss and expense under condition 24(1)(a). However the practical effect ordinarily will be that if the architect has refused an extension under the former the contractor is unlikely to be successful with the architect on an application under condition 24(1)(a).'

Comment

The problem with the dominant cause approach is that although it works well in insurance cases it is not always suited to construction cases where there are time and money claims and it seems to be at odds with other rulings on extension of time claims. See the *Wells* v. *Army & Navy* case (1902) discussed later in this chapter. Generally, however, it continues to receive judicial support – see, for example, the case of *Galoo Ltd* v. *Bright Grahame Murray* (1994) where it was said 'A plaintiff was entitled to claim damages for breach of contract by the defendant where the breach was the effective or dominant cause of loss and did not merely provide him with the opportunity to sustain loss'.

14.5 Apportionment

Although they are not an exact match, there are many similarities between global claims, on which a considerable body of case law has developed, and apportionment, on which there is little case law. On the matter of financial claims the two came together in the important decision of the Scottish courts in the case of *John Doyle* v. *Laing* (2004) discussed in Chapter 12. For extension of time claims, however, there has until recently been little or no authorative support for acceptance of apportionment in extension of time claims.

If anything, as noted in Chapter 16, the courts have generally taken a strict approach towards linkage of cause and effect.

However, something of a shift in the approach of the courts may be detected from two recent judgments.

In *London Underground Ltd* v. *Citylink Telecommunications Ltd* (2007) Mr Justice Ramsey dealing with an appeal against an architect's award on extensions of time said:

> '141. In deciding whether there was serious irregularity, I consider that the proper approach to global claims is relevant. The approach set out in the decision in [*Doyle* v. *Laing*] is not challenged on this application and I accept that approach.
> 142. The essence of a global claim is that, whilst the breaches and the relief claimed are specified, the question of causation linking the breaches and the relief claimed is based substantially on inference, usually derived from factual and expert evidence.'

And in the Scottish case of *City Inn Ltd* v. *Shepherd Construction Ltd* (2007) Lord Drummond Young explained how he arrived at the conclusion that apportionment was appropriate in a JCT type contract as follows:

> '[18] While delay for which the contractor is responsible will not preclude an extension of time based on a relevant event, the critical question will frequently, perhaps usually, be how long an extension is justified by the relevant event. In practice the various causes of delay are likely to interact in a complex manner; shortages of labour will rarely be total; some work may be possible despite inclement weather; and the degree to which work is affected by each of these causes may vary from day to day. Other more complex situations can easily be imagined. What is required by clause 25 is that the architect should exercise his judgment to determine the extent to which completion has been delayed by relevant events. The architect must make a determination on a fair and reasonable basis. Where there is true concurrency between a relevant event and a contractor default, in the sense that both existed simultaneously, regardless of which started first, it may be appropriate to apportion responsibility for the delay between the two causes; obviously, however, the basis for such apportionment must be fair and reasonable. Precisely what is fair and reasonable is likely to turn on the exact circumstances of the particular case. A procedure of that nature is in my opinion implicit in the wording of clause 25.3.1 and .3; both of these provisions direct the architect to give an extension of time by fixing a Completion Date that he considers to be fair and reasonable.
> [19] The foregoing construction of clause 25 is in my opinion supported by the approach taken to concurrent causes of delay in Federal tribunals in the United States. In *Chas. I. Cunningham Co.*, IBCA 60, 57–2 BCA P1541 (1957), the Board of Contract Appeals considered

the legal consequences where a contractor has claimed for an extension of time but is himself in default. The main opinion of the Board, delivered by one of its members, Mr Slaughter, states the law as follows:

"It is well settled that the failure of a contractor to prosecute the contract work with the efficiency and expedition requisite for its completion within the time specified by the contract does not, in and of itself, disentitle the contractor to extensions of time for such parts of the ultimate delay in completion as are attributable to events that are themselves excusable, as defined in [the relevant extension of time clause, corresponding to clause 25]. Where a contractor finishes late partly because of a cause that is excusable under this provision and partly because of a cause that is not, it is the duty of the contracting officer to make, if at all feasible, a fair apportionment of the extent to which completion of the job was delayed by each of the two causes, and to grant an extension of time commensurate with his determination of the extent to which the failure to finish on time was attributable to the excusable one. Accordingly, if an event that would constitute an excusable cause of delay in fact occurs, and if that event in fact delays the progress of the work as a whole, the contractor is entitled to an extension of time for so much of the ultimate delay in completion as was the result or consequence of that event, notwithstanding that the progress of the work may also have been slowed down or halted by a want of diligence, lack of planning, or some other inexcusable omission on the part of the contractor."

This approach recognises the fact that culpable and non-culpable causes of delay will frequently coexist and interact, and permits the contracting officer, equivalent to the architect under the JCT Forms, to apportion the delay between the culpable and non-culpable causes. That seems to me to be the only way in which a fair result can be achieved in such cases, and in my opinion such an approach is contemplated by the wording of clause 25. I should add that the decision of the Board of Contract Appeals in *Chas. I. Cunningham Co.* was followed in *Sun Shipbuilding & Drydock Co.*, ANBCA 11300, 68–1 BCA (CCH) P7054 (1968).

[20] Counsel for the pursuers founded strongly on the opinion of the court in *John Doyle Construction Ltd* v. *Laing Management (Scotland) Ltd*, 2004 SC 73. That case dealt with a claim for direct loss and expense under the equivalent of clause 26 of the JCT Standard Form 1980. It was concerned in particular with the way in which a contractor could establish a global claim, where it is impossible to demonstrate individual causal links between events for which the employer is responsible and particular items of loss and expense. Normally, when a global claim is pursued, the contractor must demonstrate that the whole of his loss and expense results from matters that are the responsibility of the employer. The court pointed out that that requirement might be mitigated in three ways. First, it may be

possible to identify a causal link between particular events for which the employer is responsible and individual items of loss. Secondly, the question of causation must be treated by the application of common sense to the logical principles of causation, and if it is possible to identify an act of the employer as the dominant cause of the loss that will suffice. Thirdly, it may in some cases be possible to apportion the loss between the causes for which the employer is responsible and other causes. In my opinion these principles have only limited application to the present case. They are concerned with claims for loss and expense, and consequently may have some bearing on the defenders' claim for prolongation costs (see below, at paragraphs [162]–[167]). They do not, however, appear directly relevant to the granting of an extension of time. The contractual wording relating to an extension of time is different from that relating to claims for loss and expense. In particular, in the form of contract that is presently under consideration, there is no reference in clause 26 to the architect's making such award as is "fair and reasonable". For the reasons discussed above, I attach considerable importance to those words in the interpretation of clause 25, especially in its practical application. In addition, the conceptual structure of the two clauses is quite different, and the events that trigger an extension of time and a claim for loss and expenses are likewise distinct. Consequently I do not think that the decision in *John Doyle Construction* is of general assistance in the construction of clause 25, subject to one exception, which is discussed in the following paragraph. Perhaps the one theme that is common to clauses 25 and 26 is that a practical common sense approach should be adopted to the interpretation of building contracts, but it is hardly necessary to refer to authority for that proposition.

[21] In the course of their submissions counsel for the pursuers advanced a number of legal propositions. First, it was said that for a contractor to establish an entitlement to an extension of time in respect of delay arising out of a relevant event he must establish that the delay was caused by the relevant event, as opposed to any other pre-existing or concurrent matter for which the contractor himself is responsible; and he must establish the extent of such delay. In my opinion that proposition is too broadly stated. It is correct that the contractor must establish that delay was caused by a relevant event, and the extent of the delay; nevertheless, I am of opinion that concurrent causes should be treated in the manner discussed in paragraph [18] above. The second proposition advanced for the pursuers was that, if a relevant event can be shown to be the "dominant or operative" cause of a delay, the party responsible for that event will be held responsible for the delay. I agree that it may be possible to show that either a relevant event or a contractor's risk event is the dominant cause of that delay, and in such a case that event should be treated as the cause of the delay. A similar principle was recognised in *Doyle*,

at paragraph [15] of the opinion of the court; the principle is derived from older case of *Leyland Shipping Company Ltd* v. *Norwich Union Fire Insurance Society Ltd* [1918] AC 350. Those cases refer to the "dominant" or "proximate" cause. The pursuers' submission went further, and referred to the "dominant or operative" cause of the delay. In my opinion this extension is not legitimate. Indeed, I have difficulty in seeing what the word "operative" adds to the notion of causation; a cause can only be relevant if it is operative, and that is as true of concurrent causes as it is of single or "dominant" causes.

[22] The pursuers' third proposition was that a variation instructed during a period when the contractor is already in delay will not absolve the contractor of responsibility for that pre-existing delay, unless it is proved that the delay resulted from the variation. As stated, this is correct. Nevertheless, the "delay" that matters is delay to the Completion Date. If the contractor is, through his own fault, in delay before a relevant event, that may explain delay that follows the Completion Date. Alternatively, it may be possible for the contractor to demonstrate that he would have made up the delay caused by his own fault, and that the delay beyond the Completion Date results from the variation. It is all a question of fact. The pursuers' fourth proposition was in two parts: first, it is a defence to a claim that a variation or late instruction caused delay to establish that the matter to which the variation or late instruction was issued was not on the critical path; secondly, it is also a defence that the claimed delay was in fact due to other events. The first of these contentions was not, I think, in dispute, although the parties were sharply in dispute as to where the critical path lay in the progress of the contractual works. The second contention, however, is perhaps stated rather simplistically. In practice causation tends to operate in a complex manner, and a delay to completion may be caused in part by relevant events and in part by contractor default, in a way that does not permit the easy separation of these causes. In such a case, the solution envisaged by clause 25 is that the architect, or in litigation the court, must apply judgment to determine the extent to which completion has been delayed by relevant events. In an appropriate case apportionment of the delay between relevant events and contractor's risk events may be appropriate. Precisely when and how that should take place is a question that turns on the precise facts of the case.'

Comment

It may be said that Lord Drummond Young's opinion is confined to construction contracts where the contract administrator's duty is to assess entitlement to extension of time on a fair and reasonable basis. And clearly it may be of limited application to contracts with precisely defined rules for assessment which differ from JCT rules. Nevertheless it provides some legal authority for apportionment in relation to extensions of time and it may prove to be an important decision in the course of time.

14.6 Rules for extensions of time

As noted in previous chapters of this book extension of time provisions are included in construction contracts primarily for the benefit of the employer in so far that they allow time to be extended for preventative acts and thereby guard against time becoming at large. But because extensions of time can also be seen as being a contractual entitlement for the contractor it is clearly necessary that assessment of entitlement in any particular case starts with examination of the applicable contractual provisions. Only thereafter is it appropriate to look for guidance from related or general legal rulings.

Accordingly it cannot be said that general rules of causation and concurrency (developed, in any event, mainly from insurance cases) are always applicable, nor can it be said rulings on particular contractual provisions are of general effect.

There is, however, an important rule of long standing relating to causa-tion and concurrency in extensions of time claims which is worth noting: where there are concurrent causes of critical delay, one of which gives rise to an extension of time then, in the absence of wording in the contract to the contrary, the contractor will be entitled to that extension even if there are other causes of delay which are the contractor's responsibility.

The rule comes from the old case of *Wells* v. *Army & Navy Co-operative Society* (1902). It was stated more recently as an agreed proposition in *Henry Boot Construction (UK) Ltd* v. *Malmaison Hotel (Manchester) Ltd* (1999) as follows:

> 'if there are two concurrent causes of delay, one of which is a relevant event, and the other is not, then the contractor is entitled to an extension of time for the period of delay caused by the relevant event notwithstanding the concurrent effect of the other event.'

Precisely how the rule is applied depends upon the wording of the particular contract.

14.7 Discussion on various approaches

Reimbursable/non-reimbursable

Consider a 10-week delay arising from prevention, overlapped by a 10-week delay for weather, making an overall 15-week delay.

The contractor might say that from the first-in-line or dominant event approach all 15 weeks should be given for prevention. The employer might argue for apportionment.

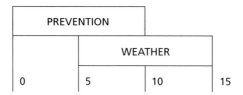

Now reverse the delays:

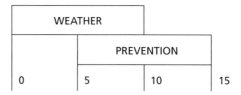

This time the employer might argue for the first-in-line approach and the contractor for apportionment or the dominant approach.

In the first case there is some logic in the first-in-line approach – the delays due to weather would not have occurred unless preceded by the prevention delays. But in the second case the logic of first-in-line breaks down – the prevention delays would have occurred whether or not preceded by the weather delays.

The probability is that in the first case the contractor would get either 10 weeks for prevention and 5 for weather; or 15 weeks for prevention. It should not make any difference financially to the contractor because he is fully covered against liquidated damages and his costs for the 5 weeks weather should, in the example shown, be recoverable irrespective of under what head the extension is granted.

In the second case the contractor would probably get 5 or 10 weeks for weather plus 10 or 5 weeks for prevention.

Reimbursable/culpable delays

```
        ┌─────────────────────────┐
        │      PREVENTION         │
        ├──────────┬──────────────┤
        │          │ CULPABLE DELAY│
        ├──────────┴──────────────┤
        0          5          10    15
```

In this case the first-in-line approach should give the contractor a 10-week extension for prevention but no more.

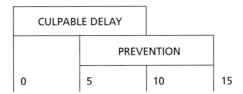

The contractor might argue for a 15-week extension for prevention but should get only 10.

Non-reimbursable/culpable delays

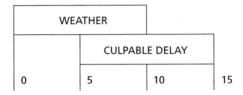

The contractor should probably get 10 weeks for weather. The argument for granting 10 rather than 5 is that the employer in strictly legal terms needs to prove breach to have damages and in concurrent situations as this, the contractor is entitled to the benefit of any doubt.

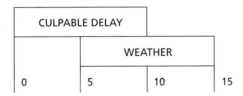

In this case the contractor is the cause of his own misfortune with weather and there appears to be no reason why he should have an extension if considerations of fairness can be invoked. In *Amalgamated Building Contractors v. Waltham Holy Cross UDC* (1952), Lord Justice Denning hypothesised on a similar situation and suggested that an extension for a neutral event should be granted to cover the actual period of delay by that event.

Defects of first-in-line

It can be seen from the above that the defect of taking the first-in-line approach is that it is not responsive to the concept of culpable delay and can only apply, if at all, to concurrent relevant events.

Delay by an employer in re-nominating, such as in *Peak v. McKinney* (1970), *Percy Bilton v. Greater London Council* (1982) and *Westminster v. Jarvis* (1970) would have little effect with first-in-line since such delay will almost invariably follow delay of the default itself for which the main contractor must, under some forms, take responsibility.

Dominant versus apportionment

The dominant cause approach holds good in all situations except where secondary considerations such as claims are relevant. Apportionment has the benefit of flexibility and offers more scope for compromise if a dispute arises than either of the other two approaches. The problem with apportionment is that however sensible an approach it may seem, it can cut across legal principles and establish incorrect liabilities in some instances.

14.8 Extensions when in culpable delay

The point has already been made that failure by the contractor to maintain progress does not disqualify him from extensions of time when he is still within the original or extended time for completion.

The question arises, and until recently was by no means fully settled, does the contractor have an entitlement to an extension if he is in culpable delay having failed to complete within the specified time? What is certain is that acts of prevention, less still, neutral events cannot always be avoided in such circumstances. The employer may not be able to issue instructions or variations on a particular matter until the works have reached a certain stage of construction and if there is no power to extend for prevention the right to liquidated damages will be lost.

However, following the decision in the case of *Balfour Beatty* v. *Chestermount* (see Chapter 5) it is now settled that it would take very clear words in a contract to produce the result that extensions of time could not be given to cover acts of prevention occurring during a period of culpable delay. As to the effect of neutral events so occurring that is less certain. But, if the test in the contract for any extension is what is fair and reasonable, then the contractor's entitlement to an extension may be dependent upon the contractor being able to show that even without his own delay the particular relevant events would have delayed completion.

Dot-on procedure

Contractors often argue that an extension for prevention should run from the original completion date to the date at which the prevention finishes – thereby conveniently absolving them from their culpable delay. This approach was rebuffed by Lord Justice Denning in *Amalgamated Building Contractors* v. *Waltham Holy Cross* (1952). Referring in that case to a neutral event he suggested adding on the time for the delay to the original time – a procedure now known as dotting-on. This leaves the contractor with liability for damages for the portion of his culpable delay. He said:

> 'where the contractors, near the end of the work, have overrun the contract time for six months without legitimate excuse . . . Now suppose . . . a strike occurs and lasts a month. The contractors can get an extension of time for that month. The architect can clearly issue a certificate which will operate retrospectively. He extends the time by one month from the original completion date, and the extended time will obviously be a date which is past.'

Current contractual position

The extent to which the dot-on approach can be taken for prevention or for that matter for neutral events depends on the wording of the extension provisions of the contract and other associated clauses. There needs to be express power for the architect or engineer to extend time after the original completion date has passed.

However, see the detailed commentary in Chapter 5 on the *Balfour Beatty* v. *Chestermount* case which approved the dot-on procedure (the net approach) and rejected the alternative gross method.

Chapter 15
Programmes, method statements and best endeavours

15.1 Status of contractor's programmes and method statements

Responsibility for programming and constructing the works rests with the contractor and under most standard forms neither the contractor's programme nor method statements are contract documents. Indeed the programme is often not published until after the contract has been signed and usually requirements for the contractor to produce a programme lie in the contract itself. This has the effect that under some forms it may be a breach of contract not to produce a programme but it is rarely a breach of contract not to proceed to programme.

The contractor's programme can be of relevance in considering rates of progress and if there is a question of determination of the contractor's employment for lack of progress, under express terms or at common law, it can be of interest as evidence. But more commonly the programme is of interest in delay and disruption claims as an aid to establishing intentions and performance.

Note, however, that it is possible for programmes and method statements to acquire contractual status without there being any direct statement to that effect in the contract.

Thus in *London Borough of Merton* v. *Stanley Hugh Leach Ltd* (1985) it was held that the contractor's programme, adequately annotated, could serve as written notice of requirements for information and drawing. And in *Howe Engineering Ltd* v. *Lindner Ceilings & Partitions Ltd* (1999) it was held that as the method statement had been referred to in both the specification and bills of quantities, both of which were contract documents, the method statement itself was of contractual effect.

15.2 Programmes

If a contractor's programme is found to be a contract document, it binds not only the contractor to perform to it but also the employer to facilitate such performance. Thus, the employer's duty to avoid prevention which in normal circumstances extends only to not obstructing the contractor in his obligation to complete on time, can be widened so that prevention can apply to programme requirements. The impact of this on the potential for claims from the contractor for delay and disruption is self-evident.

For this reason, amongst others, most employers avoid any programme, tender or otherwise, being incorporated in the contract documents.

Obligation to work to programme

A question which frequently arises is to what extent, if any, is a contractor or a sub-contractor obliged to work to a programme which is not a contract document.

The point came up in the case of *Pigott Foundations Ltd* v. *Shepherd Construction Ltd* (1993) where the sub-contractor argued that it was not under any obligation to do more than complete the sub-contract work within the time allowed and that it was not bound by any particular rate of progress. The main contractor contended that he was entitled to damages because the slow progress of the sub-contract during the contract period prevented other work from being carried out which would have been possible had the piling sub-contractor proceeded with reasonable expedition. The judge held that in the absence of any indication to the contrary (such as a contractually binding programme) the sub-contractor was entitled to plan and perform the sub-contract work as he pleased provided he finished within the time fixed for the sub-contract.

The judge said:

> 'In my judgment the obligation of the sub-contractor under clause 11.1 of DOM/1 to carry out and complete the sub-contract works "reasonably in accordance with the progress of the Works" does not upon its true construction require the sub-contractor to comply with the main contractor's programme of works nor does it entitle the main contractor to claim that the sub-contractor must finish or complete a particular part of the sub-contract works by a particular date in order to enable the main contractor to proceed with other parts of the works. The words "the progress of the Works" are in my judgment directed to requiring the sub-contractor to carry out his sub-contract works in such a manner as would not unreasonably interfere with the actual carrying out of any other works which can conveniently be carried out at the same time. The words do not however in my judgment require the sub-contractor to plan his sub-contract work so as to fit in with either any scheme of work of the main contractor or to finish any part of the sub-contract works by a particular date so as to enable the main contractor to proceed with other parts of the work.'

And, later in the judgment, the judge said:

> 'In my judgment clause 11.8 does not exclude or modify the general principle applicable to building and engineering contracts, that in the absence of any indication to the contrary, a contractor is entitled to plan and perform the work as he pleases provided that he finishes it by the time

fixed in the contract. See *Wells* v. *Army & Navy Co-operative Society* (1902) 2 Hudsons Building Contracts (Fourth Edition) 346; *GLC* v. *Cleveland Bridge & Engineering Company* (1984) 34 BLR 50.'

The *Ascon* case (1999)

In *Ascon Contracting Ltd* v. *Alfred McAlpine Construction Isle of Man Ltd* (1999) one of the matters the judge had to consider was whether a provision in a sub-contract required the sub-contractor, Ascon, to comply with the main contract programme. The relevant provision, clause 11.1, read:

> '11.1 The Sub-Contractor shall carry out and complete the Sub-Contract Works in accordance with the details in the Appendix, part 4, and reasonably in accordance with the progress of the Works . . .'

The judge concluded that the words 'reasonably in accordance with the progress of the Works' did not go so far as to require Ascon to comply with the detail of McAlpine's main contract programme but that it did impose an obligation to proceed reasonably in accordance with that programme. He said (after reciting that part of the judgment in the *Pigott Foundations* case quoted above):

> '8.8 I respectfully agree that clause 11.1 does not require the sub-contractor to comply with the detail of the main contractor's programme, either generally or in relation to the work of other specific sub-contractors. My own view, however, is that the words "reasonably in accordance with the progress of the Works" go somewhat beyond a negative duty "not unreasonably [to] interfere with the actual carrying out" of other works. The sub-contractor knows the nature of the main contract works and the place of the sub-contract works in them. As *Keating* suggests in the passage referred to by Judge Gilliland this obligation presupposes that the main contract works are proceeding regularly and diligently. The "progress" referred to is therefore, I think, that expected and observed in the light of those facts, although the obligation is only to proceed "reasonably" in accordance with that progress. It is to be noted, moreover, that in *GLC* v. *Cleveland Bridge & Engineering Co. Ltd* (1986) 8 ConLR 30, which seems to be relied upon in the *Pigott Foundations Ltd* case, there was no express term to the same or similar effect as that in question here and there had been no delay in completion of the relevant obligation. Another part of the same discussion in *Keating* (6th edn, 1995) p838 suggests that the sub-contractor is under an obligation to carry out the sub-contract works "at such a pace as will enable him to complete in accordance with the agreed [sub-contract] programme details". Here Ascon did not complete on time and it was apparent that it would not be able to do so throughout the period during which McAlpine alleges that Structal's commencement was delayed.

8.9 I do not, therefore, consider that on the point of law either party wholly succeeds. McAlpine cannot rely on any obligation by Ascon to comply with the detail of the programme for the commencement or execution of the cladding work, but is not precluded from contending that Ascon was in breach of the obligation to carry out its works "reasonably in accordance with the progress of the [main contract] works" and that in consequence commencement of cladding was delayed. The factual issue how far it has established that contention remains, but I propose to defer consideration of it until I have dealt with the second stage of causation, from that point to completion of the main contract.'

Availability of work

In *Kitsons Sheet Metal Ltd* v. *Matthew Hall Mechanical & Electrical Engineers Ltd* (1989) a sub-contractor involved in the erection of Terminal 4 at Heathrow Airport claimed that work was not made available as required by an agreed programme. The issue which came to court raised the question whether Kitsons were entitled under the contract to work to a programme and whether any written order requiring departure from it constituted a variation. It was held the parties must have recognised the likelihood of delays and of trades getting in each other's way and the prospects of working to programme were small. Provided Matthew Hall did their best to make areas available for work they were not in breach of contract even if Kitsons were brought to a complete stop.

A similar decision was given in *Martin Grant & Co. Ltd* v. *Sir Lindsay Parkinson & Co. Ltd* (1984). The main contract works were delayed by various causes and Grant, a sub-contractor, claimed that his works were rendered unprofitable by having to work over a protracted period. The court was asked to imply a term in the sub-contract:

'... That (a) the defendants would make sufficient work available to the plaintiffs to enable them to maintain reasonable progress and to execute their work in an efficient and economic manner; and (b) the defendants should not hinder or prevent the plaintiffs in the execution of the sub-contract works . . .'

The court held that, having regard to the express terms of the sub-contract, there was no room for any such implied term.

Supply of drawings, etc.

In *AMEC Process and Energy Ltd* v. *Stork Engineers & Contractors* (1999) one of the issues was whether Stork was obliged to supply drawings, design information and free issue materials so as to enable AMEC to work in accordance with its programme. The court held, having regard to the detail of the provisions in the contract, that there was such an obligation.

Instructions and variations

In *Neodox Ltd* v. *Swinton & Pendlebury Borough Council* (1958) the contractor alleged an implied term that instructions would be given to enable him to complete in an economic and expeditious manner.

It was held by Mr Justice Diplock that under the terms of the contract it was clear that instructions would be given from time to time and what was reasonable did not depend solely on the convenience and financial interests of the contractor. He said:

> 'To give business efficacy to the contract details and instructions neces-sary for the execution of the Works must be given by the Engineer from time to time in the course of the contract. If he fails to give such instruc-tions within a reasonable time the Corporation are liable in damages for breach of contract.'

In *McAlpine & Son* v. *Transvaal Provincial Administration* (1974), a South African case, a motorway contractor asked the court to define an implied term on the time for supplying information and giving instructions on variations as either:

(i)   a time convenient and profitable to himself;
(ii) a time not causing loss and expense; or
(iii) a time so that the works could be executed efficiently and economically.

The court declined on the grounds that under the contract, variations could be ordered at any time irrespective of the progress of the works, and that drawings and instructions should be given within a reasonable time after the obligation arose.

15.3 Shortened programmes

Contractors have for many years believed that by submitting programmes showing completion in a shorter time than that allowed in the contract they improve their claim prospects.

There is no doubt at all that under most standard forms contractors are entitled to finish early and are entitled to programme to finish early. That is not to say, however, that the employer's obligation is in any way changed from his original contractual obligation – not to obstruct the contractor in finishing within the full time allowed. The submission of a shortened pro-gramme which is not a contract document cannot place a greater liability on the employer than in the contract itself.

The matter did not come before the courts until the case of *Glenlion Construction Ltd* v. *The Guinness Trust* (1987). The contract allowed 114 weeks for completion and Glenlion submitted a programme showing completing in 101 weeks. On appeal from arbitration, the following points of law were raised:

'(1) . . . Whether on a true construction of clause 3.13.4 of the contract bill the contractor should provide a programme chart ("the programme") for the whole of the works showing a completion date no later than the date for completion and that agreement or approval by the architect of the programme should not relieve the contractor of his responsibility to complete the whole of the works by the date for completion.

(2) . . . Whether on a true construction of clause 21 of the conditions, namely the JCT 63 standard form of contract and clause 3.13.4 of the contract bills, if and in so far as the programme showed a completion date before the date for completion, the contractor was entitled to carry out the works in accordance with the programme and to complete the works on the said completion date.

(3) . . . Whether there was an implied term of the contract between the applicant and the respondent that, if and in so far as the programme showed a completion date before the date for completion the employer by himself, his servants or agents should so perform the said agreement as to enable the contractor to carry out the works in accordance with the programme and to complete the works on the said completion date.'

It was held that the answer to question (1) was 'Yes'. This is how Judge Fox-Andrews dealt with questions (2) and (3). He said:

'. . . As regards question (2), in the light of the wording of condition 21 it is self evident that Glenlion were entitled to complete before the date of completion. And the contractor was entitled to complete on an earlier date whether or not he produced a programme with an earlier date and whether or not he was contractually bound to produce a programme. It would follow that if he was entitled to complete before the date of completion he was entitled to carry out the works in such a way as to enable him to achieve the earlier completion date whether or not the works were programmed.

The answer to question (2) is therefore "Yes".

But in considering question (2) it becomes apparent that the answer to question (3) must be "No". It is not suggested by Glenlion that they were both entitled and obliged to finish by the earlier completion date. If there is such an implied term it imposed an obligation on the Trust but none on Glenlion. It is unclear how the variation provisions would have applied. Condition 23 operates, if at all, in relation to the date for completion stated in the appendix. A fair and reasonable extension of time for completion of the works beyond the date for completion stated in the appendix might be an unfair and unreasonable extension from an earlier date.

It is not immediately apparent why it is reasonable or equitable that a unilateral absolute obligation should be placed on an employer.

As long ago as 1970 Mr I.N. Duncan Wallace, Editor of *Hudson's Building and Engineering Contracts* 10th edn, wrote at p. 603:

"In regard to claims based on delay, litigious contractors frequently supplied to architects or engineers at an early stage in the work highly optimistic programmes showing completion a considerable time ahead of the contract date. These documents are then used (a) to justify allegations that the information or possession has been supplied late by the architect or engineer and (b) to increase the alleged period of delay, or to make a delay claim possible where the contract completion date has not in the event been extended."'

For further confirmation of the approach of the courts to shortened programmes, see also the decision in *J F Finnegan Ltd* v. *Sheffield City Council* (1988). In that case it was held that although there may be circumstances when the contractor may be entitled to recover costs incurred prior to the completion of the period contracted for, it would, at least under the JCT form, require proof of special circumstances.

15.4 *Method statements*

Method statements which become incorporated into contract documents can create the same problems as tender programmes similarly incorporated.

In *Yorkshire Water Authority* v. *Sir Alfred McAlpine & Sons (Northern) Ltd* (1985) the contractor had submitted with his tender, as instructed, a bar chart and method statement, showing he had taken note of certain specified phasing requirements providing for the construction of the works upstream. The formal contract agreement incorporated the tender and the minutes of the meeting at which the method statement was approved.

The contractor maintained that in the event it was impossible to work upstream and after some delay work proceeded downstream. The contractor then sought a variation order under clause 51(1). The court held:

(i) the tender programme / method statement was not the Clause 14 programme;

(ii) the incorporation of the method statement into the contract imposed an obligation on the contractor to follow it so far as it was legal or physically possible to do so;

(iii) the method statement, therefore, became a specified method of construction and the contractor was entitled to a variation order and payment accordingly.

Mr Justice Skinner said that:

'... In my judgment, the standard conditions recognise a clear distinction between obligations specified in the contract in detail, which both parties can take into account in agreeing a price, and those which are general and which do not have to be specified pre-contractually. In this case the applicants could have left the programme and methods as the sole responsibility of the respondents under clause 14(1) and clause 14(3). The risks inherent in such a programme or method would then have been the respondents' throughout. Instead, they decided they wanted more control over the methods and programme that clause 14 provided. Hence clause 107 of the specification; hence the method statement; hence the incorporation of the method statement into the contract imposing the obligation on the respondents to follow it save in so far as it was legally or physically impossible. It therefore became a specified method of construction by agreement between the parties . . .'

Similar decisions were reached in *Holland Dredging (UK) Ltd* v. *Dredging & Construction Co. Ltd* (1987) and *Blue Circle Industries plc* v. *Holland Dredging (UK) Ltd* (1987).

Subsequently in *Havant Borough Council* v. *South Coast Shipping Company Ltd* (1996) various issues relating to the contractor's method of working under an ICE 5th edition contract were considered. The judge, following the *Yorkshire Water* decision, distinguished between a method of working selected by the contractor and a method of working which was specified. He said:

'A method in a clause 14 Method Statement would not be specified. The method of working remains the responsibility of the contractor (Clause 14(7)).

The word "specified" I find relates to something which contractually is required to be done. With the result for example that a failure by a contractor to follow that method would be a breach of contract.'

The judge went on to hold that a specified method of working falls within the definition of temporary works and that if such temporary works prove to be impossible to undertake, within the meaning of clause 13 of the contract, then the contractor is entitled to a variation order under clause 51.

15.5 *Best endeavours and the like*

Many commercial and construction contracts use phrases such as 'best endeavours' and 'reasonable endeavours' to qualify or describe how obligations are to be performed. In construction contracts such phrases are usually attached to progress and completion obligations.

Notwithstanding the wealth of cases which have examined the meaning of the phrases, the extent to which they differ and the extent to which they are enforceable, a good deal of uncertainty remains on their application. Perhaps the best that can be said is that best endeavours means doing all that is reasonable to obtain an objective, whereas reasonable endeavours is, on the face of it, a less stringent obligation under which a party can have regard to its own interests.

Rhodia International Holdings (2007)

In *Rhodia International Holdings Ltd* v. *Huntsman International LLC* (2007) the parties were under a contractual obligation to respectively use reasonable endeavours to secure novation of an electricity supply contract. It was alleged that Huntsman had not done so. The following extracts from the High Court judgment provide a useful summary of the law and the court's reasoning:

'Reasonable endeavours
  30. Before considering in detail the parties' rival submissions as to whether on the facts Huntsman did use reasonable endeavours to obtain the consent of Cogen to the novation to HSSUK of the Energy Supply Contract, I should deal with two preliminary points.
  31. First, there was some debate at the hearing as to whether "reasonable endeavours" is to be equated with "best endeavours", a question on which there seems to be some division of judicial opinion. At the end of the day I am not convinced that it makes much difference on the facts of this case, but since the point was fully argued, I should deal with it. Mr Beazley QC for Rhodia contended that there was no difference between the two phrases. He relied upon a passage from the judgment of Buckley LJ in *IBM* v. *Rockware Glass* [1980] FSR 335 at 339:
  "in the absence of any context indicating the contrary, this [an obligation to use its best endeavours] should be understood to mean that the purchaser is to do all he reasonably can to ensure that the planning permission is granted".
There are similar statements in the judgments of Geoffrey Lane LJ at 344–5 and Goff LJ at 348.
  32. Mr Beazley also relied upon what Mustill J said in *Overseas Buyers* v. *Granadex* [1980] 2 Lloyd's Rep 608 at 613 lhc:
  "it was argued that the arbitrators can be seen to have misdirected themselves as to the law to be applied, for they have found that EIC did 'all that could reasonably be expected of them', rather than finding whether EIC used their 'best endeavours' to obtain permission to export, which is the test laid down by the decided cases. I can frankly see no substance at all in this argument. Perhaps the words 'best endeavours' in a statute or contract mean something different from doing all that can reasonably be expected-although I cannot think what the difference might be. (The unreported decision of the Court of Appeal in *IBM* v *Rockware Glass* upon which the buyers relied, does not to my mind suggest that such a difference exists . . .)."
Mr Beazley pointed out that in *Marc Rich v SOCAP* (1992) Saville J equated best endeavours with due diligence and that Rix LJ in *Galaxy Energy* v. *Bayoil* [2001] 1 Lloyd's Rep 512 at 516 equated reasonable efforts with due diligence, which suggested that best endeavours and reasonable endeavours meant the same thing. He sought to distinguish the unreported decision of Rougier J in *UBH (Mechanical Services)* v. *Standard Life* (1986) that an obligation to use reasonable endeavours was less stringent than

an obligation to use best endeavours, on the grounds that the point was not argued but conceded by Counsel.

33. I am not convinced that (apart from that decision of Rougier J) any of the judges in the cases upon which Mr Beazley relied were directing their minds specifically to the issue whether "best endeavours" and "reasonable endeavours" mean the same thing. As a matter of language and business common sense, untrammelled by authority, one would surely conclude that they did not. This is because there may be a number of reasonable courses which could be taken in a given situation to achieve a particular aim. An obligation to use reasonable endeavours to achieve the aim probably only requires a party to take one reasonable course, not all of them, whereas an obligation to use best endeavours probably requires a party to take all the reasonable courses he can. In that context, it may well be that an obligation to use <u>all</u> reasonable endeavours equates with using best endeavours and it seems to me that is essentially what Mustill J is saying in the *Overseas Buyers* case. One has a similar sense from a later passage at the end of the judgment of Buckley LJ in *IBM* v. *Rockware Glass* at 343, to which Mr Edwards-Stuart QC for Huntsman drew my attention.

34. That there is a distinction between best endeavours and reasonable endeavours and that the latter is less stringent than the former is not only supported by the decision of Rougier J in *UBH* but by the decision of Kim Lewison QC (as he then was) sitting as a Deputy High Court Judge, in *Jolley* v. *Carmel Ltd* [2000] 2 EGLR 154 upon which Mr Edwards-Stuart relied. At p. 159 the judge said:
    "Where a contract is conditional upon the grant of some permission, the courts often imply terms about obtaining it. There is a spectrum of possible implications. The implication might be one to use best endeavours to obtain it (see *Fischer* v. *Toumazos* [1991] 2 EGLR 204), to use all reasonable efforts to obtain it (see *Hargreaves Transport* v. *Lynch* [1969] 1 WLR 215) or to use reasonable efforts to do so. The term alleged in this case [to use reasonable efforts] is at the lowest end of the spectrum."

Mr Beazley sought to suggest that somehow this analysis was distinguishable because it was concerned with the implication of a term, but I cannot see any basis for such a distinction. It seems to me that the judge's analysis is equally applicable to the construction of the phrase reasonable efforts or reasonable endeavours whether it is an express or an implied term of any particular contract.

35. Accordingly, in so far as it is necessary to decide this point, I agree with Mr Edwards-Stuart that an obligation to use reasonable endeavours is less stringent than one to use best endeavours. As to what reasonable endeavours might entail, he relied upon a recent decision of Lewison J in *Yewbelle* v. *London Green Developments* [2006] EWHC 3166 (Ch) at paragraphs 122–3 where the judge said:

"... However, the essence of the obligation required Yewbelle to use reasonable endeavours to reach an agreement, not with the other party to the contract, but with a third party. To that extent it seems to me that at the very least Phillips is a useful analogy. In using reasonable endeavours towards that end, I do not consider that Yewbelle was required to sacrifice its own commercial interests.

123. I come back to the question: for how long must the seller continue to use reasonable endeavours to achieve the desired result? In his opening address, Mr Morgan said that the obligation to use reasonable endeavours requires you to go on using endeavours until the point is reached when all reasonable endeavours have been exhausted. You would simply be repeating yourself to go through the same matters again. I am prepared to accept this formulation, subject to the qualification that account must be taken of events as they unfold, including extraordinary events."

Subject to one caveat, I would agree with this analysis. The caveat is that, where the contract actually specifies certain steps have to be taken (as here the provision of a direct covenant if so required) as part of the exercise of reasonable endeavours, those steps will have to be taken, even if that could on one view be said to involve the sacrificing of a party's commercial interests.'

Chapter 16
Delay analysis

16.1 Introduction

Although the basic legal principles relating to extensions of time for completion have a long history and can be traced back a hundred or more years it is only in the last thirty or so years that detailed attention has been given as to how the detail of amounts due as extensions of time should be calculated. But out of this comparatively recent change a new industry devoted to delay analysis has developed.

The old methods were generally simplistic and, more often than not, impressionistic rather than analytical. Typically, if a contractor finished late the contract administrator would review the amount of additional work undertaken and the circumstances under which the works had been constructed and then, if the contractor was adjudged to have performed satisfactorily, an extension of time revising the date for completion to the date of actual completion would be awarded. When analysis was made of the delaying effects of particular events it was usually undertaken retrospectively by comparing the contractor's as-built progress with his planned programme. Prospective analysis was rarely undertaken. Such simple methods were not unfitting for their time – particularly when applied with motivation to achieve a result fair and reasonable to both parties. Programmes were often no more than simple bar charts and critical path analysis, when it was done, was a manual task unaided by the benefit of computers.

The main driving forces behind the development of new methods are frequently explained as being the complexity of modern construction projects and the facility of computers to generate and examine multiple critical path possibilities and the effects thereon of delaying events. This may well be correct but there are other forces in play which almost certainly have an influence on the modern approach to the assessment of extensions of time. One is that commercial pressures to impose liquidated damages for delay are greater than in bygone times – as are pressures of accountability on public bodies. Another is that the respect which was previously held for the decisions of persons holding office as contract administrators (architects, engineers and the like) has largely evaporated and anything that can be challenged will now be challenged by a dissatisfied party. Hence the present pressures for scrutiny of the decision-making process.

But, whatever the explanation for the development of delay analysis techniques, interest in the subject is now so great that it has become essential to undertake some form of delay analysis in examination of extension of time and prolongation claims. So much can be seen from the various cases mentioned later in this chapter. What can also be seen from the cases is that there is still room for debate as to which delay analysis techniques are most suitable for particular circumstances and what reliance should be placed on computer-generated results which are at odds with results obtained by more down-to-earth methods.

Terminology

One of the problems of the developing science of delay analysis is that the terminology used is still expanding and only a limited number of the terms used by those engaged in the process have a settled meaning. Only the principal terms are considered in this book but for further information see the very useful 'Delay and Disruption Protocol' published by the Society of Construction Law in 2002 which contains a glossary of the most commonly used terms.

Contractual provisions

Since, in the majority of instances, delay analysis is something undertaken in pursuance of an objective related to a contractual obligation, liability, or claim, it follows, although it is frequently overlooked, that the starting point for any analysis should be examination of the contractual provisions. Thus some contracts require extensions of time to be awarded retrospectively whereas others require a prospective approach. And whereas some contracts place emphasis on the need for fair and reasonable extensions of time, others are more prescriptive.

Such considerations have an important part to play in the selection and/or determination of which method of delay analysis should be used – whether that be by the contract administrator in making routine awards, by programming experts engaged to assist in dispute resolution processes, or by the dispute resolvers themselves.

Purposes of delay analysis

Much, if not the majority, of delay analysis work in the construction industry is undertaken to assess, to prove, or to rebut time-related claims of one sort or another. These may be extension of time claims, reasonable time claims, prolongation claims, acceleration claims or delay claims. Sometimes there is a combination of claims.

Clearly, where the focus of attention is on one particular type of claim, there is a need to utilise from the outset an appropriate method of delay analysis. However, where various types of claim are packaged together it may not be evident from the outset if there is a single appropriate method or to what extent, depending on how difficult the claims are to resolve, separate methods for separate claims are necessary. A costly mistake in such circumstances is for the claiming party to embark upon, and to soldier on with, an inappropriate or limited use delay analysis. It is far better and less costly in the long run to take legal advice to ensure that the purpose of the delay analysis is understood before it is commenced.

Concerns

Commonly expressed concerns about delay analysis are that even on a given set of facts very different conclusions can be reached depending on the type of analysis undertaken. See, for example, the Society of Construction Law publication 'The Great Delay Analysis Debate' released in 2006.

It is, of course, evident that a prospective analysis may produce a different result from a retrospective analysis. What is less evident is why on common facts, a common programme, and a common method of analysis very different results can be produced. At the best this suggests that some types of analysis are over-dependent on subjective use of data; at the worst it suggests that some are capable of being manipulated to produce whatever result is required.

Thus it was said by the judge in the *Great Eastern* case mentioned in Chapter 14:

> '223. It is evident in my judgment that Laing consistently underplayed mention of the true causes of critical delay and assert other reasons for delay that would not reflect upon them. They consistently misreported the delays actually occurring and manipulated the data in the programme update to obscure the accurate position.'

and

> '231. Because of the misreporting of progress, some of the following Trade Contractors commenced work on site before the works were ready for them, and this led to claims for extensive extensions of time together with prolongation and disruption costs. Had the true state of progress been declared, whilst it would have been necessary for Laing to have renegotiated with Trade Contractors in order to postpone their commencement on site, the cost consequences of such renegotiation would have been relatively minor, and it would have avoided the subsequent claims for extensions of time and loss and expense.'

Another commonly expressed concern about delay analysis is the high cost of engaging experts for dispute resolution procedures. Frequently it seems to be that far from simply engaging an individual to act as expert a party has, knowingly or otherwise, engaged a full supporting cast of assistants.

Similarly the costs of engaging claims consultants can result in startlingly high bills. His Honour Judge Fox-Andrews in the case of *Wessex Regional Health Authority* v. *HLM Design Ltd* (1995) even went so far as to say:

'In the arbitration proceedings only to a limited extent did Wessex use the services of counsel and solicitors. In the main the preparation of Wessex's case was put in the hands of claims consultants.

In any proceedings by Wessex against HLM where reasonableness of the preparation and settlement of the arbitration proceedings was in issue it would be likely to result in a prolonged investigation as to the conduct of the claims consultants who might be made a party by Wessex to incur costs of £1,197,363.73 in respect of the extension of time aspects alone. (Substantial other costs must have been incurred for other issues that were raised in the arbitration.)'

16.2 *Critical paths*

The Society of Construction Law Delay and Disruption Protocol defines 'critical path' and related terms as follows:

'critical path
The sequence of activities through a project network from start to finish, the sum of whose durations determines the overall project duration. There may be more than one critical path depending on workflow logic. A delay to progress of any activity on the critical path will, without acceleration or re-sequencing, cause the overall project duration to be extended, and it is therefore referred to as a "critical delay".
critical path analysis (CPA) and **critical path method** (CPM)
The critical path analysis or method is the process of deducing the critical activities in a programme by tracing the logical sequence of tasks that directly affect the date of project completion. It is a methodology or management technique that determines a project's critical path. The resulting programme may be depicted in a number of different forms, including a <u>Gantt</u> or bar chart, line-of-balance diagram, pure logic diagram, time-scaled logic diagram or as a time-chainage diagram, depending on the nature of the works represented in the programme.'

His Honour Judge Toulmin in the case of *Mirant Asia-Pacific Construction (Hong Kong) Ltd* v. *Ove Arup & Partners International Ltd* (2007) made these observations on critical paths:

'575.

1. The critical path can be defined as "the sequence of activities through a Project network from start to finish, the sum of whose durations determine the overall Project duration".
2. Duration is only the shortest time if activities on the critical path are carried out in the shortest time.
3. There may be more than one critical path.

4. It is important to look at activities at or near the critical path to understand their potential impact on the Project.

5. Windows analysis, reviewing the course of a Project month by month, provides an excellent form of analysis to inform those controlling the Project what action they need to take to prevent delay to the Project.

6. Without such analysis those controlling the Project may think they know what activities are on the critical path but it may well appear after a critical path analysis that they were mistaken.

7. A less reliable form of critical path analysis is the watershed analysis. This analyses the Project in terms of a few key events. It may be a sufficient check in the course of a Project to analyse what changes, if any, may need to be made in the Project at the time of a benchmark event.

8. Both windows analysis and watershed analysis are used frequently to analyse delays at the end of a Project. A watershed analysis will be less reliable particularly if the gaps between the watersheds are lengthy. It does not show the pattern of events between the watersheds. This may be very important where a number of activities are at or near the critical path. What the watershed analysis provides is a snapshot at the particular time when it is carried out.

9. Float in a programming sense means the length of time between when an activity is due to start and when it must start if it is to avoid being on the critical path. Float can also be used to refer to the additional time needed / allowed to complete an activity over and above the shortest time that is reasonably required.

10. It is, of course, obvious that the analysis is only valid if it is comprehensive and takes account of all activities.'

The judge went on to say:

'575.

11. As the claimant readily acknowledges, it is merely a tool which must be considered with the other evidence. The question of whether or not the failure of the Boiler foundation caused delay to the commencement of the Reliability Trials and if so what delay is a question of fact. The evidence of Programming Experts may be of persuasive assistance.

576. To these propositions I add the proposition that if a retrospective delay analysis is being conducted on a Project, the analysis must include the time to the end of the Project, otherwise activities may occur which will take them on to the (or a) critical path after the date of the final window or watershed. In this respect Mr Lechner's analysis which ends in October 1998 is seriously flawed.'

Thereafter, having considered points of criticism in respect of both experts' reports, the judge concluded 'that the Reports of the Programming Experts take me no further than the findings which I have already made'.

16.3 *Float*

In simple terms 'float' is spare time within a programme. The term can be attached to an activity or to overall completion.

The Society of Construction Law Protocol definitions are:

'**float**
The time available for an activity in addition to its planned duration. See free float and total float. Where the word "float" appears in the Protocol, it means positive not negative float, unless expressly stated otherwise.
free float
The amount of time that an activity can be delayed beyond its early start / early finish dates without delaying the early start or early finish of any immediately following activity.
total float
The amount of time that an activity may be delayed beyond its early start / early finish dates without delaying the contract completion date.'

Ownership of float

Questions and disputes as to which party owns float, or gets the benefit of it, are commonplace in construction contracts. There is no single answer since contracts differ in their provisions as to extensions of time. Thus, in a contract which allows extensions for any delay beyond planned completion end float would appear to belong to the contractor. In contrast, in a contract allowing extensions only for delay beyond the completion date any end float would appear to belong to the employer.

In the case of *Ascon Contracting Ltd* v. *Alfred McAlpine Construction Isle of Man Ltd* (1999), His Honour Judge Hicks had this to say on float:

'91. Before addressing those factual issues I must deal with the point raised by McAlpine as to the effect of its main contract "float", which would in whole or in part pre-empt them. It does not seem to be in dispute that McAlpine's programme contained a "float" of five weeks in the sense, as I understand it, that had work started on time and had all sub-programmes for sub-contract works and for elements to be carried out by McAlpine's own labour been fulfilled without slippage the main contract would have been completed five weeks early. McAlpine's argument seems to be that it is entitled to the "benefit" or "value" of this float and can therefore use it at its option to "cancel" or reduce delays for which it or other sub-contractors would be responsible in preference to those chargeable to Ascon.

92. In my judgment that argument is misconceived. The float is certainly of value to the main contractor in the sense that delays of up to that total amount, however caused, can be accommodated without involving him in liability for liquidated damages to the employer or, if he calculates his own prolongation costs from the contractual

completion date (as McAlpine has here) rather than from the earlier date which might have been achieved, in any such costs. He cannot, however, while accepting that benefit as against the employer, claim against sub-contractors as if it did not exist. That is self-evident if total delays as against sub-programmes do not exceed the float. The main contractor, not having suffered any loss of the above kinds, cannot recover from sub-contractors the hypothetical loss he would have suffered had the float not existed, and that will be so whether the delay is wholly the fault of one sub-contractor, or wholly that of the main contractor himself, or spread in varying degrees between several sub-contractors and the main contractor. No doubt those different situations can be described, in a sense, as ones in which the "benefit" of the float has accrued to the defaulting party or parties, but no-one could suppose that the main contractor has, or should have, any power to alter the result so as to shift that "benefit". The issues in any claim against a sub-contractor remain simply breach, loss and causation.

93. I do not see why that analysis should still hold good if the constituent delays more than use up the float, so that completion is late. Six sub-contractors, each responsible for a week's delay, will have caused no loss if there is a six weeks' float. They are equally at fault, and equally share in the "benefit". If the float is only five weeks, so that completion is a week late, the same principle should operate; they are equally at fault, should equally share in the reduced "benefit" and therefore equally in responsibility for the one week's loss. The allocation should not be in the gift of the main contractor.

94. I therefore reject McAlpine's "float" argument. I make it clear that I do so on the basis that it did not raise questions of concurrent liability or contribution; the contention was explicitly that the "benefit", and therefore the residual liability, fell to be allocated among the parties responsible for delay and that that allocation was entirely in the main contractor's gift as among sub-contractors, or as between them and the main contractor where the latter's own delay was in question.

In *Royal Brompton Hospital NHS Trust Ltd* v. *Hammond* (2002) His Honour Judge Lloyd said:

'. . . All activities have potential or theoretical float (even if the period is negative). What is required is to track the actual execution of the works. On a factual basis this part of the case requires no further discussion. In addition, clause 25 refers to "expected delay in the completion of the Works" and to the need for the Architect to form an opinion as to whether because of a Relevant Event "the completion of the works is likely to be delayed thereby beyond the Completion Date". Under the JCT conditions, as used here, there can be no doubt that if an architect is required to form an opinion then, if there is then unused float for the benefit of the contractor (and not for another reason such as to deal with p.c. or provisional sums or items), then the architect is bound to take it into account since an extension is only to be granted if completion would otherwise be delayed beyond the then current completion date. This may seem hard to a contractor but the objects of an extension of time clause are to avoid the contractor being liable for liquidated damages where there has been delay for which it is not responsible, and still to establish a new completion date to which the contractor should work so that both the employer and the contractor know where they stand. The architect should in such circumstances inform the contractor that, if thereafter events occur for which an extension of time cannot be granted, and if, as a result, the contractor would be liable for liquidated damages then an appropriate extension, not exceeding the float, would be given. In that way the purposes of the clause can be met: the date for completion is always known; the position on liquidated damages is clear; yet the contractor is not deprived permanently of "its" float. Under these JCT Conditions the Architect cannot revise an extension once given so as to fix an earlier date (except in the limited circumstances set out in clauses 25.3.2 and 25.3.3). Thus to grant an extension which preserved the contractor's float would not be "fair and reasonable". Under clause 23.1 the employer is entitled to completion on or before the Completion Date so the employer is ultimately entitled to the benefit of any unused float that the contractor does not need. Few contractors wish to remain on a site any longer than is needed and employers are usually happy to take possession earlier, rather than later, and, if they are not, they have to accept the risk of early completion. In practice, however, architects are not normally concerned about these points and may reasonably take the view that, unless the float is obvious, its existence need not be discovered . . .'

16.4 *Methods of delay analysis*

It is beyond the scope of this book to comment on all the various methods of delay analysis now in common use. Most methods are, however, variants of four main types:

- as-planned v. as-built
- impacted as-planned
- collapsed as-built
- time impact analysis.

As-planned v. as-built

This is a simple retrospective method of delay analysis based on factual evidence. It examines actual progress achieved against progress as planned. To be effective it requires a realistic programme and good as-built records. It can work to limited effect without a critical path analysis and without computer software.

Impacted as-planned

This is another comparatively simple method of delay analysis. It examines the impact of single events or groups of events on the planned programme. It can be used for prospective as well as for retrospective analysis. Again it can be used without a critical path analysis and computer software. However, since the effect of impacting delaying events on to a programme may well change the critical path, manual use of the method has obvious limitations.

Collapsed as-built

This is a retrospective method of analysis by which delays are extracted from the as-built programme to see when completion would have been achieved but for the delaying effects attributable to either the contractor or the employer. It requires a logic-linked programme and good as-built records. For analysis of any complexity computer software is essential.

Time impact analysis

This is a prospective method of analysis. It involves updating the contractor's programme to reflect the progress achieved at the time of a delaying event, or group of events, and then comparing the anticipated completion dates with and without the events. It is normally undertaken using computer software.

This method of analysis is selected in the Society of Construction Law Protocol as the preferred method for complex cases. Thus, at paragraph 4.8 it is said:

'Time impact analysis is based on the effect of Delay Events on the Contractor's intentions for the future conduct of the work in the light of progress actually achieved at the time of the Delay Event and can also be used to assist in resolving more complex delay scenarios involving concurrent delays, acceleration and disruption. It is also the best technique for determining the amount of EOT that a Contractor should have been granted at the time an Employer Risk Event occurred. In this situation, the amount of EOT may not precisely reflect the actual delay suffered by the Contractor. That does not mean that time impact analysis generates hypothetical results – it generates results showing entitlement. This technique is the preferred technique to resolve complex disputes related to delay and compensation for that delay.'

As a general overview of all four methods the Protocol states at paragraphs 4.14 to 4.16:

'4.14 As-planned v as-built and impacted as-planned are generally the cheapest and simplest methods of analysis.

4.15 Collapsed as-built is also an analysis simple to perform although it is often more laborious and subjective because of the inherent difficulty of establishing accurate as-built logic from records.

4.16 Time impact analysis is the most thorough method of analysis, although it is generally the most time-consuming and costly when performed forensically.'

16.5 Judicial comments on delay analysis

Prior to the judgment in the case of *John Barker Construction Ltd* v. *London Portman Hotel* (1996) there was little by way of judicial comment on delay analysis. However, as noted in Chapter 12 above, the ruling in that case made clear that the court expected extensions of time to be assessed by logical analysis. In holding that the architect's assessment was fundamentally flawed the judge said, amongst other things:

'1. [the architect] did not carry out a logical analysis in a methodical way of the impact which the relevant matters had or were likely to have on the Plaintiffs' planned programme.

2. [the architect] made an impressionistic, rather than a calculated, assessment of the time which he thought was reasonable for the various items individually and overall.

3. [the architect] misapplied the contractual provisions, as more particularly set out above.

4. Where [the architect] allowed time for relevant events, the allowance which he made in important instances . . . bore no logical or reasonable relation to the delay caused.'

This ruling that there should be logical analysis has been followed in subsequent cases – with some judges going so far as to say that critical path analysis is a necessary part of the analysis. But note the comment of Mr Justice Ramsey in *London Underground Ltd* v. *Citylink Telecommunications Ltd* (2007):

'Secondly, whilst analysis of critical delay by one of a number of well-known methods is often relied on and can assist in arriving at a conclusion of what is fair and reasonable, that analysis should not be seen as determining the answer to the question. It is at most an area of expert evidence which may assist the arbitrator or the court in arriving at the answer of what is a fair and reasonable extension of time in the circumstances.'

Moreover, in two recent cases the judges have shown, in preferring simplicity over complexity, that it is the reliability of the material put before them that counts not its apparent technical superiority, sophistication or volume.

In *Skanska Construction UK Ltd* v. *Egger (Barony) Ltd* (2004) the judge's comments included the following:

'413. [Mr S], a planning consultant originally employed by and later retained by SCL as a Consultant gave evidence at the Liability trial . . . He impressed me as someone who was objective, meticulous as to detail, and not hide bound by theory as when demonstrable fact collided with computer programme logic.

414. His analysis was accessibly depicted in a series of charts accompanying his evidence.

415. [Mr P] produced a report of some hundreds of pages supported by 240 charts. It was a work of great industry incorporating the efforts of a team of assistants in his practice.

. . .

422. It is evident that the reliability of [Mr P']s sophisticated impact analysis is only as good as the data put in. The court cannot have confidence as to the completeness and quality of the input into this complex and rushed computer project.

423. Egger submit that the software used by [Mr S] is incapable of producing a reliable analysis since Power Project is primarily a planning tool creating a graphic representation, it is a dated system and does not have the sophistication of the [P] system but I am satisfied that it also has a significant capacity for logical connections and for identifying critical paths and for re-scheduling activities to show how events change.

424. [Mr P] stated that the effective application of Power Project with its inherent limitations was also dependent upon the 'intuition' of its user. A term, it seems, that includes the power of selection of facts and interpretive judgment of them. As a criticism, it is difficult to see how this differs from the process followed and applied by [Mr P's] own team of assistants prior to input into his computer programme. [Mr S] was available to be cross-examined and his judgment and interpretation was apparent and could be tested. I was not impressed with the evidence of [Mr P] for the reasons I have set out above. It was not thorough. It was not complete. He only directly considered critical delay and did not really address disruption and he proceeded from the wrong premise in relation to sub-contract periods which proceeded on the basis of that which is agreed between SCL and the sub-contractor. I preferred the evidence of [Mr S] as to programming and planning matters to that of [Mr P].'

In *City Inn Ltd* v. *Shepherd Construction Ltd* (2007) Lord Drummond Young said in relation to the evidence of the two experts in the case:

'[27] [Mr W] was critical of the as-built critical path analysis used by [Mr L]; I deal with his specific criticisms of that analysis at paragraphs [36]–[39] below. In evidence, [Mr W] stated that he had considered undertaking a critical path analysis, but decided not to do so. He did not have access to an electronic version of the defenders' original programme for the project, and because of this it was impossible to identify the defenders' original critical path through the programme. Nevertheless, making use of his experience in programming, [Mr W] had attempted to replicate what he surmised might be the logic of the defenders' original programme; he stated, however, that he had no great confidence that his version of that programme was either correct or complete. [Mr W] stated that to continue with a critical path analysis based on logic that he knew not to be completely correct would have meant that he could not be sure of the evidence that he was giving to the court.

[28] The pursuers criticised [Mr W's] approach to the case. They referred in particular to his failure to undertake a critical path analysis of the present project. That might be explained by the fact that [Mr W] preferred to use the as-planned v as-built method. Nevertheless, the weakness of that method was that, as [Mr W] acknowledged, it does not identify the critical path and therefore needs to be used with great care and understanding of the processes in the whole of the project. The pursuers submitted that an expert could only give a meaningful opinion as to which activities in a project are critical on the basis of an as-built critical path analysis, such as that carried out by [Mr L]. For that reason it was suggested that I should treat with caution, and indeed scepticism, [Mr W's] opinion.

[29] In my opinion the pursuers clearly went too far in suggesting that an expert could only give a meaningful opinion on the basis of an as-built critical path analysis. For reasons discussed below (at paragraphs [36]-[37]) I am of the opinion that such an approach has serious dangers of its own. I further conclude, as explained in those paragraphs, that [Mr L's] own use of an as-built critical path analysis is flawed in a significant number of important respects. On that basis, I conclude that that approach to the issues in the present case is not helpful. The major difficulty, it seems to me, is that in the type of programme used to carry out a critical path analysis any significant error in the information that is fed into the programme is liable to invalidate the entire analysis. Moreover, for reasons explained by [Mr W] (paragraphs [36]–[37] below), I conclude that it is easy to make such errors. That seems to me to invalidate the use of an as-built critical path analysis to discover after the event where the critical path lay, at least in a case where full electronic records are not available from the contractor. That does not invalidate the use of a critical path analysis as a planning tool, but that is a different matter, because it is being used then for an entirely different purpose. Consequently I think it necessary to revert to the methods that were in use before computer software came to be used extensively in the programming of complex construction contracts. That is essentially what [Mr W] did in his evidence. Those older methods are still plainly valid, and if computer-based techniques cannot be used accurately there is no alternative to using older, non-computer-based techniques.'

Chapter 17
Building forms

17.1 Introduction

In the earlier editions of this book the liquidated damages and extension of time provisions from all the then widely used standard forms for building works were reproduced and briefly commented upon. That was a manageable task. However, with the publication over the last ten years or so of new model forms by various professional and commercial organisations and the expansion of others into families of forms, the number of model / standard forms of building contracts on the market precludes examination of each and every form.

There are signs that for the foreseeable future the main focus of attention in the building industry will be on the comprehensive suite of main contracts and subcontracts (the JCT 2005 forms) recently produced by the Joint Contracts Tribunal. Only time will tell whether these forms will become popular enough to eliminate the opposition but given the restricted scope of this chapter these are clearly the forms which need to be examined.

17.2 JCT 2005 contracts

The Joint Contracts Tribunal, whose members presently include representatives of professional bodies, employer's organisations and contractor's federations, was established in 1931. Its best-known productions JCT 63 and JCT 1980 were without doubt the dominant forms of building contract of their times.

In 2005 JCT launched, partly perhaps in response to competition coming from the expanding family of New Engineering Contracts, a co-ordinated suite of contracts covering various procurement, pricing and management options and including both main contracts and sub-contracts. The suite includes:

- Standard Building Contract
- Intermediate Building Contract
- Minor Works Building Contract
- Design and Build Contract

- Major Project Construction Contract
- Standard Building Sub-Contracts
- Intermediate Building Sub-Contracts
- Major Project Sub-Contract
- Design and Build Sub-Contract
- Prime Cost Building Contract
- Measured Term Contract
- Repair and Maintenance Contract

The provisions for liquidated damages and extensions of time considered in this chapter are taken from the Standard Building Contract (without quantities).

17.3 *Commencement and completion*

Within the Contract Particulars attached to the Articles of Agreement the employer is required to state dates of possession of the site and the date for completion (the Completion Date).

Clause 2.4 commences:

'On the Date of Possession possession of the site or, in the case of a Section, possession of the relevant part of the site shall be given to the Contractor who shall thereupon begin the construction of the Works or Section and regularly and diligently proceed with and complete the same on or before the relevant Completion Date.'

Completion

Completion is not a defined term. It is for the architect to form an opinion on when practical completion has been achieved. Clause 2.30 states:

'When in the opinion of the Architect / Contract Administrator practical completion of the Works or a Section is achieved and the Contractor has complied sufficiently with clauses 2.40 and 3.25.4, then:

.1 in the case of the Works, the Architect / Contract Administrator shall forthwith issue a certificate to that effect (the Practical Completion Certificate)';

.2 in the case of a Section, he shall forthwith issue a certificate of practical completion of that Section (a Section Completion Certificate)';

and practical completion of the Works or the Section shall be deemed for all the purposes of this Contract to have taken place on the date stated in that certificate.'

The references in clause 2.30 to clauses 2.40 and 3.25.4 relate to the provision of as-built drawings and information for the CDM (construction, design and management) regulations.

In properly forming his opinion on when practical completion is achieved the architect should be guided by the principles laid down by the courts. These are discussed in Chapter 10 above but, in short, practical completion is completion of all the work that has to be done save for 'de minimis' items. Or as the Hong Kong Court of Appeal in *Mariner Hotels Ltd* v. *Atlas Ltd* (2006) put it practical completion means 'a state of affairs in which the works have been completed free from patent defects other than ones to be ignored as trifling'.

But even if there is some uncertainty as to what constitutes practical completion on any particular project it is clear from clause 2.30 that it is the issue of the Practical Completion Certificate which marks completion.

17.4 *Notification of delay*

Clause 2.27.1 spells out the firm obligation on the contractor to give written notice of any delay whether or not he is seeking an extension of time and whether or not the cause of the delay is a relevant event as defined in the clause.

> 'If and whenever it becomes reasonably apparent that the progress of the Works or any Section is being or is likely to be delayed the Contractor shall forthwith give written notice to the Architect / Contract Administrator of the material circumstances, including the cause or causes of the delay, and shall identify in the notice any event which in his opinion is a Relevant Event.'

The contractor would not normally write to the architect of his own accord informing him of the problems which beset any contractor in the course of his business and for which he carries and accepts full responsibility – problems with labour, poor performance of domestic sub-contractors, late supply of materials, bad planning and faulty workmanship. Most contractors would prefer to keep such matters to themselves. Moreover, since the contractor's primary obligation is to complete the works within the time allowed, it could be argued that it is of no concern to the architect or employer unless the works are likely to be delayed beyond the completion date.

Against this, however, there are two factors to consider:

(a) by clause 2.28.6.1 the contractor is to use his best endeavours to prevent delay in the progress of the works, howsoever caused;
(b) by clause 2.28.1 the architect is to give such extension of time as he considers to be fair and reasonable.

Clearly, for an architect's decision to be fair and reasonable he must know all the facts – including those both favourable and unfavourable to the contractor's case. Thus, in the case of *Henry Boot Construction (UK) Ltd* v. *Malmaison Hotel (Manchester) Ltd* (1999) which related to a JCT 1980 contract it was held that in determining whether a relevant event was likely to have caused delay beyond the completion date the architect was entitled to consider the effect of other events.

Further requirements in respect of notices are contained in clauses 2.27.2 and 2.27.3. These read:

> '.2 In respect of each event identified in the notice the Contractor shall, if practicable in such notice or otherwise in writing as soon as possible thereafter, give particulars of its expected effects, including an estimate of any expected delay in the completion of the Works or any Section beyond the relevant Completion Date.
>
> .3 The Contractor shall forthwith notify the Architect / Contract Administrator in writing of any material change in the estimated delay or in any other particulars and supply such further information as the Architect / Contract Administrator may at any time reasonably require.'

Conditions precedent

It is doubtful if, in themselves, any of the parts of clause 2.27 contain notification requirements which would be construed by the courts as condition precedents to entitlement to extension of time. There are no clear words to the effect that they do. However, taken in conjunction with the opening sentence of clause 2.28 which makes receipt by the architect of notice and particulars under clause 2.27 the starting point for consideration of extension of time it could be argued that failure to give written notice constitutes breach of a condition precedent to entitlement. Thus in the *Steria* v. *Sigma* (2007) case discussed in Chapter 5 a requirement to give notice of delay under an MF / 1 type contract was held to be a condition precedent. However, the architect's power under clause 2.28.5.1 to take account of relevant events not notified when making a final review goes against the concept of the contractor's notice being a binding condition precedent.

17.5 Extension of time

JCT 2005 places its extension of time provisions under the heading 'Adjustment of Completion Date'. It nevertheless continues to use the phrase 'extension of time' in various clauses including clause 2.28 – the clause setting out the procedures for the adjustment of time. Clause 2.28 has six parts:

- 2.28.1 – architect's duty to consider the contractor's notices
- 2.28.2 – architect's duty to notify decision within 12 weeks
- 2.28.3 – architect's duty to itemise any extension of time
- 2.28.4 – reductions of extension of time in respect of omitted works
- 2.28.5 – architect's review after the completion date
- 2.28.6 – provisos affecting 2.28.1 to 2.28.5.

Clause 2.28.1

> 'Fixing Completion Date
> .1 If, in the opinion of the Architect / Contract Administrator, on receiving a notice and particulars under clause 2.27:
>> .1 any of the events which are stated to be a cause of delay is a Relevant Event; and
>> .2 completion of the Works or of any Section is likely to be delayed thereby beyond the relevant Completion Date,
> then, save where these Conditions expressly provide otherwise, the Architect / Contract Administrator shall give an extension of time by fixing such later date as the Completion Date for the Works or Section as he then estimates to be fair and reasonable.'

Under clause 2.28.1 the architect has two key tasks to perform – but only on receipt of the contractor's notice and particulars. The first task is to consider whether a notified event is a relevant event. The contractor should already have stated his view in compliance with clause 2.27.1. The architect may not agree. For example, the contractor may claim adverse weather conditions to be exceptional – and thereby a relevant event, whereas the architect may disagree that the weather was exceptional. The architect's second task is to consider whether delay to the completion date has been caused by a relevant event. This may involve undertaking systematic delay analysis. For more on this see Chapter 16.

The requirement in clause 2.28.1 for the architect's assessment to be fair and reasonable is important. It allows a measure of personal judgment and possibly apportionment where there are competing events – see the commentary in Chapter 14. However, it is not a licence to make an impressionistic assessment – see the case of *John Barker Construction Ltd* v. *London Portman Hotel Ltd* (1996) mentioned in Chapter 12.

Clause 2.28.2

> '.2 Whether or not an extension is given, the Architect / Contract Administrator shall notify the Contractor in writing of his decision in respect of any notice under clause 2.27 as soon as is reasonably practicable and in any event within 12 weeks of receipt of the required particulars. Where the period from receipt to the Completion Date is less than 12 weeks, he shall endeavour to do so prior to the Completion Date.'

There are sound practical and commercial reasons why the architect's decision should be given within 12 weeks. The contractor is entitled to know how long he has to complete the works. It is doubted, however, whether failure by the architect to meet the 12 week target invalidates his award. In *Temloc* v. *Errill* (1987) the 12 week requirement post-completion was held to be directory rather than mandatory (see Chapter 11). Nevertheless it is questionable whether the same rule applies to pre-completion assessments.

The requirement for the architect to notify his decision as soon as reasonably practicable leaves open the question of whether the architect's delay analysis should be prospective or retrospective. This provides a measure of flexibility for the architect in making his assessments which is lacking in some other standard forms.

Clause 2.28.3

'.3 The Architect / Contract Administrator shall in his decision state:
    .1 the extension of time that he has attributed to each Relevant Event; and
    .2 (in the case of a decision under clause 2.28.4 or 2.28.5) the reduction in time that he has attributed to each Relevant Omission.'

Clause 2.28.3 does not permit the architect to make a global award of extension of time. One reason for this is the need to be able to identify the impact of events which give rise to entitlement to recovery of loss and expense.

However, if all the delaying events under consideration are relevant events of the same standing as regards loss and expense and there is complex interaction of the type considered in *Crosby* v. *Portland* (1967) a global award may be a more accurate and realistic award than one artificially particularised.

Clause 2.28.4

'.4 After the first fixing of a later Completion Date in respect of the Works or a Section, either under clause 2.28.1 or by a Pre-agreed Adjustment, but subject to clauses 2.28.6.3 and 2.28.6.4, the Architect / Contract Administrator may by notice in writing to the Contractor, giving the details referred to in clause 2.28.3, fix a Completion Date for the Works or that Section earlier than that previously so fixed if in his opinion the fixing of such earlier Completion Date is fair and reasonable, having regard to any Relevant Omissions for which instructions have been issued after the last occasion on which a new Completion Date was fixed for the Works or for that Section.'

This clause empowers the architect to fix completion dates earlier than those previously fixed if it is fair and reasonable to do so having regard to omissions instructed after extensions of time have been granted. In effect, therefore, extensions of time previously granted can be reduced by later omissions. However the architect's power is subject to two provisos found in clauses 2.28.6.3 and 2.28.6.4 – the first that there should be no reduction of the original time for completion; the second that there should be no reduction of pre-agreed extensions of time.

Clause 2.28.5

'.5 After the Completion Date for the Works or for a Section, if this occurs before the date of practical completion, the Architect / Contract Administrator may, and not later than the expiry of 12 weeks after the date of practical completion shall, by notice in writing to the Contractor, giving the details referred to in clause 2.28.3:

.1 fix a Completion Date for the Works or for the Section later than that previously fixed if in his opinion that is fair and reasonable having regard to any Relevant Events, whether on reviewing a previous decision or otherwise and whether or not the Relevant Event has been specifically notified by the Contractor under clause 2.27.1; or

.2 subject to clauses 2.28.6.3 and 2.28.6.4, fix a Completion Date earlier than that previously fixed if in his opinion that is fair and reasonable having regard to any instructions for Relevant Omissions issued after the last occasion on which a new Completion Date was fixed for the Works or Section; or

.3 confirm the Completion Date previously fixed.'

Under this clause the architect is required to finalise his position on revisions to the completion date within 12 weeks of practical completion. He has three options: he can extend his awards of extensions of time; reduce his awards on account of later omissions; or confirm previous decisions. However, what he must do is put in writing his final decision.

Clause 2.28.6

'.6 Provided always that:

.1 the Contractor shall constantly use his best endeavours to prevent delay in the progress of the Works or any Section, however caused, and to prevent the completion of the Works or Section being delayed or further delayed beyond the relevant Completion Date;

.2 in the event of any delay the Contractor shall do all that may reasonably be required to the satisfaction of the Architect / Contract Administrator to proceed with the Works or Section;

.3 no decision of the Architect / Contract Administrator under clause 2.28.4 or 2.28.5.2 shall fix a Completion Date for the Works or any Section earlier than the relevant Date for Completion; and

.4 no decision under clause 2.28.4 or 2.28.5.2 shall alter the length of any Pre-agreed Adjustment unless the relevant Variation or other work referred to in clause 5.2.1 is itself the subject of a Relevant Omission.'

These provisos seem intended principally to operate on the preceding clauses setting out the architect's powers and duties regarding extensions of time. However, they also can be read as imposing obligations on the contractor as regards progress.

The proviso in clause 2.28.6.1 that the contractor should use his best endeavours to prevent delay seems unnecessary given the contractor's basic obligation to complete the works within the time allowed and the inclusion of the 'fair and reasonable' test in assessment of extensions of time. As to how far the contractor has to go in using his best endeavours, see the discussion in Chapter 15.

The proviso in clause 2.28.6.2 could be taken as a mildly worded obligation on the contractor to accelerate.

17.6 Relevant events

JCT 2005 retains the extensive list of relevant events found in earlier editions but it is presented in a more readable manner:

Clause 2.29

'The following are the Relevant Events referred to in clauses 2.27 and 2.28:

.1 Variations and any other matters or instructions which under these Conditions are to be treated as, or as requiring, a Variation;

.2 Instructions of the Architect / Contract Administrator:

  .1 under any of clauses 2.15, 3.15, 3.16, 3.23 or 5.3.2; or

  .2 for the opening up for inspection or testing of any work, materials or goods under clause 3.17 or 3.18.4 (including making good), unless the inspection or test shows that the work, materials or goods are not in accordance with this Contract;

  .3 deferment of the giving of possession of the site or any Section under clause 2.5;

  .4 suspension by the Contractor under clause 4.14 of the performance of his obligations under this Contract;

  .5 any impediment, prevention or default, whether by act or omission, by the Employer, the Architect, Contract Administrator, the Quantity Surveyor or any of the Employer's Persons, except to the extent caused or contributed to by any default, whether by act or omission, of the Contractor or of any of the Contractor's Persons;

  .6 the carrying out by a Statutory Undertaker of work in pursuance of its statutory obligations in relation to the Works, or the failure to carry out such work;

  .7 exceptionally adverse weather conditions;

  .8 loss or damage occasioned by any of the Specified Perils;

  .9 civil commotion or the use or threat of terrorism and / or the activities of the relevant authorities in dealing with such event or threat;

  .10 strike, lock-out or local combination of workmen affecting any of the trades employed upon the Works or any of the trades engaged in the preparation, manufacture or transportation of any of the goods or materials required for the Works or any persons engaged in the preparation of the design for the Contractor's Designed Portion;

.11 the exercise after the Base Date by the United Kingdom Government of any statutory power which directly affects the execution of the Works;

.12 force majeure.'

Instructions

The instructions referred to in clause 2.29.2.2 are:

- clause 2.15   resolution of errors, omissions etc. in documents
- clause 3.15   postponement of work
- clause 3.16   provisional sums
- clause 3.23   antiquities
- clause 5.3.2   quotations.

Statutory undertakers' works

For comment on such works see Chapter 13 and the reference there to the case of *Henry Boot* v. *Central Lancashire Development Corporation* (1980).

Specified perils

These are defined in clause 6.8 as follows:

'fire, lightning, explosion, storm, flood, escape of water from any water tank, apparatus or pipe, earthquake, aircraft and other aerial devices or articles dropped therefrom, riot and civil commotion, but excluding Excepted Risks.'

'Excepted Risks' as referred to here are risks arising from nuclear waste, sonic waves and the like.

Force majeure

JCT 2005 does not define 'force majeure'. For discussion on its meaning see Chapter 13 above.

17.7 *Non-completion certificates*

Clause 2.31

'If the Contractor fails to complete the Works or a Section by the relevant Completion Date, the Architect / Contract Administrator shall issue a certificate to that effect (a 'Non-Completion Certificate'). If a new Completion Date is fixed after the issue of such a certificate, such fixing shall cancel that certificate and the Architect / Contract Administrator shall where necessary issue a further certificate.'

In common with some other standard forms of contract, JCT 2005 makes it a condition precedent to the deduction of liquidated damages, that the architect shall have issued a certificate of non-completion. This acts as a safeguard to the contractor against premature deductions from payment certificates by the employer, and it serves as a positive reminder to the employer that the right to deduct liquidated damages has been activated.

The certificate in clause 2.31 is to be a statement of fact and not a statement of opinion. If the architect fails in his duty to issue a certificate of non-completion he will deprive the employer of his right to deduct liquidated damages.

17.8 *Payment of liquidated damages*

Clauses 2.32 and 2.37 of JCT 2005 are the principal clauses detailing the employer's rights to deduct, or require the contractor to pay, liquidated damages for late completion. Clause 2.33 which deals with partial possessions by the employer also needs to be considered.

Clause 2.32

'.1 Provided:
   .1 the Architect / Contract Administrator has issued a Non-Completion Certificate for the Works or a Section; and
   .2 the Employer has informed the Contractor in writing before the date of the Final Certificate that he may require payment of, or may withhold or deduct, liquidated damages,
the Employer may, not later than 5 days before the final date for payment of the debt due under the Final Certificate, give notice in writing to the Contractor in the terms set out in clause 2.32.2.
.2 A notice from the Employer under clause 2.32.1 shall state that for the period between the Completion Date and the date of practical completion of the Works or that Section:
   .1 he requires the Contractor to pay liquidated damages at the rate stated in the Contract Particulars, or lesser rate stated in the notice, in which event the Employer may recover the same as a debt; and / or
   .2 that he will withhold or deduct liquidated damages at the rate stated in the Contract Particulars, or at such lesser stated rate, from monies due to the Contractor.
   .3 if the Architect / Contract Administrator fixes a later Completion Date for the Works or a Section or such later Completion Date is

stated in the Confirmed Acceptance of a Schedule 2 Quotation, the Employer shall pay or repay to the Contractor any amounts recovered, allowed or paid under clause 2.32 for the period up to that later Completion Date.

.4 if the Employer in relation to the Works or a Section has informed the Contractor in writing in accordance with clause 2.32.1.2 that he may require payment of, or may withhold or deduct, liquidated damages, then, unless the Employer states otherwise in writing, clause 2.32.1.2 shall remain satisfied in relation to the Works or Section, notwithstanding the cancellation of the relevant Non-Completion Certificate and issue of any further Non-Completion Certificate.'

Conditions precedent to deduction

Clause 2.32.1 states two important general conditions precedent to the employer's rights to deduct or require payment of liquidated damages. The first is that the architect must have issued a non-completion certificate; the second is that the employer must have issued a notice of intention to deduct or require payment. Without either of these deduction of liquidated damages will be invalid.

Additionally in clause 2.32.1 there is a requirement that the employer's notice must be given not later than 5 days before the final date for payment under the final certificate.

Further conditions precedent are found in clauses 4.13.4 and 4.15.4 which deal with the service of withholding notices for interim and final payment certificates. These are additional to the clause 2.32.1 conditions.

The overall scheme therefore is that there must be a clearly stated date from which delay damages can run; the employer must have informed the contractor that he intends to exercise his rights regarding delay damages; and the employer must serve withholding notices in respect of particular deductions.

Employer's notice

At first sight the requirement in clause 2.32.2 that the employer's notice shall state the period between the completion date and the date of practical completion might be taken as suggesting that the notice, and the deduction of liquidated damages, should only follow practical completion. However, read in conjunction with the remainder of clause 2.32 it is apparent that what it means is simply that the notice shall have a statement to the effect that, for the period between the completion date and practical completion, payment or deduction of liquidated damages will be enforced.

It is, therefore, open to the employer to serve his notice under clause 2.32.2 at any time prior to the final certificate restriction.

He can, if he wishes, do so at commencement of the contract so that the contractor is under no illusion that liquidated damages are to be no more than a threat.

Partial possessions

As is common in construction contracts, sections of the works in JCT 2005 are pre-defined parts of the works with their own completion dates and rates of liquidated damages. They are covered by the phrases 'Work or any Section' found throughout the contract. In contrast, 'parts' of the work are not defined and they are not subject to provisions for completion, extension of time and liquidated damages. However, partial completions do influence such provisions for the whole of the works and for sections in so far that they have the potential to invalidate liquidated damages unless these are proportioned down.

JCT 2005 provides for partial possessions by the employer in clause 2.33:

'2.33 If at any time or times before the date of issue by the Architect / Contract Administrator of the Practical Completion Certificate or relevant Section Completion Certificate the Employer wishes to take possession of any part or parts of the Works or a Section and the consent of the Contractor has been obtained (which consent shall not be unreasonably delayed or withheld), then, notwithstanding anything expressed or implied elsewhere in this Contract, the Employer may take possession of such part or parts. The Architect / Contract Administrator shall thereupon issue to the Contractor on behalf of the Employer a written statement identifying the part or parts taken into possession and giving the date when the Employer took possession ('the Relevant Part' and 'the Relevant Date' respectively).'

17.9 *Proportioning down of liquidated damages*

The proportioning down rule of JCT 2005 is simply stated in clause 2.37 as follows:

Clause 2.37

'As from the Relevant Date, the rate of liquidated damages stated in the Contract Particulars in respect of the Works or Section containing the Relevant Part shall reduce by the same proportion as the value of the Relevant Part bears to the Contract Sum or to the relevant Section Sum, as shown in the Contract Particulars.'

Chapter 18
Civil engineering forms

18.1 NEC 3 – Engineering and Construction Contract, 2005

NEC 3 is the generic name of a family of contracts published for the Institution of Civil Engineers. NEC stands for New Engineering Contract and it is by this name that the contracts are generally known. The main contract and the subcontract were first published as consultative editions in January 1991. First formal editions followed in March 1993; second editions in November 1995; and third editions in June 2005.

Between 1991 and 2005 other contracts were produced such that by 2005 the NEC 3 family comprised:

- the NEC 3 Engineering and Construction Contract
- the NEC 3 Engineering and Construction Subcontract
- the NEC 3 Professional Services Contract
- the NEC 3 Short Contract
- the NEC 3 Short Subcontract
- the NEC 3 Adjudicator's Contract
- the NEC 3 Term Services Contract
- the NEC 3 Framework Contract.

In this chapter it is the main form NEC 3 Engineering and Construction Contract which is examined.

NEC 3 is intended for use on engineering or building projects, both at home or overseas; and to apply whether the contractor has full design responsibility or none at all; and to both ordinary contracts or management-style contracts. There are six pricing options varying from lump sum to fully cost reimbursable. The main objectives adopted in the drafting were flexibility, clarity and promotion of good project management. To achieve this NEC 3 incorporates various contractual arrangements and combinations of core clauses and optional clauses. Much of the familiar terminology of conventional construction contracts is abandoned in favour of a direct and non-legalistic style of drafting. The balance of risk between the parties can be varied according to which optional clauses are adopted, but generally the full costs of any changes fall wholly on the employer. Thus variations and the like are valued at cost including contractor's preliminaries.

Liquidated damages provisions for late completion are included in the optional clauses, as are provisions for liquidated damages for low performance which apply when the works fail to meet stipulated performance standards after hand-over. The provisions for extensions of time are included in the core clauses for compensation events. These are events which give the contractor entitlement to both extra cost and extra time. There are no provisions for events giving entitlement to extra time only.

Commencement and completion

The commencement date is a date set by the employer in the contract data. The date for completion is generally similarly set by the employer but there is provision for it to be set by the contractor if the parties so agree.

Because of the minimalistic style of wording in NEC 3 care needs to be taken in operating the contract of the distinction between the date for completion (the completion date) and the date of completion.

Completion is defined in clause 11.2(2) as follows:

'(2) Completion is when the *Contractor* has:
  – done all the work which the Works Information states he is to do by the Completion Date and
  – corrected notified Defects which would have prevented the *Employer* from using the *works* and Others from doing their work.

If the work which the *Contractor* is to do by the Completion Date is not stated in the Works Information, Completion is when the *Contractor* has done all the work necessary for the *Employer* to use the *works* and for Others to do their work.'

By clause 30.2 the project manager 'decides' the date of completion and certifies completion within one week of completion. The contractor is not required to take any steps to initiate that but in practice will generally do so. The use of the word 'decides' is interesting. It raises questions as to whether it can be said that the parties have agreed to accept the decision of the project manager such that it is not challengeable or whether via the various dispute resolution procedures in the contract, or in law, either party can seek to overturn the project manager's decision.

Sectional completion and key dates

Sectional completion requirements are set out in the contract data and are covered by optional clause X5.1 as follows:

'In these *conditions of contract*, unless stated as the whole of the *works*, each reference and clause relevant to:

  – the *works*
  – Completion and
  – Completion Date

applies, as the case may be, to either the whole of the *works* or any *section* of the *works*.'

Key dates are dates set by the employer by which the contractor has to bring a stated part of the works to a specified condition. Key dates can be extended by the same compensation event procedures as completion dates.

Extension of time

The provisions in NEC 3 for extending time are significantly different from conventional extension of time clauses. Thus the phrase 'extension of time' is never used. In its place there is only reference to changes to the completion date. Clause 11.2(3) explains this as follows 'The Completion Date is the *"completion date"* unless later changed in accordance with this contract'. More importantly the only mechanism for extending time is through the compensation event procedures of the contract. These serve two purposes – changes to the contract price and changes to completion dates and/or key dates. The listed compensation events, of which there are twenty or so, depending on which of the six main options of NEC 3 is selected, can therefore be taken as a list of relevant events covering both time and price changes.

One of the main features of the compensation event procedures is that they are intended to commence with quotations from the contractor on instruction from the project manager or by the contractor's initiative. For both time and price changes there are prescriptive rules on how the quotations should be calculated – or 'assessed' as NEC 3 puts it. Put simply, time changes are assessed by reference to the impact of the compensation event on the accepted programme and price changes are assessed by calculation of the extra cost (plus a mark-up) arising from the event. Ideally both time and price assessments are prospective calculations and as such they should include the contractor's risk allowances. The following extracts from NEC 3 illustrate the above points:

Clause 62.2 – quotations for compensation events

'Quotations for compensation events comprise proposed changes to the Prices and any delay to the Completion Date and Key Dates assessed by the *Contractor*. The *Contractor* submits details of his assessment with each quotation. If the programme for remaining work is altered by the compensation event, the *Contractor* includes the alterations to the Accepted Programme in his quotation.'

Clause 63.3 – assessment of delay

'A delay to the Completion Date is assessed as the length of time that, due to the compensation event, planned Completion is later than planned Completion as shown on the Accepted Programme. A delay to a Key Date is assessed as the length of time that, due to a compensation event, the planned date when the Condition stated for a Key Date will be met is later than the date shown on the Accepted Programme.'

Clause 63.6 – risk allowances

'Assessment of the effect of a compensation event includes risk allowances for cost and time for matters which have a significant chance of occurring and are at the *Contractor's* risk under this contract.'

Other particular points of interest are:

(a) Early warnings
By clause 16.1 the contractor and the project manager are required to give early warning notice of any matters which could (amongst other things) delay completion. Clause 63.5 states that if the contractor did not give an early warning the event is assessed as if it had been given.
(b) Time-bars / condition precedent
Clause 61.3 states:

'The *Contractor* notifies the *Project Manager* of an event which has happened or which he expects to happen as a compensation event if:

– the *Contractor* believes that the event is a compensation event and
– the *Project Manager* has not notified the event to the *Contractor*.

If the *Contractor* does not notify a compensation event within eight weeks of becoming aware of the event, he is not entitled to a change in the Prices, the Completion Date or a Key Date unless the *Project Manager* should have notified the event to the *Contractor* but did not.'
(c) Float
As can be seen from the above-quoted extract from clause 63.3, a delay to completion is assessed by reference to planned completion not by reference to delay beyond the date for completion. In effect, therefore, the contractor owns (or gets the benefit of) any terminal float in the accepted programme.

June 2006 amendment

Regarding the accepted programme, users of NEC 3 should be aware of an amendment made to the contract in June 2006. That deletes from clause 32.1 the words "and of notified early warning matters" from the part of the clause which reads:

'The Contractor shows on each revised programme:
. . .
the effects of implemented compensation events and of notified early warning matters'
. . .

The effect of this is to prevent the contractor benefiting in his time and cost assessments from events which might not happen. This is an important point given that clause 65.2 gives a measure of finality to all implemented compensation events in stating 'The assessment of a compensation event is not revised if a forecast upon which it is based is shown by later recorded information to have been wrong'.

Listed compensation events

Clause 60.1 contains nineteen items stated to be compensation events. These are:

(1) instructions changing the works information
(2) late access given by employer
(3) things not provided
(4) suspensions
(5) works by the employer and others
(6) late replies to communications
(7) fossils, antiquities, etc.
(8) changed decisions
(9) withholding acceptances
(10) searches for defects
(11) inspections
(12) physical conditions
(13) weather
(14) employer's risks
(15) early take-over
(16) late provision of testing facilities
(17) correction of assumptions
(18) breach by the employer
(19) prevention events.

In addition to the above, there are three further compensation events in main options B and D covering changes of quantities and the like arising from the use of bills of quantities.

Some employers prune the above list to achieve what they consider to be a more acceptable balance of risk. The prime candidate for deletion is weather, but physical conditions usually comes out where the contractor is responsible for design and the prevention events item which is close to being 'force majeure' certainly requires attention.

In NEC 3 contracts where the physical conditions event remains included it needs to be noted that the test is not the usual 'foreseeability' test but "what an experienced contractor would have judged at the Contract Date to have had such a small chance of occurring that it would have been unreasonable for him to have allowed for them". Similarly where the prevention event remains included it needs to be noted that not only is its scope potentially wider than force majeure but it also is tested by consideration of what it would have been unreasonable to have allowed for.

Delay damages

The only provisions for delay damages in NEC 3 are in optional clause X7. From the reference there to delay damages at rates stated in the contract data this is evidently intended to act as a liquidated damages clause. In the absence of clause X7 the employer's remedy for late completion is to sue for general damages.

Clause X7 (delay damages) reads:

'X7

X7.1 The *Contractor* pays delay damages at the rate stated in the Contract Data from the Completion Date for each day until the earlier of
– Completion and
– the date on which the *Employer* takes over the *works*.

X7.2 If the Completion Date is changed to a later date after delay damages have been paid, the *Employer* repays the overpayment of damages with interest. Interest is assessed from the date of payment to the date of repayment and the date of repayment is an assessment date.

X7.3 If the *Employer* takes over a part of the *works* before Completion, the delay damages are reduced from the date on which the part is taken over. The *Project Manager* assesses the benefit to the *Employer* of taking over the part of the *works* as a proportion of the benefit to the *Employer* of taking over the whole of the *works* not previously taken over. The delay damages are reduced in this proportion.'

An important point to note about clauses X7.1 and X7.3 is that there are two possible end dates for conclusion of liability for delay damages – completion and take-over. Whichever of these is the earlier is the effective end date. However, under clause 35 whenever the employer uses any part of the works before completion the contractor becomes entitled to a take-over certificate for that part. At that point in time the contractor's liability for delay damages starts to reduce and clause X7.3 comes into play. In short, the rule is that if the employer uses any part of the works, delay damages start to reduce.

Another point to note about clause X7.1 is that it only covers damages for delay to 'the Completion Date'. Such delay can include for sectional completions providing that secondary option X5 is also included in the contract but, on the face of it, clause X7 is not intended to have any application to delays in meeting key dates. That raises interesting questions on the exclusivity effect of clause X7 and on the employer's rights (if any) to recover as general damages costs resulting from late achievement of key dates and from late completions of sections for which no stipulated rates for delay damages are stated. For comment on suchlike questions see Chapter

3 above and, in particular, the extracts there from the judgment in the *Biffa Waste* case.

Yet a further point of note on clause X7.1 is that it has little to say on the procedures for deduction of delay damages. What seems to be intended is that for payments due after the completion date has passed the project manager assesses the amounts of any delay damages and allows for these in payment certificates issued under clause 50.

Proportioning down

The usual rule is that delay damages are reduced in proportion to the value of the works taken over. This is largely an arithmetic or quantity surveying exercise. The benefit rule in clause X7.3 of NEC 3 is an interesting departure from the usual rule. The clause gives no guidance as to how the employer's benefit is to be assessed and it is not difficult to visualise endless argument as to how it should be assessed. The principles of assessment are likely to be contentious and similarly the facts.

A basic point which needs to be considered is that stipulated rates of delay damages have to be taken as genuine pre-estimates of loss if they are to stand as valid liquidated damages. Any adjustments to the stipulated rates need therefore to follow some logical and identifiable process to avoid voiding the rates. It is therefore arguable that in assessing the employer's benefit the project manager should only take into account circumstances anticipated at the time the contract was made. However, the Guidance Notes to NEC 3 take the opposite view suggesting that only benefits qualifying at the time of calculation of proportioning down should be considered.

Given the potential in the present drafting of clause X7.3 for disputes and differences in applying its 'benefit' rule, many employers may be disposed to amend the clause to bring it into line with the conventional 'value' rule for proportioning down.

18.2 ICE Conditions of Contract – 7th edition, 1999

The first edition of ICE Conditions of Contract was published in 1945; the second in 1950; the third in 1951; the fourth in 1955; the fifth in 1973; the sixth in 1991; and the seventh in September 1999. For fifty years or so these were the principal standard forms for civil engineering works but with the growing popularity of the New Engineering Contract it is doubtful if they still hold that position.

The following commentary relates to the ICE 7th edition.

Commencement and completion

By tradition ICE contracts state times for completion rather than completion dates. Accordingly it is essential that the starting date (or Works

Commencement Date as it is called) is fixed with certainty. This is dealt with in clause 41 as follows:

'41 (1) The Works Commencement Date shall be
  (a) the date specified in the Appendix to the Form of Tender or if no date is specified
  (b) a date between 14 and 28 days of the award of the Contract to be notified to the Contractor by the Engineer in writing or
  (c) such other date as may be agreed between the parties.
  (2) The Contractor shall start the Works on or as soon as is reasonably practicable after the Works Commencement Date. Thereafter the Contractor shall proceed with the Works with due expedition and without delay in accordance with the Contract.'

The contractor's obligations with regard to completion are stated in clause 43 (time for completion):

'43 The whole of the Works and any Section required to be completed within a particular time as stated in the Appendix to the Form of Tender shall be substantially completed within the time so stated (or such extended time as may be allowed under Clause 44 or revised time agreed under Clause 46(3)) calculated from the Works Commencement Date.'

Completion is covered in clause 48, the first part of which deals with notification of substantial completion:

'48 (1) When the Contractor considers that
  (a) the whole of the Works or
  (b) any Section in respect of which a separate time for completion is provided in the Appendix to the Form of Tender
has been substantially completed and has satisfactorily passed any final test that may be prescribed by the Contract he may give notice in writing to that effect to the Engineer or to the Engineer's Representative. Such notice shall be accompanied by an undertaking to finish any outstanding work in accordance with the provisions of Clause 49(1).'

The second part of clause 48 makes clear that it is the engineer's duty to form an opinion as to whether the works have been substantially completed and, if not, to state what remains to be done:

'48 (2) The Engineer shall within 21 days of the date of delivery of such notice either
  (a) issue to the Contractor (with a copy to the Employer) a Certificate of Substantial Completion stating the date on which in his opinion the Works were or the Section was substantially completed in accordance with the Contract or
  (b) give instructions in writing to the Contractor specifying all the work which in the Engineer's opinion requires to be done by the Contractor before the issue of such certificate.'

Sectional completion

ICE 7th edition distinguishes between parts and sections of the works. Normally only sections have stipulated times for completion. A section is defined in clause 1(1)(u) which states 'Section means a part of the Works separately identified in the Appendix to the Form of Tender'.

Extension of time

The procedures and relevant events for extending time under ICE 7th edition are found in clause 44 which, in its entirety, reads as follows:

'44 (1) Should the Contractor consider that:

(a) any variation ordered under Clause 51(1) or

(b) increased quantities referred to in Clause 51(4) or

(c) any cause of delay referred to in these Conditions or

(d) exceptional adverse weather conditions or

(e) any delay impediment prevention or default by the Employer or

(f) other special circumstances of any kind whatsoever which may occur

be such as to entitle him to an extension of time for the substantial completion of the Works or any Section thereof he shall within 28 days after the cause of any delay has arisen or as soon thereafter as is reasonable deliver to the Engineer full and detailed particulars in justification of the period of extension claimed in order that the claim may be investigated at the time.

(2)

(a) The Engineer shall upon receipt of such particulars consider all the circumstances known to him at that time and make an assessment of the delay (if any) that has been suffered by the Contractor as a result of the alleged cause and shall so notify the Contractor in writing.

(b) The Engineer may in the absence of any claim make an assessment of the delay that he considers has been suffered by the Contractor as a result of any of the circumstances listed in sub-clause (1) of this Clause and shall so notify the Contractor in writing.

(3) Should the Engineer consider that the delay suffered fairly entitles the Contractor to an extension of the time for the substantial completion of the Works or any Section thereof such interim extension shall be granted forthwith and be notified to the Contractor in writing with a copy to the Employer. In the event that the Contractor has made a claim for an extension of time but the Engineer does not consider the Contractor entitled to an extension of time he shall so inform the Contractor without delay.

(4) The Engineer shall not later than 14 days after the due date or extended date for completion of the Works or any Section thereof (and whether or not the Contractor shall have made any claim for an extension of time) consider all the circumstances known to him at that time and take actions similar to that provided for in sub-clause (3) of this Clause. Should the Engineer consider that the Contractor is not entitled to an extension of time he shall so notify the Employer and the Contractor.

(5) The Engineer shall within 28 days of the issue of the Certificate of Substantial Completion for the Works or for any Section thereof review all the circumstances of the kind referred to in sub-clause (1) of this Clause and shall finally determine and certify to the Contractor with a copy to the Employer the overall extension of time (if any) to which he considers the Contractor entitled in respect of the Works or the relevant Section. No such final review of the circumstances shall result in a decrease in any extension of time already granted by the Engineer pursuant to sub-clauses (3) or (4) of this Clause.'

Relevant events

The causes of delay referred to in clause 44(1)(c) are for the most part found in the clauses listed below. These are clauses which make specific references to Clause 44:

- clause 7(4)   late drawings and instructions
- clause 12(2)   adverse physical conditions or artificial obstructions
- clause 13(3)   instructions causing delay
- clause 14(8)   delay in engineer's consent to contractor's methods or because of engineer's requirements
- clause 27(4)   variations relating to public utilities
- clause 31(2)   facilities for other contractors
- clause 40(1)   suspension of work
- clause 42(3)   failure to give possession.

Additionally, by cross-references to clause 13, clause 5 (documents mutually explanatory) and clause 59(4)(b) (expulsion of nominated sub-contractor) include relevant events.

The relevant event stated in clause 44(1)(e) for prevention or default by the employer was new to ICE 7th edition – previous editions having relied on such events being covered by the 'special circumstances' event. For the dangers of that see the comment on 'catch all' phrases in Chapter 6 above.

Notice requirements

The requirement in clause 44(1) for the contractor to claim extension of time within 28 days of the cause of delay is probably not worded strongly enough to serve as a condition precedent to entitlement. But, in any event, the requirements later in clause 44 for the engineer to make assessments of the extensions of time due whether or not the contractor has made any claim, dispose of the proposition that timely notice is a condition precedent to entitlement.

Method of assessment

ICE 7th edition does not prescribe any particular method of assessment of extension of time. The procedures for interim assessments under clause 44(3) seem to suggest prospective assessments by the requirement that they be made 'forthwith'. However, the requirements for reviews in clauses 44(4) and 44(5) suggest retrospective assessments.

The underlying requirement that the engineer shall make assessments which are fair in all the circumstances known to him appears to offer plenty of scope for choice in the method of assessment. However against that the specific references in clauses 44(2)(b) and 44(5) to the circumstances referred to in sub clause (1) – i.e. to the relevant events – might be taken as indicative of the need for time-impact analysis.

Liquidated damages for delay

The provisions in ICE conditions of contract for liquidated damages for delay have traditionally been expressed with a level of detail which makes clear both their purpose and procedures. In full the provisions in ICE 7th edition as set out in clause 47 are as follows:

'47 (1)

    (a) Where the whole of the Works is not divided into Sections the Appendix to the Form of Tender shall include a sum which represents the Employer's genuine pre-estimate (expressed per week or per day as the case may be) of the damages likely to be suffered by him if the whole of the Works is not substantially completed within the time prescribed by Clause 43 or by any extension thereof granted under Clause 44 or by any revision thereof agreed under Clause 46(3) as the case may be.

    (b) If the Contractor fails to achieve substantial completion of the whole of the Works within the time so prescribed he shall pay to the Employer the said sum for every week or day (as the case may be) which shall elapse between the date on which the prescribed time expired and the date the whole of the Works is substantially completed.

Provided that if any part of the Works is certified as substantially complete pursuant to Clause 48 before the completion of the whole of the Works the said sum shall be reduced by the proportion which the value of the part so completed bears to the value of the whole of the Works.

(2)

   (a) Where the Works is divided into Sections (together comprising the whole of the Works) which are required to be completed within particular times as stated in the Appendix to the Form of Tender sub-clause (1) of this Clause shall not apply and the said Appendix shall include a sum in respect of each Section which represents the Employer's genuine pre-estimate (expressed per week or per day as the case may be) of the damages likely to be suffered by him if that Section is not substantially completed within the time prescribed by Clause 43 or by any extension thereof granted under Clause 44 or by any revision thereof agreed under Clause 46(3) as the case may be.

   (b) If the Contractor fails to achieve substantial completion of any Section within the time so prescribed he shall pay to the Employer the appropriate stated sum for every week or day (as the case may be) which shall elapse between the date on which the prescribed time expired and the date of substantial completion of that Section.

Provided that if any part of that Section is certified as substantially complete pursuant to Clause 48 before the completion of the whole thereof the appropriate stated sum shall be reduced by the proportion which the value of the part so completed bears to the value of the whole of that Section.

   (c) Liquidated damages in respect of two or more Sections may where circumstances so dictate run concurrently.

(3) All sums payable by the Contractor to the Employer pursuant to this Clause shall be paid as liquidated damages for delay and not as a penalty.

(4)

   (a) The total amount of liquidated damages in respect of the whole of the Works or any Section thereof shall be limited to the appropriate sum stated in the Appendix to the Form of Tender. If no such limit is stated therein then liquidated damages without limit shall apply.

   (b) Should there be omitted from the Appendix to the Form of Tender any sum required to be inserted therein either by sub-clause (1)(a) or by sub-clause (2)(a) of this Clause as the case may be or if any such sum is stated to be 'nil' then to that extent damages shall not be payable.

(5) The Employer may

   (a) deduct and retain the amount of any liquidated damages becoming due under the provision of this Clause from any sums due or which become due to the Contractor or

   (b) require the Contractor to pay such amount to the Employer forthwith.

If upon a subsequent or final review of the circumstances causing delay the Engineer grants a relevant extension or further extension of time the Employer shall no longer be entitled to liquidated damages in respect of the period of such extension.

Any sum in respect of such period which may already have been recovered under this Clause shall be reimbursed forthwith to the Contractor together with interest compounded monthly at the rate provided for in Clause 60(7) from the date on which such sums were recovered from the Contractor.

(6) If after liquidated damages have become payable in respect of any part of the Works the Engineer issues a variation order under Clause 51 or adverse physical conditions or artificial obstructions within the meaning of Clause 12 are encountered or any other situation outside the Contractor's control arises any of which in the Engineer's opinion results in further delay to that part of the Works

   (a) the Engineer shall so notify the Contractor and the Employer in writing and

   (b) the Employer's further entitlement to liquidated damages in respect of that part of the Works shall be suspended until the Engineer notifies the Contractor and the Employer in writing that the further delay has come to an end.

Such suspension shall not invalidate any entitlement to liquidated damages which accrued before the period of further delay started to run and subject to any subsequent or final review of the circumstances causing delay any monies already deducted or paid as liquidated damages under the provision of this Clause may be retained by the Employer.'

For the most part the above provisions are self-explanatory and require no further comment. Clause 47(6) on the intervention of variations is the exception.

Intervention of variations

Clause 47(6) deals in an unusual way with the question of what extension, if any, is due when delay for which the contractor is not responsible occurs after the due completion date. Instead of attempting to resolve whether an extension is due up to the date at which the delay ends, or whether it is due only for the period of delay itself, this provision avoids the argument by suspending damages for the period of delay. In principle this seems a good approach but the drafting of the clause is likely to lead to difficulties.

The first problem is that the grounds for suspending liquidated damages are arguably too wide and it is the engineer's 'opinion' not a test of fairness which applies. The stated grounds are:

(i)   a variation order under clause 51;
(ii) adverse physical conditions or artificial obstructions within clause 12;
(iii) any other situation outside the contractor's control.

The last item goes well beyond the relevant events for extensions of time. Thus any delay by weather would seem to qualify, whether exceptionally adverse or not, and it does not seem to be open to the engineer to consider whether it is fair for suspension of damages to be made: he can only form an opinion on whether there has been delay. Contractors in culpable delay and paying liquidated damages will readily find a range of matters outside their control to offer as excuses for continuing delay – not least, as discussed in Chapter 13, lack of performance by their sub-contractors or suppliers.

A second problem lies in the use of the phrase 'that part of the works' with regard to both delay and suspension of damages. It is not obvious why delay to part of the works, particularly if non-critical, should lead to suspension of damages. Nor is it clear how the level of suspended damages should be calculated. Perhaps the intended meaning is that damages should be suspended if there is delay to a section or the whole of the works, but by definition in clause 1(1)(c) a section has its own identity whereas a part is undefined.

A third problem is interpretation of the final paragraph of the sub-clause. The wording throws into question the whole purpose of the suspension of damages provision. Clearly the suspension does not invalidate entitlement to damages before the period of delay.

But, in any event, following the decision in the *Balfour Beatty* case mentioned in Chapter 5 of this book there is little doubt that the contractors' entitlement to extension of time for acts of prevention during a period of culpable delay is on a net basis rather than a gross basis. Consequently, it is arguable whether clause 47(6) serves much useful purpose.

18.3 ICE Conditions of Contract for Minor Works – 3rd edition, 2001

The ICE minor works form, first published in 1988, is intended for works of a simple and straightforward nature where the risks are small, the period for completion does not exceed six months and the contract value does not exceed £500,000.

As with all briefly drafted and simply worded forms of contract on a subject as complex as works of construction there is some lack of legal precision. The price of brevity is paid with uncertainty and the minor works form does leave a few questions unanswered. Thus, one notable omission is the absence of any provision for the employer to deduct liquidated damages from amounts due to the contractor. This may be an unintended omission although the wording of the payment clauses suggests that the point might have been considered and rejected by the draftsmen. Also there are references throughout the form to 'parts of the works' which are confusing since the phrase seems to apply to both parts defined in the schedule and to other parts not so defined but which are taken into use or possession independently of the remainder.

Commencement and completion

Clause 4.1 (starting date)

> 'The starting date shall be the date specified in the Appendix or if no date
> is specified a date to be notified by the Engineer in writing being within
> a reasonable time and in any event within 28 days after the date of accep-
> tance of the Tender. The Contractor shall begin the Works at or as soon
> as reasonably possible after the starting date.'

Clause 4.2 (period for completion)

> 'The period or periods for completion shall be as stated in the Appendix
> or such extended time as may be granted under Clause 4.4 and shall
> commence on the starting date.'

Completion is defined in clause 4.5(1) by reference to the term 'practical
completion'. Partial completion is covered in clause 45(2). Those clauses read
as follows:

> '4.5(1) Practical completion of the whole of the Works shall occur when
> the Works reach a state when notwithstanding any defect or out-
> standing items therein, they are taken or are fit to be taken into
> use or possession by the Employer.'
> '4.5(2) Similarly practical completion of part of the Works may also occur
> but only if it is fit for such part to be taken into use or possession
> independently of the remainder.'

Clause 4.4 – Extensions of time

> 'If the progress of the Works or any part thereof shall be delayed for any
> of the following reasons:
>
> (a) an instruction given under Clause 2.3(a)(c) or (d);
> (b) an instruction given under Clause 2.3(b) where the test or investiga-
> tion fails to disclose non-compliance with the Contract;
> (c) encountering an obstruction or condition falling within Clause 3.8
> and/or an instruction given under Clause 2.3(e);
> (d) delay in receipt by the Contractor of necessary instructions, drawings
> or other information;
> (e) failure by the Employer to give adequate access to the Works or
> possession of land required to perform the Works;
> (f) delay in receipt by the Contractor of materials to be provided by the
> Employer under the Contract;
> (g) exceptional adverse weather;
> (h) any delay impediment prevention or default by the Employer;
> (i) other special circumstances of any kind whatsoever outside the
> control of the Contractor

then provided that the Contractor has taken all reasonable steps to avoid or minimise the delay the Engineer shall upon a written request by the Contractor promptly by notice in writing grant such extension of the period for completion of the whole or part of the Works as may in his opinion be reasonable. The extended period or periods for completion shall be subject to regular review provided that no such review shall result in a decrease in any extension of time already granted by the Engineer.'

'Any part of the works'

The reference to 'any part' in the first sentence should logically apply only to those parts of the works detailed in the Appendix and which have their own periods for completion. If the phrase is given a wider meaning it brings in any delay to any part, whether critical or not. This may be intended since there is a cross reference from the additional payments provisions, clause 6.1, to delays in clause 4.4 but clearly although additional payment may be due for a non-critical delay, an extension of time will not be due.

Support for the view that 'any part of the works' refers only to the parts detailed in the Appendix comes later in the clause where the engineer is empowered to grant extensions of time for 'the whole or part of the works'. It would seem unlikely that the engineer is intended to grant extensions of time for unspecified parts of the works since they have no stipulated time for completion in the first place, other than the time for the whole of the works, and because liquidated damages do not apply to unspecified parts independently of the whole.

This latter point reveals a major difficulty in the clause whatever interpretation is taken. Although there can be specified parts of the works with their own times for completion entered in the Appendix, there is no provision for separate liquidated damages for each part. Consequently late completion of a part, whether specified or not, cannot attract liquidated damages independently of the whole since there is no mechanism for arriving at the damages due. There is a procedure in clause 4.6 similar to the 'scaling down in proportion to work completed' provisions of most forms which may be an attempt to proportion the stipulated damages down to sections, but trial insertion of figures suggests that it does not work as such.

The argument against 'any part of the works' being only a specified part is that in other clauses the phrase is used in a general sense. Thus in clause 8.2, the contractor shall not sub-let 'any part of the works'; and in clause 2.6(2), a suspension of work affecting 'part of the works' can be treated as an omission variation.

Clause 4.6 Liquidated damages for late completion

'4.6 If by the end of the period or extended period or periods for completion the Works have not reached practical completion the Contractor shall be liable to the Employer in the sum stated in the Appendix as liquidated damages for every week (or pro rata for part of a week) during which the Works so remain uncompleted up to the limit stated in the Appendix. Similarly where part or parts of the Works so remain uncompleted the Contractor shall be liable to the Employer in the sum stated in the Appendix reduced in proportion to the value of those parts which have been certified as complete provided that the said limit shall not be reduced. Provided that if after liquidated damages have become payable in respect of any part of the Works the Engineer issues a variation order under Clause 2.3(a) or an artificial obstruction or physical condition within the meaning of Clause 3.8 is encountered or any other situation outside the Contractor's control arises any of which in the opinion of the Engineer results in further delay to that part of the Works

(a) the Engineer shall so inform the contractor and the Employer in writing

and

(b) the Employer's further entitlement to liquidated damages in respect of that part of the Works shall be suspended until the Engineer notifies the Contractor and the Employer in writing that the further delay has come to an end.

Such suspension shall not invalidate any entitlement to liquidated damages which accrued before the period of delay started to run and any monies deducted or paid in accordance with this Clause may be retained by the Employer without incurring interest thereon under Clause 7.8.'

18.4 CECA Form of sub-contract, 2008

A model form of sub-contract for civil engineering works has been in use for many years. Originally produced by what was then the Federation of Civil Engineering Contractors it was known as the FCEC Blue Form. It is currently published as the CECA Form of Sub-Contract by the Civil Engineering Contractors Association. The July 1998 version was reprinted with amendments in February 2008.

Few changes have been made to the Blue Form since the 1984 edition was produced for use on sub-contracts associated with main contracts under ICE 5th edition conditions of contract. And although there is a reference to ICE 6th edition conditions on the cover of the present version it remains effective for use with ICE 7th edition conditions.

Because of its simplicity, the Blue Form does not fit in well with main contracts under NEC conditions but nevertheless it is widely used with such conditions. There are various reasons for this – tradition; concern over using the NEC sub-contract which simply steps-down with all its complexities and risk-sharing arrangements the provisions of the main NEC contract, and also concern that the stepped-down administrative and procedural arrangements of the NEC sub-contract are too weighty for small sub-contracts and for small sub-contractors.

Commencement and completion

Clause 6.1 of the Blue Form deals with commencement and completion. It states:

'Within 10 days, or such other period as may be agreed in writing, of receipt of the Contractor's written instructions so to do, the Sub-Contractor shall enter upon the Site and commence the execution of the Sub-Contract Works and shall thereafter proceed with the same with due diligence and without any delay, except such as may be expressly sanctioned or ordered by the Contractor or be wholly beyond the control of the Sub-Contractor. Subject to the provisions of this Clause, the Sub-Contractor shall complete the Sub-Contract Works within the Period for Completion specified in the Third Schedule hereto.'

Extension of time

Although there are no provisions for liquidated damages for late completion in the Blue Form, it does have provisions for extensions of time matching those in the main contract and covering additionally variations of the sub-contract works and delays caused by the main contractor's breaches of the sub-contract. Such extensions are granted by the main contractor on the basis of what is 'fair and reasonable' and they serve to formalise the completion obligations of the sub-contract in claims for general damages. It is an express condition precedent to gaining an extension under the matching provisions of the main contract that the sub-contractor gives notice within 14 days of any delay first occurring. This is to ensure that the sub-contractor only acquires the benefits which should with adequate notice also be available to the main contractor. This is reinforced by limitation of the sub-contractor's entitlement to that of the main contractor.

Clauses 6.2 to 6.5 which deal with extensions of time read:

'(2) If the Sub-Contractor shall be delayed in the execution of the Sub-Contract Works:
    (a) by any circumstances or occurrence (other than a breach of this Sub-Contract by the Sub-Contractor) entitling the Contractor to an extension of his time for completion of the Main Works under the Main Contract; or
    (b) by the ordering of any variation of the Sub-Contract Works to which paragraph (a) of this sub-clause does not apply; or
    (c) by any breach of this Sub-Contract by the Contractor;
then in any such event the Sub-Contractor shall be entitled to such extension of the Period for Completion as may in all the circumstances be fair and reasonable.

Provided always that in any case to which paragraph (a) of this sub-clause applies it shall be a condition precedent to the Sub-Contractor's right to an extension of the Period for Completion that he shall have given written notice to the Contractor of the circumstances or occurrence which is delaying him within 14 days of such delay first occurring and in any such case the extension shall not in any event exceed the extension of time to which the Contractor is properly entitled under the Main Contract.

(3) Where differing Periods of Completion are specified in the Third Schedule for different parts of the Sub-Contract Works, then for the purposes of the preceding provisions of this clause each such part shall be treated separately in accordance with sub-clause (2) above.

(4) Nothing in this clause shall be construed as preventing the Sub-Contractor from commencing off the Site any work necessary for the execution of the Sub-Contract Works at any time before receipt of the Contractor's written instructions under sub-clause (1) of this clause.

(5) The contractor shall notify the Sub-Contractor in writing of all extensions of time obtained under the provisions of the Main Contract which affect the Sub-Contract.'

Delay damages

The Blue Form does not provide for liquidated damages for delay. As explained in Chapter 9, there are good reasons why it is more appropriate to leave damages for delay unliquidated in sub-contracts. However, since one element of such unliquidated damages may be liquidated damages imposed on the main contractor it is commonplace to include details of main contract rates of liquidated damages in the sub-contract schedules. These can then be included in claims for general damages for delay if it is the sub-contractor's delay which has caused main contract delay.

This is confirmed by clauses 3.3 and 3.4 of the Blue Form which deal with the sub-contractor's liability for general damages:

'(3) The Sub-Contractor shall indemnify the Contractor against every liability which the Contractor may incur to any other person whatsoever and against all claims, demands, proceedings, damages, costs and expenses made against or incurred by the Contractor by reason of any breach by the Sub-Contractor of the Sub-Contract.

(4) The Sub-Contractor hereby acknowledges that any breach by him of the Sub-Contract may result in the Contractor's committing breaches of and becoming liable in damages under the Main Contract and other contracts made by him in connection with the Main Works and may occasion further loss or expense to the Contractor in connection with the Main Works and all such damages loss and expense are hereby agreed to be within the contemplation of the parties as being probable results of any such breach by the Sub-Contractor.'

Chapter 19
Process and plant forms

19.1 The IChemE Red Book – 4th edition, 2001

The Institution of Chemical Engineers published its first model form for lump sum process plant contracts in 1968. It rapidly became known as 'the Red Book' from the colour of its cover. The present version of the form is the fourth edition, published in 2001. It is titled simply Form of Contract, Lump Sum Contracts, The Red Book, Fourth edition 2001.

The omission of reference to process and / or plant in the title of the fourth edition is perhaps a reflection of the wider usage of the Red Book – particularly in civil engineering. This arose from the trend towards design and build lump sum contracts in the 1980s at a time when there were few other established model forms of that type. Moreover until comparatively recently the Red Book enjoyed the remarkable record of having avoided the attention of the courts.

The current provisions in the Red Book for completion, extensions of time for completion and delay damages are, save for a few exceptions, not far removed from standard provisions in construction contracts.

Completion

Completion under the Red Book can be a more complex concept than completion as understood in construction contracts. It depends on the amount of testing required. Thus completion of construction of the plant may be followed by take-over tests and then performance tests – with each stage having its own certification. Additionally the Red Book provides for sectional completions and allows other unspecified things, examples of which could be the provisions of programmes or manuals, to be given completion dates. However, as can be seen from clause 13.1, the completion provisions concentrate principally on completion of construction of the plant. This is marked by the issue of a Construction Completion Certificate under clause 32.3.

Clause 13.1 reads:

'13.1 Subject to the provisions of Clause 14 (Delays), the Contractor shall complete the construction of the Plant ready for the carrying out of the take-over procedures on or before the date, or within the period, specified in Schedule 11 and shall also complete any specified section of the Plant and do any other thing in the performance of the Contract on or before the dates, or within the periods, specified in the said Schedule.'

A well-recognised problem with clause 13.1 is that it seems to require only readiness for take-over tests rather than satisfactory passing of take-over tests to establish completion. But that can readily be rectified, and frequently is, by including in Schedule 11 the passing of take-over tests as one of the things with a stipulated completion date.

Sectional completion

The Red Book provides for sectional completion, at least as far as completion of construction of the plant is concerned in clause 32.1. This reads:

'32.1 If the Contract provides for the completion of construction of the Plant to be by specified sections, the provisions of Sub-clauses 32.2 to 32.5 shall apply as if the reference therein to Plant were a reference to a specified section.'

Extensions of time for completion

The principal clause in the Red Book dealing with extension of time for completion is clause 14.1 which reads:

'14.1 If the Contractor is delayed in the performance of any of his obligations under the Contract by any of the matters specified below, or if either party is delayed by Force Majeure in the performance of any of his obligations under the Contract, the relevant party shall forthwith give notice to the Project Manager and as appropriate to the Contractor.

As soon as reasonably possible, the Contractor shall advise the Project Manager of the extension of any date or period specified in the Contract for the completion of such obligations which he considers would be fair and reasonable in the circumstances. The Contractor shall keep contemporaneous records of the circumstances, extent and effect of such delay. The Project Manager shall, within fourteen days of the time that the extent and consequences of any such delay are known, issue a Variation Order both to the Purchaser and to the Contractor stating the appropriate extension to the Approved Programme and to Schedule 11 (Times of completion) or, if appropriate, to the period in Schedule 16 (Performance tests and procedures) by the end of which the Plant should have passed all its performance tests. If either party does not agree with such extension and such disagreement is not settled in accordance with Clause 45 (Disputes) then the matter may be referred to an Expert in accordance with Clause 47 (Reference to an Expert).

The matters entitling the Contractor to an extension under this Sub-clause are delays caused by:

(a) the occurrence of conditions to which the provisions of Sub-clause 6.3 apply;

(b) a Variation ordered by the Project Manager (other than a Variation Order given by reason of the Contractor's default) except where the delay is already covered in a Variation Order issued by the Project Manager under Sub-clause 16.3;

(c) the giving of any Suspension Order by the Project Manager, except where given by reason of the Contractor's default;

(d) a breach of the Contract by the Purchaser; or

(e) the failure of any Subcontractor nominated by the Project Manager in accordance with Clause 10 (Nominated Subcontractors) to perform such Subcontractor's obligations despite all due supervision by the Contractor.'

Clause 6.3 mentioned in clause 14.1(a) above is an unforeseen physical conditions clause, the main body of which reads:

'6.3 If during the carrying out of the Works the Contractor encounters on the Site any physical condition which at the date of tender as stated in the Agreement could not reasonably have been foreseen by an experienced contractor possessed of all the information which the Contractor then had or could have obtained by visual inspection of the Site or by reasonable enquiry, and if the Contractor considers that he will in consequence of such condition incur an increase in the cost of performing his obligation under the Contract, he shall give the Project Manager a notice under this Sub clause within fourteen days of becoming aware of such unforeseen condition and otherwise shall comply with the requirements of Sub-clause 18.1. Any such notice shall . . .'

In summary therefore the relevant events for extensions of time are:

• unforeseen physical conditions
• variations
• suspensions
• breach by the purchaser
• failures of nominated sub-contractors
• force majeure.

Clause 14.1 has some interesting features. Firstly, the onus is on the contractor not only to initiate any claim for extension of time but also to state what he considers to be a fair and reasonable extension. The project manager has no express function to perform other than to issue a variation order stating 'the appropriate extension' – this presumably being the project manager's assessment. In the event that either party disagrees with this assessment that matter proceeds to dispute resolution.

Another point of interest is that in so far as delays are caused by force majeure the clause applies even-handedly to both parties. That is to say, the purchaser has the same entitlement as the contractor to apply for an extension of time in which to perform his obligations. At first sight this would appear to be an arrangement which could, in some circumstances, relieve the purchaser of his financial liabilities to the contractor for delay claims. However, on close reading of the clause it can be seen that only specified times can be extended. It will not often be the case that the contract imposes specified times for performance on the purchaser and it is doubted if the clause should be operated so as to grant extensions of time to the purchaser for delays which are no more than ordinary acts of prevention.

Also of interest is that extensions of time are formalised by the issue of a variation. This has the effect of bringing extensions of time within the scope of clause 16 (variations) of the Red Book. Consequently, financial claims from the contractor for delay should be normally evaluated under the rules for the valuation of variations. Note, however, that clause 14.4 specifically requires delay claims arising from the purchaser's breach or failures of nominated sub-contractors to be valued on a cost plus profit basis and clause 14.5 requires the parties to bear their own costs if the delay is caused by force majeure.

Finally the facility in clause 14.1 for disputes on extension of time to be referred to an expert needs to be considered with caution. Under clause 47 of the Red Book the parties agree to be bound by decisions of an expert and that his decisions shall be final and binding. For most Red Book contracts this will be the end of the matter but not necessarily so for contracts falling within the scope of statutory adjudication laws.

Force majeure

Force majeure, as referred to in clause 14.1, is defined in clause 14.2 as follows:

'14.2 In the context of this Clause, 'Force Majeure' shall mean any circumstance beyond the reasonable control of either party which prevents or impedes the due performance of the Contract by that party including, but not limited to, the following:

(a) government action or trade embargo;

(b) war, hostilities or acts of terrorism;

(c) riot or civil commotion;

(d) epidemic;

(e) earthquake, flood, fire or other natural physical disaster;

(f) exceptionally severe weather conditions or the consequences thereof;

(g) denial of the use of any railway, port, airport, shipping service or other means of public transport; or

(h) industrial disputes, other than any solely confined to the Contractor and / or his Subcontractors or their employees including employees of any Affiliate of the Contractor or Subcontractor.

The mere shortage of labour, materials or utilities shall not constitute Force Majeure unless caused by circumstances which are themselves Force Majeure.'

Damages for delay

The Red Book's provisions for damages for delay are set out principally in clauses 15.1 and 15.2. These read:

'15.1 If the Contractor fails to complete the Plant or any specified section thereof or to do any other thing in accordance with Schedule 11 (Times of completion), the Contractor shall pay the Purchaser liquidated damages as prescribed in Schedule 12, but shall have no liability to pay damages in excess of the maximum (if any) stated in Schedule 12.

15.2 If after liquidated damages for delay have become payable in respect of any part of the Plant the Project Manager issues a Variation Order or a physical condition is encountered as envisaged in Sub-clause 6.3, either of which delays the Contractor and in the opinion of the Project Manager properly entitles the Contractor to an extension of time in respect of such further delay to that part of the Plant, the Project Manager shall forthwith so inform the Contractor and the Purchaser in writing.

The Purchaser's further entitlement to liquidated damages in respect of that part of the Plant shall thereupon be suspended until the Project Manager notifies the Contractor and Purchaser in writing that such further delay has come to an end.

Such suspension shall not invalidate any entitlement to liquidated damages which accrued before the period of further delay started to run and (subject to any final review of the circumstances) any monies already deducted or paid as liquidated damages for delay may be retained by the Purchaser.'

Provided that the data entered into Schedules 11 and 12 is consistent and complete, the operation of clause 15.1 is reasonably straightforward. However, there is no express provision for the purchaser to deduct liquidated damages nor is there any provision for proportioning down liquidated damages for partial completions.

Clause 15.2 is less clear in its intent and operation. It relates to delays occurring when the contractor is in a period of culpable delay but it apparently restricts the contractor's entitlement to any further extension of time to delays caused by variations and unforeseen physical conditions.

19.2 MF / 1 (Rev 4) 2000 edition

Since 1903 there has been a model form of conditions of contract for plant and electrical works. Originally known as MF'A', it was re-named MF / 1 in 1988, since when it has been published in various editions – the latest being MF / 1 (Rev 4) 2000 edition. Its full title is 'Model form of General Conditions of Contract for use in connection with home or overseas contracts for the supply of electrical, electronic or mechanical plant – with erection'. It is published for a joint committee of the Institutions of Electrical and Mechanical Engineers.

The provisions in MF / 1 for completion, extensions of time for completion, and delay damages are significantly different from those in standard building and civil engineering forms. Most notably, liquidated damages can become due even if there is no stipulated time for completion and in the event of prolonged delay time can be made of the essence.

Completion

Completion in MF / 1 is when the works pass their tests on completion. This can be seen from the following clauses:

Clause 13.1 (contractor's general obligations)

'13.1 The Contractor shall, subject to the provisions of the Contract, with due care and diligence, design, manufacture, deliver to Site, erect and test the Plant, execute the Works and carry out the Tests on Completion within the Time for Completion.'

Clause 29.2 (taking-over certificate)

'29.2 When the Works have passed the Tests on Completion and are complete (except in minor respects that do not affect their use for the purpose for which they are intended) the Engineer shall issue a certificate to the Contractor and to the Purchaser (herein called a 'Taking-Over Certificate'). The Engineer shall in the Taking-Over Certificate certify the date upon which the Works passed the Tests on Completion and were so complete.'

Clause 32.1

32.1 Subject to any requirement under the Contract for the completion of any Section before the completion of the whole of the Works, the Contractor shall so execute the Works that they shall be complete and pass the Tests on Completion (but not the Performance Tests, if any be included) within the time for Completion.'

Note that there is no obligation or need for the engineer to issue a certificate of completion – the taking-over certificate serves for this purpose.

Sectional completions

MF / 1 allows for sectional completions by stating in clause 29.1: 'If the Contract provides for the Works to be taken over by Sections the provisions of this clause shall apply to each such Section as it applies to the Works.'

Extension of time for completion

Clause 33.1 (extension of time for completion) states

'If, by reason of any variation ordered pursuant to clause 27 (Variations) or of any act or omission on the part of the Purchaser or the Engineer or of any industrial dispute or by reason of circumstances beyond the reasonable control of the Contractor arising after the acceptance of the Tender, the Contractor shall have been delayed in the completion of the Works, whether such delay occurs before or after the Time for Completion, then provided that the Contractor shall as soon as reasonably practicable have given to the Purchaser or the Engineer notice of his claim for an extension of time with full supporting details, the Engineer shall on receipt of such notice grant the Contractor from time to time in writing either prospectively or retrospectively such extension of the Time for Completion as may be reasonable.'

As can be seen, relevant events are grouped into four categories:

- variations
- acts or omissions of the purchaser or the engineer
- industrial disputes
- circumstances beyond the reasonable control of the contractor.

Act or omission

The phrase 'act or omission on the part of the Purchaser' may not be wide enough to cover all delays for which the purchaser could be liable – for example, delays caused by the purchaser's other contractors. However, since any such delay could be covered by the category of circumstances 'beyond the control of the contractor', the omission is probably not serious.

Industrial disputes

'Any industrial dispute' is wide enough to cover disputes both within and without the contractor's organisation. To the extent that the contractor's own management practices may have contributed to an industrial dispute, the wording seems generous to the contractor.

Circumstances beyond the control of the Contractor

'Circumstances beyond the reasonable control of the Contractor' are limited in clause 33.1 to those arising after acceptance of the tender. This is not as straightforward a limitation as it might appear. Its interpretation turns on what is meant by circumstances. For example, on one interpretation the contractor would have no entitlement to an extension of time for unexpected site conditions which existed prior to acceptance of the tender if the conditions themselves rather than the discovery of the conditions are to be regarded as the 'circumstances'. Another aspect of the phrase 'circumstances beyond the reasonable control of the Contractor' to consider is how wide is its scope. A similar phrase was given an unexpectedly wide meaning by the House of Lords in *Scott Lithgow Ltd* v. *Secretary of State for Defence* (1989). See Chapter 6 for comment on this case.

As to delays caused by force majeure as defined in clause 46.1 of MF / 1 that clause clearly contemplates that extensions of time will be awarded and presumably, since clause 33.1 does not expressly mention force majeure, such delays come under circumstances beyond the reasonable control of the contractor.

Adverse weather

There is no specific relevant event in clause 33.1 for delays caused by adverse weather but it would certainly be open to the contractor to argue that these were beyond his reasonable control.

Delays after the time for completion

Clause 33.1 is more explicit than the extension of time provisions in most other contracts in expressly stating that it applies to delays which occur after the time for completion. By that is meant, after the time the contractor should have completed and the contractor is proceeding in what is sometimes known as culpable delay and is liable for damages for late completion.

The legal position in respect of delays after the time for completion was examined, albeit in the context of the standard building form of contract, JCT 1980, in the case of *Balfour Beatty Building Ltd* v. *Chestermount Properties Ltd* (1993). It was held that the extension of time provisions did apply to such delays and that the extensions granted should be on a net basis rather than on a gross basis. That decision is thought to be of general application and is likely to be applicable to plant and process contracts as well as to construction contracts. Another point to note is that for MF / 1 the provisions for delays after the time for completion are so worded that they extend to neutral delays as well as to delays caused by the purchaser. The

Balfour Beatty case left open the question of whether the contractor had an entitlement to an extension of time for delays caused by such neutral events as strikes and floods which would have been avoided if the contractor had completed on time.

Notice of claim

Clause 33.1 contains a proviso on the duty of the engineer to grant such extensions of time as are due. It reads: 'then providing that the Contractor shall as soon as reasonably practicable have given . . . notice of his claim for an extension of time . . . the Engineer shall on receipt of such notice . . . grant . . . such extension . . . as may be reasonable'.

This may be intended to act as a condition precedent to the granting of any extension of time and it can be argued that if the contractor fails to give notice he loses his entitlement. See, for example, the *Steria* v. *Sigma* (2007) case discussed in Chapter 5 where a similar provision in a MF / 1-based contract was held to be a condition precedent.

'Prospectively or retrospectively'

These words are used to overcome difficulties which have arisen in some contracts on the retrospective granting of extensions of time.

Clearly it is desirable that whenever possible the contractor should be granted extensions of time prospectively so that he has a completion date to aim for and can plan accordingly. It can even be argued that the words 'on receipt of such notice the Engineer shall' imposes a duty on the engineer to deal with claims for any extension with reasonable promptitude – if perhaps, not 'forthwith' as required by some contracts.

'As may be reasonable'

This phrase apparently gives the engineer some flexibility in determining the amount of an extension of time. In circumstances such as concurrent delay where the contractor's entitlement may not match exactly the recorded delay, apportionment would appear to be in order.

Delay by sub-contractors

Clause 33.2 reads:

> '33.2 Any delay on the part of a Sub-Contractor which prevents the Contractor from completing the Works within the Time for Completion shall entitle the Contractor to an extension thereof provided

such delay is due to a cause for which the Contractor himself would have been entitled to an extension of time under sub-clause 33.1 (Extension of time for completion).'

This is not it is suggested a general provision entitling the contractor to claim an extension of time for any delay caused by a sub-contractor. Such a provision would cut right across the basic principle that the contractor is responsible for the performance of his sub-contractors. More likely, clause 33.2 merely provides that the contractor retains his entitlement to extensions of time even though the relevant events apply directly to sub-contractors. In most contracts this is to be implied.

Mitigation of consequences of delay

Clause 33.3 reads:

'In all cases where the Contractor has given notice under Sub-Clause 33.1 (Extension of Time for Completion) the Contractor shall consult with the Engineer in order to determine the steps (if any) which can be taken to overcome or minimise the actual or anticipated delay. The Contractor shall thereafter comply with all reasonable instructions which the Engineer shall give in order to overcome or minimise such delay. If compliance with any such instruction shall cause the Contractor to incur additional Cost and the Contractor is entitled to an extension of time, the amount thereof shall be added to the Contract Price.'

There is much good sense in this provision to mitigate delay but the wording could be clearer. The question is – does the contractor get payment instead of an extension of time or does he get both? The point is important in determining whether there is any transfer of risk to the contractor in respect of attempts to mitigate delay. If there is not then the purchaser risks the expenditure of the extra costs without any guarantee of a corresponding benefit.

On strict interpretation of the wording of clause 33.3 it is suggested that the contractor is entitled to receive both an extension of time and payment of extra costs. However, there must be some doubt as to whether this is intended. The procedural arrangements of the clause place the initiative with the contractor and this seems odd if he is simply spending the purchaser's money.

'The Contractor shall consult'

Under the wording of clause 33.3 the contractor is obliged to consult with the engineer whenever an application is made for an extension of time. This applies both to delays for which the purchaser is responsible and to neutral delays. Although the contractor does not have discretion on whether or not to consult he must in practice have discretion on what he proposes.

'Shall comply with all reasonable instructions'

In so far that the engineer gives instructions which match or are developed from the contractor's proposals they would no doubt be deemed to be reasonable instructions. For instructions not agreed in the consultation process these might well be deemed to be unreasonable.

Note that the contractor's obligation is only to comply with reasonable instructions – not all instructions. But in any event the obligation may be more theoretical than real because if the contractor refuses to comply with instructions, whether reasonably given or unreasonably given, there is no obvious contractual remedy short of invoking the termination procedures for his default.

The key question for the contractor in deciding whether or not to comply with the engineer's instructions is likely to be – will payment be made for the extra costs? For the answer to this the contractor needs to know whether or not the engineer accepts that there is an entitlement to extension of time. Unfortunately for the contractor the wording of clause 33.3 does not require the engineer to state his position in advance of giving instructions – although since it is the purchaser's money which is at stake the engineer would be well advised to know his position and to have obtained the approval of the purchaser before giving any instructions.

As a general point it is questionable whether instructions on acceleration type matters should ever be left to the engineer. Usually any payments to the contractor for attempting to finish before the due date are agreed directly between the purchaser and the contractor and both parties know where they stand before acceleration measures are instructed and commenced. And to complicate matters, clause 33.3 also requires the contractor to comply with instructions to overcome or minimise delay for which there is no entitlement to payment – a matter which is already covered in the contract in clause 14.6 which relates to rate of progress.

To alleviate the more obvious difficulties of clause 33.3 it is suggested that the following procedure might be adopted:

- contractor applies for an extension of time giving appropriate details;
- contractor indicates to the engineer what steps could be taken to overcome or minimise the delay;
- engineer forms a view on whether any extension of time is due;
- if no extension of time considered due the engineer informs the contractor accordingly and asks for the contractor's proposals under clause 14.6;
- if an extension considered due the engineer consults with the purchaser on the financial implications of taking acceleration measures;
- if the purchaser consents to additional expenditure the engineer instructs the contractor under clause 33.3.

Delay in completion and delay damages

Clause 34.1 deals with delay in completion and liquidated damages for late completion. The clause reads:

> '34.1 If the Contractor fails to complete the Works in accordance with the Contract, save as regards his obligations under Clauses 35 (Performance Tests) and 36 (Defects Liability), within the Time for Completion, or if no time be fixed, within a reasonable time, there shall be deducted from the Contract Price or paid to the Purchaser by the Contractor the percentage stated in the Appendix of the Contract Value of such parts of the Works as cannot in consequence of the said failure be put to the use intended for each week between the Time for Completion and the actual date of completion. The amount so deducted or paid shall not exceed the maximum percentage stated in the Appendix of the Contract Value of such parts of the Works, and such deduction or payment shall subject to Sub-Clause 34.2 (Prolonged Delay) be in full satisfaction of the Contractor's liability for the said failure.'

'If no time be fixed'

The provision in clause 34.1 that liquidated damages become due if the contractor fails to complete within a reasonable time 'if no time be fixed' is most unusual. Normal convention is that liquidated damages apply only to fixed times and unliquidated damages apply when time is at large. But even if legal obstacles are not fatal to the application of liquidated damages to unspecified times there is the practical point to resolve of who decides whether or not the contractor has failed to complete within a reasonable time. This alone should be sufficient to make the point that entering into a contract with no fixed time for completion will rarely be in the purchaser's interests.

'The percentage stated in the Appendix'

The appendix of MF / 1, in common with other plant and process model forms, requires the rate of liquidated damages to be stated as a percentage of the contract value for each week of delay. The rate suggested in the official 'Commentary' on MF / 1 published by the sponsors of the model form is between ¼% and 1% per week.

If the rate selected is equivalent to, or less than, the purchaser's genuine pre-estimate of loss then it can genuinely stand as liquidated damages. But otherwise the rate may be challengeable as a penalty. In practice the application of a financing formula of the type approved in the *Finnegan* case gives a rate in the order of ½% per week when modest supervision charges are included. And for the majority of contracts the purchaser's true and full pre-estimate of loss would probably greatly exceed 1% per week.

'Such part of the Works'

These words suggest the intention that the rate of liquidated damages should be applied not so much to the contract price of the whole of the works as to the contract value of parts of the works.

Two construction cases, *Bruno Zornow (Builders) Ltd* v. *Beechcroft Developments Ltd* (1989) and *Turner* v. *Mathind* (1986), illustrate the difficulty of applying liquidated damages to parts of the works which are not specified as sections with their own rates of damages. Those cases did admittedly deal with specified sums rather than percentages as liquidated damages but they show that the courts may not enforce liquidated damages provisions which are not fully specified. The problem with MF / 1 in referring to parts of the works is who is to specify the value of the parts in question and how is the difficulty that there is no provision for extension of time for parts to be overcome.

'As cannot in consequence be put to the use intended'

The words 'cannot in consequence of the said failure be put to the use intended' introduce more uncertainty into clause 34.1. The said 'failure' is the failure of the contractor to complete the works. That in itself is all that is needed to activate the contractor's liability for late completion. Either the contractor has got a taking-over certificate or he has not. It is therefore difficult to see why failure to complete is qualified by consideration of whether or not the works, or parts of the works, cannot in consequence be put to their intended use. On one view the clause seems to suggest that the contractor can challenge his liability for liquidated damages on the grounds that although he failed to complete in time the purchaser was not ready for them. Again this cuts across the basic legal principle that with liquidated damages the purchaser is not required to prove his loss.

'The amount . . . shall not exceed the maximum percentage'

The appendix to MF / 1 requires a maximum percentage to be stated for liquidated damages – thus putting a gross limit on the contractor's liability. The official Commentary on MF / 1 suggests a maximum of between 5% and 15%.

This is purely a commercial arrangement and there is no legal requirement for any maximum limit. So the question of what maximum percentage should be used has to be decided on the commercial circumstances of the parties in each contract. The purchaser may get the benefit of better competitive prices if the maximum percentage is kept low but that needs to be balanced against the potential problems of the purchaser if the contractor is under no financial pressure to complete on time.

Prolonged delay

Clause 34.2 deals with prolonged delay. The clause states:

'34.2 If any part of the Works in respect of which the Purchaser has become entitled to the maximum amount provided under Sub-Clause 34.1 (Delay in Completion) remains uncompleted the Purchaser may by notice to the Contractor require him to complete. Such notice shall fix a final Time for Completion which shall be reasonable having regard to such delay as has already occurred and to the extent of the work required for completion. If for any reason, other than one for which the Purchaser or some other contractor employed by him is responsible, the Contractor fails to complete within such time, the Purchaser may by further notice to the Contract elect either:

(a) to require the Contractor to complete, or

(b) to terminate the Contract in respect of such part of the Works,

and recover from the Contractor any loss suffered by the Purchaser by reason of the said failure up to an amount not exceeding the sum stated in the Appendix or, if no sum be stated, that part of the Contract Price that is properly apportionable to such part of the Works as cannot by reason of the Contractor's failure be put to the use intended.'

Commentary on clause 34.2 – prolonged delay

The purpose of clause 34.2 is clearly to keep the contractor under pressure to complete when the maximum amount of liquidated damages for late completion has been reached. The purchaser is, in effect, given power to make time of the essence (for which the remedy for default is termination of the contract) by serving notice on the contractor.

'Any part of the Works'

These words reintroduce the complications in clause 34.1 on the application of the provisions to parts of the works. But the concept of termination of the contract in respect of parts is even more legally dubious than that of applying liquidated damages to parts.

'A final time for completion'

It is apparently left to the purchaser (not the engineer) to fix the final time for completion. The words used suggest that this time cannot later be extended. However, it is far from clear how clause 34.2 operates if the purchaser or some other contractor employed by him is responsible for delay after the notice fixing a final time for completion is served. The possibility is that the provisions for termination lapse and the purchaser has no further remedy. It would also seem from the words 'Purchaser or some other contractor' that the risk of all other causes of delay rests with the contractor and that no relief is given after the fixing of the final time for completion for the effects of neutral events.

'The purchaser may by further notice ... elect'

By further notice, after the final time for completion has elapsed, the purchaser may either require the contractor to complete, or terminate the contract in respect of parts. However, the purchaser is not under a duty to serve this further notice – the clause reads 'the Purchaser may'. The contractual position if the purchaser fails to serve any further notice would seem to be the same as if the purchaser serves notice requiring the contractor to complete. Namely the purchaser is prepared to accept completion whenever it is achieved. There is nothing in the clause to suggest that the purchaser has the power to fix a second final time for completion thereby re-activating his option to terminate the contract.

'Terminate the contract'

There is no reference in clause 34.2 linking termination under the clause with the provisions for termination under clause 49 (contractor's default). Nor is there anything in clause 49 referring back to clause 34.2. This suggests that termination under clause 34.2 is wholly independent of the principal termination provisions of MF / 1 as set out in clause 49. The consequences of termination under clause 34.2 are therefore purely those described in the clause. These consequences are that the purchaser may terminate the contract in respect of the delayed parts, and recover unliquidated damages up to the amount stated in the appendix. This, of course, is after the contractor has already exhausted the limit of liquidated damages stated in the contract.

'Any loss suffered by the purchaser'

This phrase may not mean what it says. Clause 44.2 of the contract prevents recovery of indirect and consequential damage except in relation to clauses 34.1 and 35.8.

'An amount not exceeding the sum in the appendix'

The appendix requires the maximum loss recoverable by the purchaser to
be stated as a sum of money. There is nothing to stop this sum exceeding
the contract price but the intention is probably that if a sum is stated it is
less than the contract price.

'If no sum be stated'

Where no sum is stated in the appendix the maximum liability of the contractor under clause 34.2 is the contract price. But this applies only when
the whole of the works cannot be put into use as intended. Where only part
of the works cannot be put into use an apportionment must be made.

In other words, if no maximum sum is stated in the appendix, the contractor's maximum liability for any part of the works which is still not
capable of being used after prolonged delay is the proportion of the contract
price that the price of the part has to the whole.

As a general rule damages for late completion are deductible from the
contract price. Thus the contractor retains his entitlement to payment for
work completed notwithstanding that he completes late and is liable for a
deduction for damages. In the case of damages payable for prolonged delay
where the termination option is exercised under MF / 1 the contractor may
not have entitlement to payment of any of the contract price but he remains
liable nevertheless for the purchaser's loss.

Chapter 20
FIDIC conditions of contract 1999

20.1 FIDIC contracts

FIDIC conditions of contract (Fédération Internationale Des Ingénieurs – Conseils) were first published in 1957 based on ICE conditions, 4th edition. A second edition was published in 1969; a third in 1977; and a fourth in 1987 – all reflecting developments in ICE conditions of contract and including much the same text and clause numbering of ICE conditions. Throughout that time, FIDIC conditions, commonly known as the Red Book because of the cover, were widely used for international construction projects of a civil engineering nature.

In 1999 FIDIC introduced first editions of a new suite of contracts:

Conditions of contract for construction

Recommended for building or engineering works designed in whole or in part by the Employer and / or his Engineer

Conditions of contract for plant and design-build

Recommended for the provision of electrical and / or mechanical plant

Conditions of contract for EPC / Turnkey projects

Recommended for power plants, infrastructure projects and the like where the contractor takes full responsibility for design and construction.

Short form of contract

Recommended for works of small capital value.

Note

The provisions for liquidated damages and extensions of time considered in this chapter are taken from the Conditions of Contract for Construction. For convenience it is referred to as the new Red Book.

20.2 *Commencement and completion*

Clause 8.1 (commencement of works) of the new Red Book states:

'The Engineer shall give the Contractor not less than 7 days' notice of the Commencement Date. Unless otherwise stated in the Particular Conditions, the Commencement Date shall be within 42 days after the Contractor receives the Letter of Acceptance.'

This is a significant improvement on the wording of the corresponding provision in the old Red Book which left much to be desired by way of certainty in saying 'The Contractor shall commence the Works as soon as is reasonably possible after the receipt by him of a notice to this effect from the Engineer'.

Completion

Clauses 8.2 and 10.1 make clear that where tests on completion are specified these have to be passed before completion.

Clause 8.2 (time for completion) reads:

'The Contractor shall complete the whole of the Works, and each Section (if any), within the Time for Completion for the Works or Section (as the case may be), including:

(a) achieving the passing of the Tests on Completion, and

(b) completing all work which is stated in the Contract as being required for the Works or Section to be considered to be completed for the purposes of taking-over under Sub-Clause 10.1 [*Taking Over of the Works and Sections*].'

Claim 10.1 (taking over of the works and sections) reads:

'Except as stated in Sub-Clause 9.4 [*Failure to Pass Tests on Completion*], the Works shall be taken over by the Employer when (i) the Works have been completed in accordance with the Contract, including the matters described in Sub-Clause 8.2 [*Time for Completion*] and except as allowed in sub-paragraph (a) below, and (ii) a Taking-Over Certificate for the Works has been issued, or is deemed to have been issued in accordance with this Sub-Clause.

The Contractor may apply by notice to the Engineer for a Taking-Over Certificate not earlier than 14 days before the Works will, in the Contractor's opinion, be complete and ready for taking over. If the Works are divided into Sections, the Contractor may similarly apply for a Taking-Over Certificate for each Section.

The Engineer shall, within 28 days after receiving the Contractor's application:

(a) issue the Taking-Over Certificate to the Contractor, stating the date on which the Works or Section were completed in accordance with the Contract, except for any minor outstanding work and defects which will not substantially affect the use of the Works or Section for their intended purpose (either untiI or whilst this work is completed and these defects are remedied); or

(b) reject the application, giving reasons and specifying the work required to be done by the Contractor to enable the Taking-Over Certificate to be issued. The Contractor shall then complete this work before issuing a further notice under this Sub-Clause.'

In short it is the issue of the taking-over certificate which defines completion.

20.3 Extension of time

Clause 8.4 (extension of time for completion) contains the principal provisions in the new Red Book for extensions of time. However, it needs to be read in conjunction with other clauses which detail particular relevant events and clause 20.1 (contractor's claims) which includes an important condition precedent to the contractor's entitlement to extension of time.

Clause 8.4

In full clause 8.4 states:

'The Contractor shall be entitled subject to Sub-Clause 20.1 [*Contractor's Claims*] to an extension of the Time for Completion if and to the extent that completion for the purposes of Sub-Clause 10.1 [*Taking Over of the Works and Sections*] is or will be delayed by any of the following causes:

(a) a Variation (unless an adjustment to the Time for Completion has been agreed under Sub-Clause 13.3 [*Variation Procedure*]) or other substantial change in the quantity of an item of work included in the Contract,

(b) a cause of delay giving an entitlement to extension of time under a Sub-Clause of these Conditions,

(c) exceptionally adverse climatic conditions,

(d) unforeseeable shortages in the availability of personnel or Goods caused by epidemic or governmental actions, or

(e) any delay, impediment or prevention caused by or attributable to the Employer, the Employer's Personnel, or the Employer's other contractors on the Site.

If the Contractor considers himself to be entitled to an extension of the Time for Completion, the Contractor shall give notice to the Engineer in accordance with Sub-Clause 20.1 [*Contractor's Claims*]. When determining each extension of time under Sub-Clause 20.1, the Engineer shall review previous determinations and may increase, but shall not decrease, the total extension of time.'

Variations

The reference in clause 8.4(a) to adjustment to the time for completion under clause 13.3 seems to concern the procedural aspects of extending time. It creates a distinction between agreed extensions of time for variations (recognised by changing the time for completion) and extensions claimed for variations.

Changes in quantity

The new Red Book is a re-measurement contract as confirmed in clause 12.1. Hence the provision in clause 8.4(a) for extending time for substantial changes in quantities.

Particular entitlements

Clause 8.4(b) effectively collects together the various entitlements to extension of time stated elsewhere in the conditions. These are:

- clause 1.9 – delayed drawings or instructions
- clause 2.1 – failure to give possession
- clause 4.7 – errors in setting-out information
- clause 4.12 – unforeseeable physical conditions
- clause 4.24 – fossils
- clause 7.4 – testing instructions
- clause 8.5 – delays caused by authorities
- clause 8.9 – suspensions of works
- clause 10.3 – interference with tests on completion
- clause 13.7 – changes in legislation
- clause 19.4 – force majeure.

Clause 8.5 (delays caused by authorities)

Clause 8.5 clarifies a matter which in many construction contracts is left unanswered – namely which party carries the risk of delays attributable to the performance or attitude of public bodies and authorities? The clause reads:

'If the following conditions apply, namely

(a) the Contractor has diligently followed the procedures laid down by the relevant legally constituted public authorities in the Country,
(b) these authorities delay or disrupt the Contractor's work, and
(c) the delay or disruption was Unforeseeable, then this delay or disruption will be considered as a cause of delay under sub-paragraph (b) of Sub-Clause 8.4 [*Extension of Time for Completion*].'

Clause 19.1 (definition of force majeure)

Clause 19 deals in considerable detail with circumstances arising from 'force majeure'. The clause commences with the following definition:

'In this Clause, "Force Majeure" means an exceptional event or circumstance:

(a) which is beyond a Party's control,
(b) which such Party could not reasonably have provided against before entering into the Contract,
(c) which, having arisen, such Party could not reasonably have avoided or overcome, and
(d) which is not substantially attributable to the other Party.'

Climatic conditions

To qualify for extension of time climatic conditions must be 'exceptionally adverse'. This can be a controversial matter although it can be avoided by stating applicable parameters in the contract documents.

Unforeseeable shortages

The provision in clause 8.4(d) for extending time for unforeseeable shortages is another risk sharing arrangement of the new Red Book. On its wording it seems to apply to the actions of any government but that may not be its intention.

Prevention

Clause 8.4(e) covers acts of prevention by the employer, his personnel and his other contractors. It does not expressly extend to the engineer and his staff but it could arguably do so.

Clause 20 (contractor's claims)

Clause 8.4 commences with a reference to clause 20.1. That clause reads:

'If the Contractor considers himself to be entitled to any extension of the Time for Completion and/or any additional payment, under any Clause of these Conditions or otherwise in connection with the Contract, the Contractor shall give notice to the Engineer, describing the event or circumstance giving rise to the claim. The notice shall be given as soon as practicable, and not later than 28 days after the Contractor became aware, or should have become aware, of the event or circumstance.

If the Contractor fails to give notice of a claim within such period of 28 days, the Time for Completion shall not be extended, the Contractor shall not be entitled to additional payment, and the Employer shall be discharged from all liability in connection with the claim. Otherwise, the following provisions of this Sub-Clause shall apply.'

The probability is that the clearly worded condition precedent to extension of time stated in clause 20.1 would be held to be effective under English law – albeit that there remains some debate on whether such conditions should apply to delays caused by prevention. Under some civil and equity based laws it may not be so effective.

20.4 Delay damages

Clause 8.7 deals in straightforward terms with liquidated damages for delay. It states:

'If the Contractor fails to comply with Sub-Clause 8.2 [*Time for Completion*], the Contractor shall subject to Sub-Clause 2.5 [*Employer's Claims*] pay delay damages to the Employer for this default. These delay damages shall be the sum stated in the Appendix to Tender, which shall be paid for every day which shall elapse between the relevant Time for Completion and the date stated in the Taking-Over Certificate. However, the total amount due under this Sub-Clause shall not exceed the maximum amount of delay damages (if any) stated in the Appendix to Tender.

These delay damages shall be the only damages due from the Contractor for such default, other than in the event of termination under Sub-Clause 15.2 [*Termination by Employer*] prior to completion of the Works. These damages shall not relieve the Contractor from his obligation to complete the Works, or from any other duties, obligations or responsibilities which he may have under the Contract.'

Proportioning down

The provisions for proportioning down in the event of partial completions or occupations are found in clause 10.2. They state:

'If a Taking-Over Certificate has been issued for a part of the Works (other than a Section), the delay damages thereafter for completion of the remainder of the Works shall be reduced. Similarly, the delay damages for the remainder of the Section (if any) in which this part is included shall also be reduced. For any period of delay after the date stated in this Taking-Over Certificate, the proportional reduction in these delay damages shall be calculated as the proportion which the value of the part so certified bears to the value of the Works or Section (as the case may be) as a whole. The Engineer shall proceed in accordance with Sub-Clause

3.5 [*Determinations*] to agree or determine these proportions. The provisions of this paragraph shall only apply to the daily rate of delay damages under Sub-Clause 8.7 [*Delay Damages*], and shall not affect the maximum amount of these damages.'

Because the new Red Book is a re-measurement contract it should be a relatively easy task to value parts of the works and to establish reduced rates of liquidated damages for the remainder of the works. As might be expected the lump sum Conditions of Contract for EPC / Turnkey Projects do not permit the taking-over of parts except by agreement of the parties and correspondingly they do not contain a proportioning down clause.

Table of Cases

Abbreviations of Law Reports

The following abbreviations of Reports are used:

AC	Law Reports Appeal Cases Series
ALJ	Australian Law Journal
ALJR	Australian Law Journal Reports
All ER	All England Law Reports
App Cas	Law Reports, Appeal Cases
BING	Bingham's Reports, Common Pleas
BLR	Building Law Reports
CAMP	Campbell's Reports
CH D	Law Reports, Chancery Division
CILL	Construction Industry Law Letter
CLR	Commonwealth Law Reports
CONST LJ	Construction Law Journal
C & P	Carrington & Payne's Reports
DLR	Dominion Law Reports
EG	Estates Gazette
EXCH	Exchequer Reports
EWCA	England and Wales Court of Appeal
EWHC	England and Wales High Court
FLR	Federal Law Reports (Australia)
GIFF	Giffard's Reports
HBC	Hudson's Building Contracts
HLC	House of Lords Cases
KB	Law Reports Kings Bench Division
LJEX	Law Journal, Exchequer
LJQB	Law Journal, Queens Bench
Lloyds Rep	Lloyd's List Law Reports
LR/CP	Law Reports, Common Pleas
LT	Law Times Reports
M & W	Meeson & Welsby's Reports
NSWLR	New South Wales Law Reports
NZLR	New Zealand Law Reports
QB	Queens Bench Reports
SALR	South African Law Reports
SLT	Scots Law Times
TLR	Times Law Reports
TR	Term Reports
WLR	Weekly Law Reports
WR	Weekly Reporter

Index